The Serpin Family

Margarethe Geiger • Felix Wahlmüller •
Margareta Furtmüller
Editors

The Serpin Family

Proteins with Multiple Functions in Health and Disease

Springer

Editors
Margarethe Geiger
Department of Vascular Biology and Thrombosis Research, Center of Physiology and Pharmacology
Medical University of Vienna
Vienna, Austria

Felix Wahlmüller
Department of Vascular Biology and Thrombosis Research, Center of Physiology and Pharmacology
Medical University of Vienna
Vienna, Austria

Margareta Furtmüller
Department of Vascular Biology and Thrombosis Research, Center of Physiology and Pharmacology
Medical University of Vienna
Vienna, Austria

ISBN 978-3-319-22710-8 ISBN 978-3-319-22711-5 (eBook)
DOI 10.1007/978-3-319-22711-5

Library of Congress Control Number: 2015955182

Springer Cham Heidelberg New York Dordrecht London

Printed on acid-free paper

Springer International Publishing AG Switzerland is part of Springer Science+Business Media (www.springer.com)

Preface

We were just about to organize the 7th International Symposium on Serpin Biology, Structure, and Function (Serpins 2014) when Springer approached us with the idea to edit a book on topics of "vascular biology." Since the meeting, which then took place in March/April 2014 in Leogang, Austria, was a great success, it was clear to us that a book reviewing recent advances in the field of serpins in connection with the serpin meeting is needed by the scientific community. We therefore invited speakers of that meeting to contribute a chapter to a book about serpins and were very pleased that most of them agreed. In addition we could get on board some renowned scientists who were not able to attend the conference but who nevertheless agreed to contribute a chapter.

For those who are not members of the serpin community, the term "serpin" in an acronym for **ser**ine **p**rotease **in**hibitor. In fact, the first identified and characterized members of the serpin family were inhibitors of serine proteases present in plasma (e.g. antithrombin or α1-antitrypsin). However, the common feature of serpins is not so much the inhibition of serine proteases but their common structure. There are serpin family members that not only inhibit serine proteases but also cysteine proteinases (cross class inhibitors) and other so-called non-inhibitory members lacking any protease inhibitory activity. These have acquired other functions, e.g., hormone carriers, chaperones, or storage proteins. We are aware of the fact that our book does neither cover all serpins nor all aspects of serpin biology. With this book we rather tried to address some recent developments in serpin evolution, functions of nonmammalian serpins, new roles of serpins in diseases, or mechanisms regulating the intracellular maturation and processing of secreted serpins. The book starts with chapters dealing with the evolution of serpins and the mechanism of serpin action. These are followed by chapters on specific intracellular and secreted serpins. Finally there are chapters focusing on serpin-related diseases, disease mechanisms, and disease models, as well as the potential therapeutic use of virus-derived serpins.

We would like to thank all authors, who contributed to the book for their excellent chapters and also for their patience with the editors.

We hope that this book will also attract readers not so familiar with the serpin field and stimulate their curiosity for this exciting protein family.

Vienna, Austria
June 2015

Margarethe Geiger
Felix Wahlmüller
Margareta Furtmüller

Contents

The Human SERPIN Repertoire and the Evolution of 14q32.1 and 18q21.3 Gene Clusters

1

Susana Seixas

Contents

Abstract

The human genome comprises two large clusters of serine protease–inhibitor genes (*SERPIN*s) originated by duplication events that occurred at different moments of vertebrate evolution. The 14q32.1 cluster includes 11 members, all sharing a similar gene structure to alpha-1-antitrypsin, and the 18q21.3 cluster comprises 10 members, characterized by their homology to chicken ovalbumin. Although the majority of these genes are widespread across mammalian species, some are restrained to certain phylogenetic groups, making the repertoire of each species unique. In primates, events of gene duplication and divergence were associated to the origin of *SERPINA2* and *SERPINB3*. Evolutionary processes specific to the human lineage included the loss of *SERPINA13*, an ancient gene only kept in primates, and the pseudogenization of *SERPINB11*, a gene under strong constrains in other species. More recently in humans, natural

S. Seixas (✉)
Institute of Molecular Pathology and Immunology of the University of Porto (IPATIMUP), Rua Dr. Roberto Frias s/n, 4200-465 Porto, Portugal
e-mail: sseixas@ipatimup.pt

M. Geiger et al. (eds.), *The Serpin Family*, DOI 10.1007/978-3-319-22711-5_1

selection acted in *SERPINA2* and *SERPINB11* favoring, on one hand, a non-functional allele carrying a 2 kb deletion and, on the other, a resurrected gene linked to a novel non-inhibitory function. Considering a possible role of SERPINs in inflammation and immunity, together with the perception of the impact of infectious diseases in the natural history of human populations and other species, raised the hypothesis of an evolution driven by host–pathogens interactions. Overall, gains and losses of genes seem to have had an important adaptive value in the long-term evolution of 14q32.1 and 18q21.3 clusters with current implications in SERPIN activities and effects in human diseases.

1.1 Introduction

The superfamily of serine protease inhibitors (serpins) comprises a large number of proteins widely distributed among animals, plants, viruses, and bacteria, characterized by a highly order tertiary structure conserved throughout evolution. The prototypical serpin structure has three β-sheets, nine α-helices, and an exposed flexible stretch of approximately 20 residues named as reactive center loop (RCL). In most cases, serpins neutralize serine or cysteine proteases by a unique suicide substrate-like mechanism that entails a dramatic rearrangement in protein folding. Briefly, serpins are able to entrap proteases by acting as a pseudo-substrate. Specificity is determined by the RCL residue composition, in particular those located at the P1–P1′ residues (scissile bond). Once cleaved by the target protease, the RCL moves through a functional domain—the shutter—to the opposite pole of the molecule causing the distortion of protease structure and its loss of catalytic activity. As protease inhibitors, serpins play key roles in diverse vertebrate processes such as coagulation, fibrinolysis, angiogenesis, tissue repair, inflammation, and apoptosis. However, a small fraction of these proteins developed other functions outside of proteolysis regulation and exert activities as hormone carriers, chaperones, or storage proteins (Irving et al. 2000; Stein and Carrell 1995; Law et al. 2006).

Currently, in the human genome, there are more than 40 annotated *SERPINs*, among them several active genes and a few pseudogenes (Heit et al. 2013). These are classified into nine phylogenetic clades (A–I) defined by similarities in protein sequence, gene structure, and/or chromosomal position. Interestingly, all these clades are restricted to the vertebrate lineage placing the emergence of such serpins within a maximum time frame of 800 million years (Irving et al. 2000, 2007). Clade B members, also called ov-serpins due to their homology to chicken ovalbumin, occupy a more ancestral node in the SERPIN phylogenetic tree and encode mostly intracellular proteins, while the remaining clades include with few exceptions extracellular SERPINs (Irving et al. 2000, 2007). The largest *SERPIN* groups are clade A (13 members) and clade B (15 members), and these are organized into three major gene clusters in chromosome 14q32.1 and in chromosomes 6p25 and 18q21.3, respectively. The other groups include unique (clades C, D, and G) or a low number (clades E, F, H, and I) of sequences and are more dispersed along the genome (Law et al. 2006; Heit et al. 2013).

1.2 *SERPINA* Cluster at Chromosome 14q32.1

The clade A cluster located at chromosome 14q32.1 spans over ~370 kilobases (kb) and contains 11 *SERPIN* members arranged in the following centromere–telomere orientation: protein Z inhibitor (*SERPINA10*), corticosteroid-binding globulin (*SERPINA6*), alpha1-antitrypsin-related protein (*SERPINA2*), alpha1-antitrypsin (*SERPINA1*), antiproteinase-like 2 (*SERPINA11*), centerin (*SERPINA9*), vaspin (*SERPINA12*), kallistatin (*SERPINA4*), protein C inhibitor (*SERPINA5*), alpha 1-antichymotrypsin (*SERPINA3*), and kallistatin-like (*SERPINA13*) (Fig. 1.1) (Namciu et al. 2004; Marsden and Fournier 2005). Most of these genes share a typical structure comprising at least an untranslated exon and four coding exons, and accordingly, these were proposed to have evolved from a common ancestral gene through a series of duplication events (van Gent et al. 2003; Atchley et al. 2001). However, the vast majority of those events are likely to represent ancient duplications given the overall low level of protein sequence identity (average 37 %) and their occurrence in ortholog regions in other annotated mammalian genomes (Fig. 1.1). The exception to this common pattern is the case of *SERPINA1* and *A2*, which are recent duplicates originated approximately 90 million years ago (MYA) in the crown of primates and that still display identity of 60 % and 80 % at protein and gene levels, respectively (Marques et al. 2013).

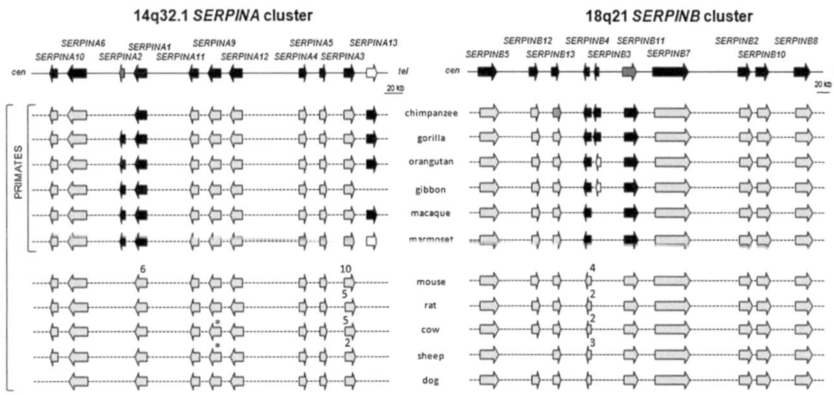

Fig. 1.1 Schematic representation of 14q32.1 and 18q21 *SERPIN* gene clusters. *Upper diagram* shows the relative position of human genes in the two clusters. *Lower diagrams* illustrate the presence/absence of ortholog sequences in annotated mammalian genomes (NCBI: http://www.ncbi.nlm.nih.gov/ and ENSEMBL: http://www.ensembl.org/index.html databases). Functional genes are shown in *black*, polymorphic gene/pseudogenes in *dark gray*, annotated functional genes in *light gray*, and known pseudogenes in *white*. *Numbers* indicate gene copy number without predicted pseudogenes. Genes labeled with (*asterisk*) are annotated as *SERPINA14*

1.2.1 *SERPINA1* and *SERPINA2* Duplicates

SERPINA1 is the archetypical protein and probably the most studied member of the superfamily. This molecule is the major protease inhibitor in the human serum and an important acute phase protein, which underlies a rare genetic disorder prevalent among individuals of European descent—the alpha1-antitrypsin deficiency (AATD). The main clinical manifestations of AATD are emphysema and hepatic disease due to mutations in *SERPINA1* gene which may lead to the unopposed activity of neutrophil elastase released in the lower respiratory tract during inflammation (loss-of-function mutations) and/or to the polymerization of misfolded proteins in the endoplasmic reticulum of hepatocytes causing cellular damage and death (gain-of-function mutations) (Gooptu et al. 2014; Janciauskiene et al. 2011). Conversely, *SERPINA2* is still mostly unknown in spite of its sequence identification several decades ago (Bao et al. 1988; Hofker et al. 1988). Originally, *SERPINA2* was thought to be a pseudogene because no promoter region or liver expression was detected, and a significant level of sequence degeneration was observed, including a disrupted starting codon (ATG to ATA) and a 2 kb deletion encompassing exon IV and part of exon V (Bao et al. 1988; Hofker et al. 1988). More recently, a study centered in the characterization of the sequence variation in the proximal region of the *SERPINA* cluster confirmed the existence of two distinct *SERPINA2* isoforms segregating within human populations: an inactive *SERPINA2* or pseudogene, carrying two disrupting mutations in strong linkage disequilibrium (ATA and 2 kb deletion) and without any recognizable transcript, and an active *SERPINA2* or gene, mostly expressed in testis and leukocytes that if translated would encode a regular SERPIN with a distinct inhibitory activity from SERPINA1—P1–P1′ residues composed by tryptophan and serine instead of methionine and serine (Seixas et al. 2007). Later on, the in vitro expression of SERPINA2 and A1 uncovered additional differences between these close homologs; while SERPINA1 was processed into a fully glycosylated protein and secreted, SERPINA2 was only partially glycosylated and was localized inside the cells in the endoplasmic reticulum (ER) (Marques et al. 2013). Nevertheless, SERPINA2 did not show any evidence of polymerization in the endoplasmic reticulum in contrast to the common misfolded variants of SERPINA1 (Z allele, Glu342Lys, and S allele, Glu264Val) suggesting that SERPINA2 is a correctly folded protein with intracellular activity (Marques et al. 2013). So far, all clade A SERPINs were found to be extracellular molecules, and only SERPINH1 (also known as HSP47) was identified in the ER as a chaperone with a role in the biosynthesis of collagen (Christiansen et al. 2010; Mala and Rose 2010).

The divergence of *SERPINA2* is also supported by phylogenetic studies. Overall, the evolution of *SERPINA2* and *A1* in primates has been shaped by functional constrains, as it could be expected for two active genes; however, *SERPINA2* has experienced higher evolutionary rates than *SERPINA1* with several residues evolving under positive selection (Marques et al. 2013). One of those residues was the reactive site (P1: Met358Trp), which—based on previous assays done in SERPINA1 scaffold—is predicted to change the inhibitory affinity toward chymotrypsin and to slower the formation of protease–inhibitor complexes (Futamura et al. 1998). Other selected residues (Arg196Asp, Pro197Lys, Leu241Ala, and Met242Gln) were located in the breach, a crucial structural domain where significant residue changes are known to affect SERPIN folding rearrangements and to increase protein stability with a concomitant decrease in the inhibitory activity (Seo et al. 2002; Askew et al. 2007a). According to this evidence, while one duplicate has maintained its original function (*SERPINA1*), the other copy has undergone an accelerated divergence possibly associated with the emergence of novel advantageous function in primate evolution (*SERPINA2*) (Marques et al. 2013). Similarly, other examples of gene duplication and divergence occurred in rodents, after the mouse–rat split approximately 25 MYA; in the mouse genome, *Serpina1* repertoire has expanded into six members (a–f; Fig. 1.1) with diverse activities as indicated by the composition of their reactive sites (P1–P1′: Met-Ser, Tyr-Ser, Leu-Ser, or Ser-Thr).

Conversely, *SERPINA2* has also suffered two independent events of gene loss; one in the chimpanzee lineage caused by a 7.5 kb deletion, which removed the entire *SERPINA2* sequence, and another in humans caused by the polymorphic 2 kb deletion, as previously mentioned. Interestingly, a remarkable discrepancy in the 2 kb deletion frequency was observed between European and African populations (18 % vs. 58 %), where the finding of an unusual homogenous haplotype has raised the hypothesis of a selective advantage through *SERPINA2* pseudogenization (Seixas et al. 2007). Here, the adaptive value of *SERPINA2* divergence and pseudogenization in primate evolution may sound contradictory, yet such events of gene gains and losses are common within gene families and often correlated with significant differences in species immunity and reproductive biology (Puente et al. 2003, 2005).

1.2.2 *SERPINA13* Pseudogenization

SERPINA13 is another example of gene loss in the 14q32.1 *SERPIN* cluster (Fig. 1.1). In this case, *SERPINA13* pseudogenization occurred through the insertion of a premature stop in the third coding exon, which is already fixated in modern humans and also present in Neanderthal and Denisova genomes but absent in nonhuman primates. Indeed, the lack of additional inactivating mutations in the closest primate species suggests *SERPINA13* as a possible functional gene silenced in the human lineage after the split from chimpanzee 6 MYA (Hedges et al. 2006). *SERPINA13* has been reported as a sequence related to *SERPINA4* (Namciu et al. 2004); however, the low level of protein identity (33 %) presupposes that a long time has elapsed since their origin in a common ancestral sequence. Accordingly, the RCL has diverged from all other members of the cluster indicating that active *SERPINA13* will encode a molecule with an unknown activity (P1–P1′: Gly-Pro). It is worth to note that among annotated mammalian genomes, only primates seem to have kept *SERPINA13* orthologs (Fig. 1.1), which may imply the conservation of an ancient function with some relevance to those lineages.

1.3 *SERPINB* Cluster at Chromosome 18q21.3

The clade B cluster located at chromosome 18q21.3 spans over ~ 513 kb and contains 10 *SERPIN* members arranged in the following centromere–telomere orientation: maspin (*SERPINB5*), yukopin (*SERPINB12*), hurpin (*SERPINB12*), squamous cell carcinoma antigen-2 (*SERPINB4*), squamous cell carcinoma antigen-1 (*SERPINB3*), epipin (*SERPINB11*), megsin (*SERPINB7*), plasminogen activator inhibitor-2 (*SERPINB2*), bomapin (*SERPINB10*), and protease inhibitor 8 (*SERPINB8*) (Fig. 1.1). The typical structure of clade B genes comprises seven or eight exons, in which the eight-exon genes encode an additional protein structure of variable length—the CD loop (Silverman et al. 2004). This structure found among SERPINB12, B13, B7, B4, B3, B11, B7, B2, and B10 does not affect protein folding, and it may inclusively contribute to protein function as it happens with SERPINB2, which is able to crosslink with fibrin and other proteins (Silverman et al. 2004; Benarafa and Remold-O'Donnell 2005). The acquisition of the CD loop occurred by the insertion of extra genetic material into the third exon of an ancestral gene, dating backwards into a tetrapod lineage (450 MYA) from which all eight-exon genes originated by serial events of duplication (Benarafa and Remold-O'Donnell 2005). Most of clade B genes located in 18q21.3 cluster are also widespread across mammalian genomes in ortholog regions, and they tend to show low rates of identity at the protein level (average 44 %). Here, the exception to the rule are *SERPINB3* and *SERPINB4*, which result from a recent duplication originating approximately 30 MYA after the split from Old Word monkeys and which still preserve 92 % and 95 % identity at protein and gene level, respectively (Gomes et al. 2014).

1.3.1 *SERPINB3* and *SERPINB4* Duplicates

Originally described as important cancer biomarkers, SERPINB3 and B4 were found to be upregulated in several squamous epithelial carcinomas (e.g., the uterine cervix, lung, esophagus, and head and neck) and in diverse inflammatory conditions (e.g., pneumonia, psoriasis, scleroderma), but their role outside of pathological circumstances is far from being fully understood, and like other members of clade B, SERPINB3 and B4 are thought to be implicated in the control of lysosomal enzymes and apoptosis (Silverman et al. 2004; Vidalino et al. 2009). Remarkably, these molecules share only 54 % identity at the RCL and present quite different protease affinities. SERPINB3 has two serine residues at the reactive site, and it is able to regulate the activity of cysteine proteases such as the cathepsins L1, L2, K, and S (CTSL1, CTSL2, CTSK, and CTSS). On the other hand, SERPINB4 has leucine and serine residues at the reactive site, and it is a potent inhibitor of the serine proteases like cathepsin G (CTSG) and mast cell chymase (CMA1) and a poor inhibitor of CTSS (Schick et al. 1997, 1998a, b).

The recent analysis of nonhuman primate sequences has demonstrated that SERPINB3 inhibitory activity has arisen only approximately 9 MYA in a common ancestor to humans, chimpanzees, and gorillas, whereas SERPINB4 activity is likely to have been maintained throughout primate evolution merely alternating the residue allocated to P1 position of the RCL as either leucine or threonine (Gomes et al. 2014). Furthermore, the analysis of the orangutan and gibbon genomes indicates that these species have accumulated in independent ways several disruptive mutations leading in both cases to the inactivation of one of the duplicates (Fig. 1.1) (Gomes et al. 2014). Altogether, these findings support contrasting evolutionary fates of *SERPINB4* duplication, while in some primate lineages one of the copies has undergone pseudogenization, in others it appear to have experienced an accelerated divergence to accommodate a novel function as *SERPINB3*. Indeed, this assumption is supported by phylogenetic studies, which confirmed a general conservation of SERPINB4/B3 structures with several residues at critical positions evolving under positive selection. Among these seven are located at the RCL and all of them differ between SERPINB4 and B3 molecules (Val351Gly, Val352Phe, Glu353Gly, Leu354Ser, Ser356Pro, Pro357Thr, and Cys364His). Moreover, the amino acid replacements Val351Gly and Val352Phe located at positions P3 and P2 of the RCL, respectively, were also pinpointed as major drivers of *SERPINB3* divergence in the human, chimpanzee, and gorilla clade (Gomes et al. 2014). Accordingly, key position replacements in the RCL of SERPINB4 including several targeted by natural selection (P2, P2′, P3′, and P10′) were shown by biochemical studies to affect significantly the affinity toward CTSS and to turn the molecule more alike to SERPINB3 (Luke et al. 2000). A second cluster of positively selected residues was located at the opposite pole of the molecule with potential implications in the RCL insertion and the stability of protease–inhibitor complexes (Gomes et al. 2014).

Conversely, studies done in target proteases have shown that CTSL2, which is regulated by SERPINB3 activity, has several residues favored by natural selection during primate evolution (Gomes et al. 2014), while CTSG and CMA1, both inhibited by SERPINB4, were identified as fast-evolving molecules in primates and other mammals (Gomes et al. 2014; da Fonseca et al. 2010). Furthermore, in the human, chimpanzee, and gorilla clade, CTSS was the single protease identified as a potential target of natural selection, a protease that is more efficiently inhibited by SERPINB3 than by B4, and thus it might be more closely correlated with the origin of SERPINB3. Importantly, many of the protease residues identified as positively selected are located near the protease catalytic pocket and, as predicted by tridimensional models of protease–inhibitor complexes, are placed in the inhibitor/protease interface in close proximity to the fast-evolving residues of the RCL (Gomes et al. 2014). These findings give support to a divergence of *SERPINB3* and *B4* possibly correlated with an improved inhibitory repertoire, in part driven by the adaptive forces acting in their target proteases. In this context, the fine balance between pathogens permanently changing in both time and space and the host immune response has already been hypothesized to have affected *CMA1* and *CTSG* evolution in mammals (da Fonseca et al. 2010). However, in the particular case of *SERPINB3* evolution, a direct role of exogenous proteases cannot be ruled out since this molecule has been implicated in the inhibition of diverse cysteine proteases released by different infectious agents (Kanaji et al. 2007; Kantyka et al. 2011). A direct activity in innate immunity is also plausible given that SERPINB3 has been localized on the surface of B-lymphocytes (Vidalino et al. 2012), and it is co-expressed with SERPINB4 in squamous epithelium of mucous membranes, skin, and the respiratory system, where they may act as primary defense mechanisms by preventing pathogens to cross epithelial barriers. Furthermore, the duplication event has also encompassed the upstream promoter region, and thus both *SERPINB3* and *B4* still share several transcription factors (e.g., STAT3) enrolled in modulation of cytokine receptors, inflammatory response, and immune system (Gomes et al. 2014; Ahmed and Darnell 2009).

Interestingly, in mammalian genomes, the number of *SERPINB4* homologs appears to be variable and correlated with several events of gene duplication and loss (Fig. 1.1). For example, the mouse genome has four gene copies and probably several pseudogenes (Heit et al. 2013); the rat has only two copies, as the cow, which is likely to include several pseudogene sequences as well; the sheep has three copies; and the dog has a single copy annotated in the genome. The analysis of residues located at the RCL (P1–P1′) shows that leucine–serine or threonine–serine is the most prevalent composition; nevertheless several other configurations are allowed which seems to imply a large functional divergence of these genes between and within mammalian species (Askew et al. 2004). Even so, the serine–serine bond is only observed in humans, chimpanzees, and gorilla, thus confirming the novelty of SERPINB3 activity.

1.3.2 *SERPINB11* Resurrection

SERPINB11 is a member of the 18q21.3 cluster common to different annotated mammalian genomes, which has probably arisen from an ancient duplication together with *SERPINB7* (protein identity 43 %) (Benarafa and Remold-O'Donnell 2005; Heit et al. 2013). In humans, *SERPINB11* has two major isoforms, one corresponding to a full-length transcript and coding for a regular SERPIN, and another carrying a premature stop codon at position 90, which results in a nonfunctional variant or pseudogene (Askew et al. 2007a). In addition, biochemical assays in which the RCL was swapped between human and mouse orthologs showed that SERPINB11 lost the ability to inhibit trypsin-like proteases due to the accumulation of a series of amino acid replacements outside of the RCL region (Askew et al. 2007a).

The comparison of *SERPINB11* sequences across multiple primate species disclosed an initial process of *SERPINB11* pseudogenization followed later by its gene resurrection, as a more parsimonious hypothesis. While all nonhuman primates have a leucine residue at position 90 generally encoded by a TTA codon, humans carry either a TAA termination codon or a GAA glutamic acid codon, pointing out a likely TTA-TAA-GAA mutational pathway (Seixas et al. 2012). In addition, the discovery of a TAA codon in both Neanderthal and Denisova genomes strengthens the assumption of an early loss of function that occurred some time before their divergence from modern humans, approximately 600,000–800,000 years ago (Reich et al. 2011; Green et al. 2008).

The loss of *SERPINB11* contrasts with the strong evolutionary constrains registered in the primate phylogeny and the high evolutionary rates observed in the human linage associated to the active *SERPINB11*. There, several residues under selection were detected including two fixed positions (Val194Glu and Thr253Ile) and five polymorphic sites (Leu90Glu, Thr148Met, Ala181Thr, Trp188Arg, and Ser303Pro). According to three-dimensional models, all these residues gained in the course of human evolution were predicted to interfere directly or indirectly with SERPIN functional domains (breach, shutter, and gate regions) possibly influencing the formation of structural rearrangements and SERPINB11 shifting of into a non-inhibitory molecule (Seixas et al. 2012; Askew et al. 2007a). Interestingly, the active *SERPINB11* (Glu90, Thr181, Arg188, Pro303) was also found to be associated with hallmarks of selection in Africa because it could define a common long-range haplotype even in the presence of a recombination hotspot. However, the worldwide distribution of the active haplotype and the decoupling of age estimates obtained for the amino acids substitutions alone (>240,000 years) and for the entire haplotype (16,500 years) suggested that the active *SERPINB11* isoform might have become beneficial, only recently under a novel set of environmental conditions. The findings of a slight overrepresentation of the active *SERPINB11* in geographic areas with a pastoral mode of subsistence, non-humid temperate climate, and greater diversity of pathogens appear to support such hypothesis (Seixas et al. 2012). Indeed, the major ecological changes that took place during the Holocene including the end of the glacial period, the onset of agriculture and animal husbandry, and the social modification from small groups of hunter-gatherers to more densely settled communities are known to have had a great impact on the pathogen burden in human populations and on a faster dispersal of infectious agents, leading to more severe outbreaks of diseases. Hence, the emergence of an active *SERPINB11* could have been triggered by major environmental changes initiated about 12,000 years ago and by a selective advantage related to host–pathogen interactions (Seixas et al. 2012). Although this hypothesis has been built based strictly on genetic data, the recent evidence from a chicken homolog lacking inhibitory activity and also displaying an exposed CD loop with strong antimicrobial potential may argue in favor of a role of SERPINB11 in innate immunity and host defense (Rehault-Godbert et al. 2013).

1.4 *SERPIN* Repertoire in Human Health and Disease

As previously described, the number of active *SERPIN* genes in 14q32.1 and 18q21.3 clusters may vary within human populations and between humans and other primate or mammalian species. In addition and aside from the copy number variation, serpins may as well exert different activities across diverse phylogenetic groups, in result of their functional divergence and organism's adaption to different environments. Even though these events of gain and loss of function are expected to have contributed to the overall fitness of individuals, nowadays these may have repercussions in the patterns of human disease and implications in the usage of animal models to access SERPIN function and dysfunction. Another central issue about copy number variation within gene families is the possible existence of some level of redundancy, which in the case of serpins may be correlated with their ability to inhibit a myriad of different proteases although with variable level of efficiency.

AATD, a Mendelian disorder affecting *SERPINA1* gene located at 14q32.1 cluster, has more severe effects when linked to gain-of-function mutations, which increase significantly the ability of protein to aggregate and cause cellular stress, than perhaps those associated only to loss-of-function mutations and reduced SERPINA1 serum levels. For instance, subjects homozygous for the Glu342Lys allele may in a low percentage of cases exhibit severe hepatic disease in childhood or, as it happens more frequently, remain as nearly asymptomatic until later in adulthood when they start to have respiratory complaints (ATS/ERS 2003). Conversely, AATD patients lacking SERPINA1 (null homozygous) do not show any other phenotype beside the lung disease in fourth decade of life (Vaz Rodrigues et al. 2012; Fregonese et al. 2008). The late onset of lung disease and the relevance of patients' smoking history in its development and progression place AATD as a health problem possibly correlated with a recent increase in life expectancy. In accordance, animal models of the human disease were capable of reproducing liver disease phenotypes through transgenic mouse expressing the Glu342Lys misfolded allele, but fail in mimicking emphysema, probably due to the multiple copies of *Serpina1* (Askew et al. 2007b).

Conversely, several SERPINs seem to have roles in the inflammatory response, which is an important underlying factor in complex diseases such as cancer or autoimmunity disorders. For example, SERPINA1 has been shown to be an anti-inflammatory and immunomodulatory molecule capable of lowering cytokines and chemokines released by macrophages and of inducing changes in the expression of important immune markers in pancreatic islet cells (Jonigk et al. 2013). In this context, SERPINA1 was also found to protect pancreatic beta-cells from the immunological attack in type 1 diabetes; to drastically reduce adipose inflammation, insulin resistance, and body weight gain in overfed mice; and to be a possible therapeutic strategy to increase insulin secretion in type 2 diabetes patients (Kalis et al. 2010; Mansuy-Aubert et al. 2013). Moreover, in systemic lupus erythematosus, the loss of SERPINB3 expression on the surface of B-lymphocytes was proposed to impair B-cell maturation (Vidalino et al. 2012), whereas in psoriasis, the principal antigen underlying immune reactions was confirmed to be an end product of SERPINB3 enzymatic cleavage by mast cell chymase (Lysvand et al. 2014). Worth to note, some serpins were also found to accumulate functions as antimicrobial and antiviral molecules when cleaved by nontarget proteases (Malmstrom et al. 2009; Munch et al. 2007; Rehault-Godbert et al. 2013; Pellegrini et al. 2004). Despite not exhaustively reviewed, these examples seem to indicate clade A and clade B SERPINs as pleiotropic molecules that may be implicated in many different activities with possible effects in human health. Even though SERPIN losses do appear to cause strong deleterious effects in humans, especially before reaching reproductive age, this does not exclude a possible contribution of the most recent evolutionary events of serpin gains and losses in the susceptibility to complex disease in particular when taking into account the new set of environmental conditions of the western societies.

1.5 Concluding Remarks

The human SERPIN repertoire results from the combination of old functions shared across different mammalian species, which result from early events of gene duplication in the evolution of vertebrates and novel functions restricted to some phylogenetic groups, probably associated to adaptive responses not fulfilled by other activities of standing members. Considering other examples of the evolution of gene families including those of proteases and the multiple SERPIN functions, it seems plausible a recent evolution driven by innate immunity with concomitant effects in current patterns of human health and disease.

References

Ahmed ST, Darnell JE Jr (2009) Serpin B3/B4, activated by STAT3, promote survival of squamous carcinoma cells. Biochem Biophys Res Commun 378(4):821–825. doi:10.1016/j.bbrc.2008.11.147

Askew DJ, Askew YS, Kato Y, Turner RF, Dewar K, Lehoczky J, Silverman GA (2004) Comparative genomic analysis of the clade B serpin cluster at human chromosome 18q21: amplification within the mouse squamous cell carcinoma antigen gene locus. Genomics 84 (1):176–184. doi:10.1016/j.ygeno.2004.01.015, S0888754304000412 [pii]

Askew DJ, Cataltepe S, Kumar V, Edwards C, Pace SM, Howarth RN, Pak SC, Askew YS, Bromme D, Luke CJ, Whisstock JC, Silverman GA (2007a) SERPINB11 is a new noninhibitory intracellular serpin. Common single nucleotide polymorphisms in the scaffold impair conformational change. J Biol Chem 282(34):24948–24960. doi:10.1074/jbc.M703182200, M703182200 [pii]

Askew DJ, Coughlin P, Bird PI (2007b) Mouse serpins and transgenic studies. In: Silverman GA, Lomas DA (eds) Molecular and cellular aspects of the serpinopathies and disorders in serpin activity. World Scientific Publishing, Singapore, pp 101–129

Atchley WR, Lokot T, Wollenberg K, Dress A, Ragg H (2001) Phylogenetic analyses of amino acid variation in the serpin proteins. Mol Biol Evol 18(8):1502–1511

ATS/ERS (2003) American Thoracic Society/European Respiratory Society statement: standards for the diagnosis and management of individuals with alpha-1 antitrypsin deficiency. Am J Respir Crit Care Med 168(7):818–900. doi:10.1164/rccm.168.7.818, 168/7/818 [pii]

Bao JJ, Reed-Fourquet L, Sifers RN, Kidd VJ, Woo SL (1988) Molecular structure and sequence homology of a gene related to alpha 1-antitrypsin in the human genome. Genomics 2 (2):165–173, 0888-7543(88)90099-7 [pii]

Benarafa C, Remold-O'Donnell E (2005) The ovalbumin serpins revisited: perspective from the chicken genome of clade B serpin evolution in vertebrates. Proc Natl Acad Sci USA 102 (32):11367–11372. doi:10.1073/pnas.0502934102, 0502934102 [pii]

Christiansen HE, Schwarze U, Pyott SM, AlSwaid A, Al Balwi M, Alrasheed S, Pepin MG, Weis MA, Eyre DR, Byers PH (2010) Homozygosity for a missense mutation in SERPINH1, which encodes the collagen chaperone protein HSP47, results in severe recessive osteogenesis imperfecta. Am J Hum Genet 86(3):389–398. doi:10.1016/j.ajhg.2010.01.034, S0002-9297 (10)00077-7 [pii]

da Fonseca RR, Kosiol C, Vinar T, Siepel A, Nielsen R (2010) Positive selection on apoptosis related genes. FEBS Lett 584(3):469–476. doi:10.1016/j.febslet.2009.12.022

Fregonese L, Stolk J, Frants RR, Veldhuisen B (2008) Alpha-1 antitrypsin Null mutations and severity of emphysema. Respir Med 102(6):876–884. doi:10.1016/j.rmed.2008.01.009

Futamura A, Stratikos E, Olson ST, Gettins PG (1998) Change in environment of the P1 side chain upon progression from the Michaelis complex to the covalent serpin-proteinase complex. Biochemistry 37(38):13110–13119. doi:10.1021/bi981234m, bi981234m [pii]

Gomes S, Marques PI, Matthiesen R, Seixas S (2014) Adaptive evolution and divergence of SERPINB3: a young duplicate in great Apes. PLoS One 9(8), e104935. doi:10.1371/journal.pone.0104935

Gooptu B, Dickens JA, Lomas DA (2014) The molecular and cellular pathology of alpha(1)-antitrypsin deficiency. Trends Mol Med 20(2):116–127. doi:10.1016/j.molmed.2013.10.007

Green RE, Malaspinas AS, Krause J, Briggs AW, Johnson PL, Uhler C, Meyer M, Good JM, Maricic T, Stenzel U, Prufer K, Siebauer M, Burbano HA, Ronan M, Rothberg JM, Egholm M, Rudan P, Brajkovic D, Kucan Z, Gusic I, Wikstrom M, Laakkonen L, Kelso J, Slatkin M, Paabo S (2008) A complete Neandertal mitochondrial genome sequence determined by high-throughput sequencing. Cell 134(3):416–426. doi:10.1016/j.cell.2008.06.021, S0092-8674(08) 00773-3 [pii]

Hedges SB, Dudley J, Kumar S (2006) TimeTree: a public knowledge-base of divergence times among organisms. Bioinformatics 22(23):2971–2972. doi:10.1093/bioinformatics/btl505, btl505 [pii]

Heit C, Jackson BC, McAndrews M, Wright MW, Thompson DC, Silverman GA, Nebert DW, Vasiliou V (2013) Update of the human and mouse SERPIN gene superfamily. Hum Genomics 7:22. doi:10.1186/1479-7364-7-22

Hofker MH, Nelen M, Klasen EC, Nukiwa T, Curiel D, Crystal RG, Frants RR (1988) Cloning and characterization of an alpha 1-antitrypsin like gene 12 KB downstream of the genuine alpha 1-antitrypsin gene. Biochem Biophys Res Commun 155(2):634–642

Irving JA, Pike RN, Lesk AM, Whisstock JC (2000) Phylogeny of the serpin superfamily: implications of patterns of amino acid conservation for structure and function. Genome Res 10(12):1845–1864

Irving JA, Cabrita LD, Kaiserman D, Worrall MM, Whisstock JC (2007) Evolution and classification of the serpin superfamily. In: Silverman GA, Lomas DA (eds) Molecular and cellular aspects of the serpinopathies and disorders in serpin activity. World Scientific Publishing, Singapore, pp 1–33

Janciauskiene SM, Bals R, Koczulla R, Vogelmeier C, Kohnlein T, Welte T (2011) The discovery of alpha1-antitrypsin and its role in health and disease. Respir Med 105(8):1129–1139. doi:10.1016/j.rmed.2011.02.002

Jonigk D, Al-Omari M, Maegel L, Muller M, Izykowski N, Hong J, Hong K, Kim SH, Dorsch M, Mahadeva R, Laenger F, Kreipe H, Braun A, Shahaf G, Lewis EC, Welte T, Dinarello CA, Janciauskiene S (2013) Anti-inflammatory and immunomodulatory properties of alpha1-antitrypsin without inhibition of elastase. Proc Natl Acad Sci USA 110(37):15007–15012. doi:10.1073/pnas.1309648110

Kalis M, Kumar R, Janciauskiene S, Salehi A, Cilio CM (2010) alpha 1-antitrypsin enhances insulin secretion and prevents cytokine-mediated apoptosis in pancreatic beta-cells. Islets 2 (3):185–189. doi:10.4161/isl.2.3.11654

Kanaji S, Tanaka Y, Sakata Y, Takeshita K, Arima K, Ohta S, Hansell EJ, Caffrey C, Mottram JC, Lowther J, Donnelly S, Stack C, Kadowaki T, Yamamoto K, McKerrow JH, Dalton JP, Coombs GH, Izuhara K (2007) Squamous cell carcinoma antigen 1 is an inhibitor of parasite-derived cysteine proteases. FEBS Lett 581(22):4260–4264. doi:10.1016/j.febslet.2007.07.072

Kantyka T, Plaza K, Koziel J, Florczyk D, Stennicke HR, Thogersen IB, Enghild JJ, Silverman GA, Pak SC, Potempa J (2011) Inhibition of Staphylococcus aureus cysteine proteases by human serpin potentially limits staphylococcal virulence. Biol Chem 392(5):483–489. doi:10.1515/BC.2011.044

Law RH, Zhang Q, McGowan S, Buckle AM, Silverman GA, Wong W, Rosado CJ, Langendorf CG, Pike RN, Bird PI, Whisstock JC (2006) An overview of the serpin superfamily. Genome Biol 7(5):216. doi:10.1186/gb-2006-7-5-216, gb-2006-7-5-216 [pii]

Luke C, Schick C, Tsu C, Whisstock JC, Irving JA, Bromme D, Juliano L, Shi GP, Chapman HA, Silverman GA (2000) Simple modifications of the serpin reactive site loop convert SCCA2 into a cysteine proteinase inhibitor: a critical role for the P3' proline in facilitating RSL cleavage. Biochemistry 39(24):7081–7091

Lysvand H, Hagen L, Klubicka L, Slupphaug G, Iversen OJ (2014) Psoriasis pathogenesis - Pso p27 is generated from SCCA1 with chymase. Biochim Biophys Acta 1842(5):734–738. doi:10.1016/j.bbadis.2014.02.005

Mala JG, Rose C (2010) Interactions of heat shock protein 47 with collagen and the stress response: an unconventional chaperone model? Life Sci 87(19-22):579–586. doi:10.1016/j.lfs.2010.09.024, S0024-3205(10)00412-1 [pii]

Malmstrom E, Morgelin M, Malmsten M, Johansson L, Norrby-Teglund A, Shannon O, Schmidtchen A, Meijers JC, Herwald H (2009) Protein C inhibitor--a novel antimicrobial agent. PLoS Pathog 5(12), e1000698. doi:10.1371/journal.ppat.1000698

Mansuy-Aubert V, Zhou QL, Xie X, Gong Z, Huang JY, Khan AR, Aubert G, Candelaria K, Thomas S, Shin DJ, Booth S, Baig SM, Bilal A, Hwang D, Zhang H, Lovell-Badge R, Smith SR, Awan FR, Jiang ZY (2013) Imbalance between neutrophil elastase and its inhibitor alpha1-antitrypsin in obesity alters insulin sensitivity, inflammation, and energy expenditure. Cell Metab 17(4):534–548. doi:10.1016/j.cmet.2013.03.005

Marques PI, Ferreira Z, Martins M, Figueiredo J, Silva DI, Castro P, Morales-Hojas R, Simoes-Correia J, Seixas S (2013) SERPINA2 is a novel gene with a divergent function from SERPINA1. PLoS One 8(6), e66889. doi:10.1371/journal.pone.0066889

Marsden MD, Fournier RE (2005) Organization and expression of the human serpin gene cluster at 14q32.1. Front Biosci 10:1768–1778, 1660 [pii]

Munch J, Standker L, Adermann K, Schulz A, Schindler M, Chinnadurai R, Pohlmann S, Chaipan C, Biet T, Peters T, Meyer B, Wilhelm D, Lu H, Jing W, Jiang S, Forssmann WG, Kirchhoff F (2007) Discovery and optimization of a natural HIV-1 entry inhibitor targeting the gp41 fusion peptide. Cell 129(2):263–275. doi:10.1016/j.cell.2007.02.042

Namciu SJ, Friedman RD, Marsden MD, Sarausad LM, Jasoni CL, Fournier RE (2004) Sequence organization and matrix attachment regions of the human serine protease inhibitor gene cluster at 14q32.1. Mamm Genome 15(3):162–178. doi:10.1007/s00335-003-2311-y

Pellegrini A, Hulsmeier AJ, Hunziker P, Thomas U (2004) Proteolytic fragments of ovalbumin display antimicrobial activity. Biochim Biophys Acta 1672(2):76–85. doi:10.1016/j.bbagen.2004.02.010, S0304416504000443 [pii]

Puente XS, Sanchez LM, Overall CM, Lopez-Otin C (2003) Human and mouse proteases: a comparative genomic approach. Nat Rev Genet 4(7):544–558. doi:10.1038/nrg1111, nrg1111 [pii]

Puente XS, Sanchez LM, Gutierrez-Fernandez A, Velasco G, Lopez-Otin C (2005) A genomic view of the complexity of mammalian proteolytic systems. Biochem Soc Trans 33 (Pt 2):331–334. doi:10.1042/BST0330331, BST0330331 [pii]

Rehault-Godbert S, Labas V, Helloin E, Herve-Grepinet V, Slugocki C, Berges M, Bourin MC, Brionne A, Poirier JC, Gautron J, Coste F, Nys Y (2013) Ovalbumin-related protein X is a heparin-binding ov-serpin exhibiting antimicrobial activities. J Biol Chem 288 (24):17285–17295. doi:10.1074/jbc.M113.469759

Reich D, Patterson N, Kircher M, Delfin F, Nandineni MR, Pugach I, Ko AM, Ko YC, Jinam TA, Phipps ME, Saitou N, Wollstein A, Kayser M, Paabo S, Stoneking M (2011) Denisova admixture and the first modern human dispersals into southeast Asia and oceania. Am J Hum Genet 89(4):516–528. doi:10.1016/j.ajhg.2011.09.005, S0002-9297(11)00395-8 [pii]

Schick C, Kamachi Y, Bartuski AJ, Cataltepe S, Schechter NM, Pemberton PA, Silverman GA (1997) Squamous cell carcinoma antigen 2 is a novel serpin that inhibits the chymotrypsin-like proteinases cathepsin G and mast cell chymase. J Biol Chem 272(3):1849–1855
Schick C, Bromme D, Bartuski AJ, Uemura Y, Schechter NM, Silverman GA (1998a) The reactive site loop of the serpin SCCA1 is essential for cysteine proteinase inhibition. Proc Natl Acad Sci USA 95(23):13465–13470
Schick C, Pemberton PA, Shi GP, Kamachi Y, Cataltepe S, Bartuski AJ, Gornstein ER, Bromme D, Chapman HA, Silverman GA (1998b) Cross-class inhibition of the cysteine proteinases cathepsins K, L, and S by the serpin squamous cell carcinoma antigen 1: a kinetic analysis. Biochemistry 37(15):5258–5266. doi:10.1021/bi972521d
Seixas S, Suriano G, Carvalho F, Seruca R, Rocha J, Di Rienzo A (2007) Sequence diversity at the proximal 14q32.1 SERPIN subcluster: evidence for natural selection favoring the pseudogenization of SERPINA2. Mol Biol Evol 24(2):587–598. doi:10.1093/molbev/msl187, msl187 [pii]
Seixas S, Ivanova N, Ferreira Z, Rocha J, Victor BL (2012) Loss and gain of function in human SERPINB11: an example of a gene under selection on standing variation, with implications for host-pathogen interactions. PLoS One 7, e32518
Seo EJ, Lee C, Yu MH (2002) Concerted regulation of inhibitory activity of alpha 1-antitrypsin by the native strain distributed throughout the molecule. J Biol Chem 277(16):14216–14220. doi:10.1074/jbc.M110272200, M110272200 [pii]
Silverman GA, Whisstock JC, Askew DJ, Pak SC, Luke CJ, Cataltepe S, Irving JA, Bird PI (2004) Human clade B serpins (ov-serpins) belong to a cohort of evolutionarily dispersed intracellular proteinase inhibitor clades that protect cells from promiscuous proteolysis. Cell Mol Life Sci 61(3):301–325. doi:10.1007/s00018-003-3240-3
Stein PE, Carrell RW (1995) What do dysfunctional serpins tell us about molecular mobility and disease? Nat Struct Biol 2(2):96–113
van Gent D, Sharp P, Morgan K, Kalsheker N (2003) Serpins: structure, function and molecular evolution. Int J Biochem Cell Biol 35(11):1536–1547, 10.1016/s1357-2725(03)00134-1
Vaz Rodrigues L, Costa F, Marques P, Mendonca C, Rocha J, Seixas S (2012) Severe alpha-1 antitrypsin deficiency caused by Q0(Ourem) allele: clinical features, haplotype characterization and history. Clin Genet 81(5):462–469. doi:10.1111/j.1399-0004.2011.01670.x
Vidalino L, Doria A, Quarta S, Zen M, Gatta A, Pontisso P (2009) SERPINB3, apoptosis and autoimmunity. Autoimmun Rev 9(2):108–112. doi:10.1016/j.autrev.2009.03.011
Vidalino L, Doria A, Quarta SM, Crescenzi M, Ruvoletto M, Frezzato F, Trentin L, Turato C, Parolin MC, Ghirardello A, Iaccarino L, Cavalletto L, Chemello L, Gatta A, Pontisso P (2012) SERPINB3 expression on B-cell surface in autoimmune diseases and hepatitis C virus-related chronic liver infection. Exp Biol Med 237(7):793–802. doi:10.1258/ebm.2012.012024

Serpins in Plants

2

Maja Cohen, Thomas H. Roberts, and Robert Fluhr

Contents

Abstract

The serpins are a family of structurally conserved protease inhibitors found mainly in the animal and plant kingdoms. This chapter will focus on the evolution of plant serpins brought to light by mining of the growing number of whole plant genomes. Our analysis shows remarkable diversification of plant serpins as compared to those in animals. Indeed, a degree of conservation between species is noted only at the level of comparison within plant orders, whereas in animals conservation of structure and function is noted at the phylum level. Within plants one serpin, containing the amino acids Leu-Arg in the P2–P1 position of the reactive center loop, is exceptional in that it is conserved in all

M. Cohen • R. Fluhr (✉)
Department of Plant and Environmental Sciences, Weizmann Institute of Science, Rehovot, Israel
e-mail: robert.fluhr@weizmann.ac.il

T.H. Roberts
Department of Plant and Food Sciences, University of Sydney, NSW 2006, Australia

M. Geiger et al. (eds.), *The Serpin Family*, DOI 10.1007/978-3-319-22711-5_2

genomes examined. The recently discovered function in programmed cell death is described for this serpin. We will further review the knowledge recently acquired on other plant serpins and their possible functions in cellular processes.

2.1 Introduction: Plant Serpins, an Evolutionary Perspective

The comparative analysis of serpin evolution throughout plant and animal phyla, including close unicellular relatives, offers insights into their structure and function. Almost all animal serpins are known to inhibit serine proteases, but the best-studied plant serpin targets cysteine proteases. Serpins of the animal phyla Chordata and Arthropoda show remarkable islands of conservation, an observation that facilitates their comparative study. The conservation among gene families of plant serpins is much less evident except for one group of serpins that contains an LR amino acid motif in the critical P2–P1 region of the reactive center loop (RCL). This chapter will highlight insights from new genomic data brought to light by the recent whole genome sequencing of scores of plant species and will describe the functionality of the most highly conserved and ubiquitous representative, AtSerpin1, from the model plant *Arabidopsis thaliana*.

2.2 Rapid Evolution of Plant Serpins as Compared to Animal Serpins

Genome families of plant serpins tend to be small, with an average 5–15 members. This small but highly variable number may indicate specialized conserved functions. The last review of plant serpins featuring integrative genomics was completed in 2007 (Roberts and Hejgaard 2008) and was based mainly on information from expressed sequence tags (ESTs) and cDNA libraries. The recent addition of numerous whole genomes made it worthwhile to reexamine the plant serpins for new insights. Complete genomes, as compared to EST and cDNA data, enabled us to emphasize evolutionary conservation irrespective of gene expression levels. Below we analyzed the complete genomes of 31 higher plant species (Fig. 2.1).

In Fig. 2.1a genomes of different species are organized by the orders to which they belong and for clarity are denoted by the same color. Clearly at this evolutionary level of diversification, serpins tend to form order-specific clusters. This clustering implies that the serpins have rapidly diversified to serve functions that are specific to the needs of each species. For example, the divergence between monocots (which include the grasses) and eudicots occurred about 200 million years (Myr) ago (Wolfe et al. 1989). The lack of obvious homology between serpins from monocots (Poales in pink) and eudicots (all other colors) observed in the evolutionary tree (Fig. 2.1a) suggests that serpins acquired their main diversity after this speciation. Consequently, plant serpins are specifically adapted to the needs of

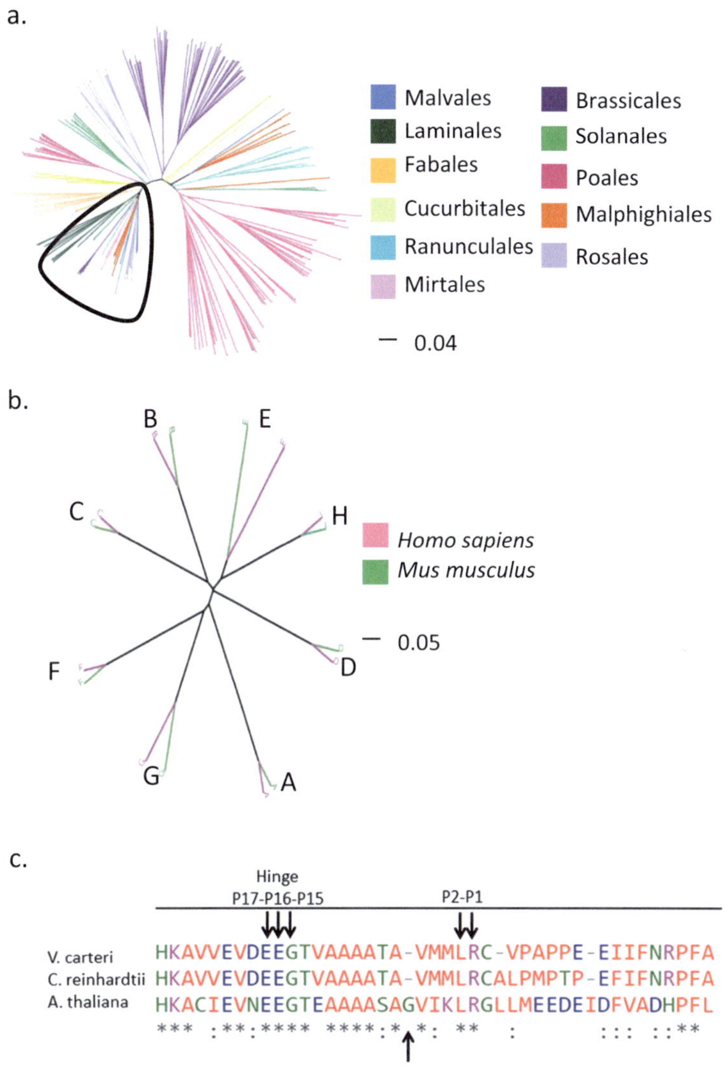

Fig. 2.1 Phylogenetic relationships among plant and animal serpins. (**a**) Evolutionary tree of the serpins from the plant kingdom. The tree is based on a multiple sequence alignment using ClustalW2. Sequences within the same length range (400–450 amino acid residues) were selected in order to avoid intron-exon prediction issues. The longer the branch, the more distant are the proteins from each other. Scale is in amino acid substitutions. Branches are *colored* by their specific order as shown in the tree legend. The following species were included (grouped according to order): Poales, *Oryza sativa*, *Zea mays*, *Sorghum bicolor*, *Hordeum vulgare*, *Setaria italica*, and *Brachypodium distachyon*. Rosales: *Malus domestica*, *Prunus persica*, and *Fragaria vesca*. Myrtales: *Eucalyptus grandis*. Ranunculales: *Aquilegia caerulea*. Cucurbitales: *Cucumis sativus*. Malpighiales: *Populus trichocarpa*, *Linum usitatissimum*, *Manihot esculenta*, and *Ricinus communis*. Fabales: *Glycine max*, *Medicago truncatula*, and *Phaseolus vulgaris*. Brassicales: *Arabidopsis thaliana*, *Arabidopsis lyrata*, *Brassica rapa*, *Capsella rubella*, and *Thellungiella halophila*. Laminales: *Mimulus guttatus*. Solanales: *Solanum lycopersicum* and *Solanum tuberosum*. Sapindales: *Citrus sinensis* and *Citrus clementina*. Malvales: *Theobroma cacao* and

the organism, and their function might not necessarily be related to basic plant machinery.

Unlike plant serpins, which show species-specific radiation at the level of order, animal serpins display conserved serpin identity at the phylum, the ancient level above the order. This is exemplified by a phylogenetic comparison of 16 serpins from two *Chordata* species, human and mouse (Fig. 2.1b). Each animal serpin type has been classified previously (serpin clades A to H; Irving et al. 2000). They form distinct evolutionary units that are well conserved between human and mouse, a pattern that likely attests to conservation of function throughout Chordata evolution. Remarkably, there are almost twice as many serpins in the mouse genome than in the human genome (60 and 36, respectively) (Heit et al. 2013). This difference might have arisen from independent gene duplications in order to fit the need to regulate the larger number of proteinase present in the mouse genome (Kaiserman et al. 2002). Therefore, some mouse serpins are encoded by gene paralogs that have human orthologs. It is tempting to hypothesize that several serpin orthologs were present before the speciation of the Chordata phylum some 1000 Myr ago (Wang et al. 1999). Thus, serpin functions in these taxons are linked to traits common to all the Chordata species. For example, all antithrombins (serpin clade C) from each of the terrestrial vertebrate groups share a similar heparin target and anticoagulant properties (Jordan 1983). In contrast to these ancient animal lineages, the differentiation of plant monocots and eudicots and within eudicot species occurred more recently, some 200 Myr ago (Wolfe et al. 1989). Yet, compared to animals, they have rapidly diversified in their serpin sequences. This may indicate that they have adopted different functions or, if they are inhibitory-type serpins, that they have rapidly adapted to new target proteases. Thus, plants and animals have taken very different strategies in pacing the evolutionary development of serpins. The common phenomenon of polyploidization in plants and the abundance of transposable elements in their genomes may explain the rapidity of serpin evolution.

Fig. 2.1 (continued) Gossypium raimondii. (**b**) Evolutionary tree of serpins from the Chordata. Serpin sequences from mouse and human belonging to clades (A to H; Irving et al. 2000) were selected and aligned using ClustalW. The *colors* represent the species as noted in the tree legend. The leaf labels (A to H) are for the serpin clades. Scale is in amino acid substitutions. (**c**) Multiple sequence alignment of the RCL regions of serpins from *Chlamydomonas reinhardtii* (Locus: XP_001695981), *Volvox carteri* (Locus: XP_002949428), and *Arabidopsis thaliana* LR serpin (AT1G47710). The *arrows* point to specific position with the sequences: P17–P15 (hinge region) and P2–P1 (Leu, Arg residues). All the trees were built based on the alignment files performed using ClustalW2 with default parameters. The phylogenetic trees were uploaded to the ITOL software that allowed their editing (Letunic and Bork 2011)

2.3 The LR Serpins Represent an Ancient Lineage

One group of plant serpins stands out in its conservation and contains representatives from all sequenced genomes (Fig. 2.1a, framed in black). This group has been noted before (Roberts and Hejgaard 2008) and is made up of the so-called LR serpins, i.e., the serpins that present a conserved Leu-Arg motif at the P2–P1 cleavage site. The ancient origin of this lineage can be established by examining *Chlamydomonas reinhardtii* and *Volvox carteri* as examples of single and multicellular volvocine-type green algae for which genome sequences are available. The last common ancestor diverged from the ancestors of flowering land plants at least 450 Myr ago and from each other about 200 Myr (Wolfe et al. 1989; Herron et al. 2009). The alignment of the RCL regions of the algae serpins and the *Arabidopsis* LR serpin is shown in Fig. 2.1c. The hinge region seems rather conserved with the P17–P15 positions similar to that found in *Arabidopsis* LR serpins. However, as reported earlier (Roberts and Hejgaard 2008), the critical P2–P1 site in the green algal serpins appears 16 residues distant from the start of the hinge region, falling one residue short of the canonical 17 aa length normally required for protease inhibition (Zhou et al. 1997). Shortening of the RCL seems to be less critical than adding residues: crmA, C1 inhibitor, and alpha-2-antiplasmin are examples of inhibitory serpins with an RCL length of 16 residues (Gettins 2002). Considering the fact that the LR serpins are conserved throughout the evolution of the plant kingdom, including in close relatives such as single-celled green algae, it is conceivable that the LR serpin is the ancestral progenitor of all serpins in plants and that its function has been retained as well as its conserved sequence. Functions of the *Arabidopsis* LR serpin, AtSerpin1, will be discussed below.

Apart from the LR serpins, the arrangement of most plant serpin clades in a species-specific manner strongly suggests that extant plant serpins evolved during plant speciation. It was therefore of interest to examine if and when serpin gene members exhibit conservation of sequence, which may also allude to conservation of function. To this end, sequences were compared for species within the Poales and Brassicales as these orders are represented by numerous species.

2.4 Conservation of Serpin Sequence Within Orders

The order Poales, for which complete genome sequence of six species is available, contains the grasses, including some of the important cereals that sustain humankind, including barley, maize, and rice. In all, 55 genes encoding full-length serpins are distributed between the six species. At this level of comparison, distinct groups that contain representatives of at least 5–6 species are evident (marked as clades 1–3 in Fig. 2.2a). Residues P17–P15 are normally highly conserved in the hinge region as they facilitate rapid insertion of the RCL into the β-sheet A (Huntington et al. 1997). The hinge of the serpins in group 1 of the Poales species presents an unusual amino acid composition; e.g., they lack a glutamic acid residue (E) at P17,

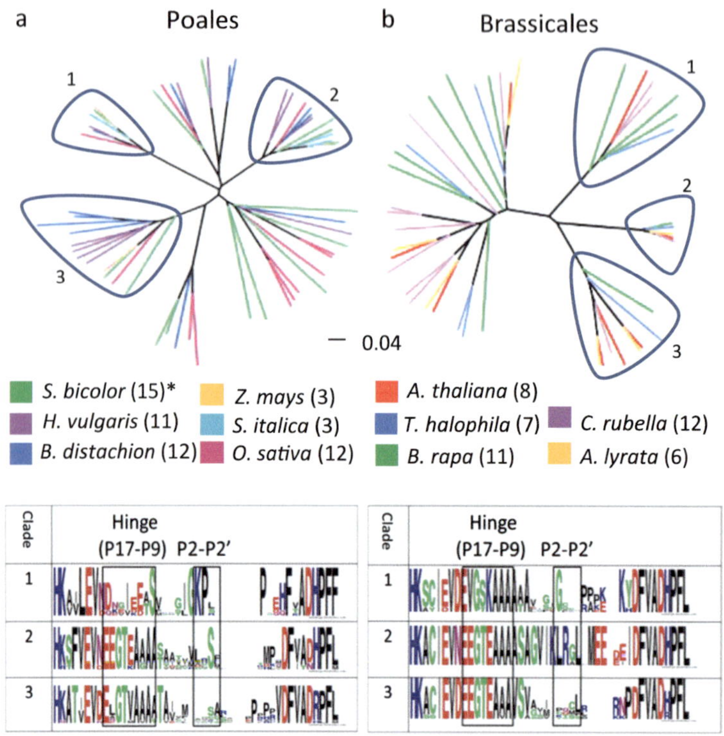

c

Locus	Protein name(s)	P4-P4′ and variant residues	No. accessions (of 91) with variant residues
At1g47710	AtSerpin1 (ArathZx)	IKLRGLLM	None
At1g62170	(ArathZ5)	SAYLGCAL T-------	6 (P4 = T)
At1g64030	AtSRP3 (ArathZ1)	DCGCSLDF YV-F-H-L -----?--	14 (P4 = Y); 1 (P3 = V); 3 (P1 = F); ? (P2′ = H); ? (P2′ = ?); ? (P4′ = L)
At2g14540	AtSRP2 (ArathZ2)	VVTGSCLW -ID-----	14 (P3 = I); 1 (P2 = D)
At2g25240	(ArathZ10)	VSCTSFRR	None
At2g26390	AtSRP4	MMPQCLMR	None
At2g35580		GIGCRRPP	None
At3g45220	(ArathZ4)	MTKDMLLM --NI----	4 (P2 = N) 4 (P1 = I)

Fig. 2.2 Phylogenetic analysis of serpins in plant orders and ecotypes. (**a**, **b**) Evolutionary tree and sequence logo of the serpins belonging to species in the orders Poales (*Oryza sativa*, *Zea mays*, *Sorghum bicolor*, *Hordeum vulgaris*, *Setaria italica*, and *Brachypodium distachyon*) and Brassicales (*Arabidopsis thaliana*, *Arabidopsis lyrata*, *Brassica rapa*, *Capsella rubella*, and *Thellungiella halophile*), respectively. Each *color* represents a different species as mentioned in the tree legends. The sequence logo represents the 40-amino acid sequence that surrounds the RCL. The hinge and the P2–P2′ regions are framed in the logo. The sequence logo was performed using the alignment data of serpins from the same order; therefore, the presence of gaps can be detected. Clades 1, 2, and 3 from both Poales and Brassicales are framed in the trees and have been selected due to their relatively long and significant branch length. (*Asterisk*) The numbers in

which would affect the ability of the serpin to inhibit proteases. Moreover, the residues Gly, Lys, and Pro are highly conserved around the P1–P1′ region (Fig. 2.2a, bottom). Those serpins may perform non-inhibitory functions and are conserved among the Poales species (Francis et al. 2012). The second clade includes the monocot LR serpins, which are in a group that also contains non-LR serpins and in which the hinge has the conserved sequence of inhibitory serpins. A serine residue at the P1′ position is highly conserved in all of the sequences of this clade and might be related to the target proteases that could also be conserved. The third group in Fig. 2.2a is characterized by the presence of unusual threonine residues appearing on either side of the hinge region and a valine residue in the C-terminus of the sequence. In contrast, the position of the cleavage P1–P1′ appears to be highly variable. Functions have yet to be associated with any of these monocot serpins.

The analysis of the order Brassicales also confirms clustering into clades at this level. The first group shows a hydrophobic Val instead of the conserved negatively charged Glu at the P16 position of the hinge. The P13 position is usually occupied by a Glu; however, this group presents a positively charged Lys instead. This serpin group has other noncanonical residues in the hinge region, and therefore these proteins might be candidates for non-inhibitory serpins. The second group consists of the LR serpins. The RCL region of these serpins is almost entirely conserved in the order Brassicales, suggesting that they are the targets of conserved proteases. The third group displays characteristics of inhibitory serpins, containing all the conserved residues in the hinge domain. However, the P1–P1′ amino acids are highly variable.

Fig. 2.2 (continued) parentheses represent the total number of sequences from the same species. (**c**) Amino acid variation in the reactive centers of *Arabidopsis* serpins. Loci for genes encoding full-length serpins in Columbia (Col-0) were examined at the 1001 Proteomes portal: http://1001proteomes.masc-proteomics.org (Cao et al. 2011). For each serpin, amino acid variations (non-synonymous single nucleotide polymorphisms, nsSNPs) for the P4–P4′ peptide region were assessed among the *Arabidopsis* accessions from MPI (Max Planck Institute for Developmental Biology; 83 accessions) and JGI (Joint Genome Institute; eight accessions) available on 27 November 2014. Loci are listed in chromosome order. Amino acid residues (one-letter abbreviations) are *colored* according to physicochemical properties: *black*, small; *green*, medium-sized and large hydrophobic; *pink*, polar; *red*, negatively charged; and *blue*, positively charged. The P1 residue is shown in *bold*. Residues given below the P4–P4′ peptide sequence are those recorded as different in at least one of the 91 MPI or JGI accessions (a *dash* indicates no residues were variant at that position). For sites within the P4–P4′ peptide that had more than one variant residue, the residues are displayed in an additional row. The number of accessions corresponding to each specific substitution is also given. A question mark indicates that data could not be mined due to multiple variant residues at single sites creating overlapping images. Note that the accession "proteomes" used at the 1001 Proteomes portal are based on SNP calls from various re-sequencing projects and analysis procedures and are completely reliant on the analysis and SNP calls by the original data providers. Consequently, all SNPs should be treated as putative until demonstrated experimentally through targeted sequence validation or mass spectrometry

2.5 Conservation of Genes at the Level of the Species: Analysis of Arabidopsis Ecotypes

Arabidopsis thaliana, or thale cress, is a small flowering plant of the mustard family (Brassicaceae or Cruciferae) related to agricultural plants such as cabbage, cauliflower, radish, and rapeseed. Its small genome, size, and easy manipulation in the laboratory have made it a model organism in plant biology. Its broad distribution in nature has resulted in a high degree of variation among ecotypes. These have been collected around the world as they exhibit remarkable morphological differences; yet they all belong to the same species. In this context, it was of interest to analyze the amino acid variation in the RCL region of 91 ecotypes of *Arabidopsis* (Fig. 2.2c). The eight-member gene family is conserved; however, a number of variations occur among the ecotypes. It is of interest that all the serpins that show no variation in their P4–P4′ peptide sequence within all the accessions are also the same ecotypes that belong to clades 1–3 shown in Fig. 2.2b. This suggests evolutionary pressure to conserve this sequence. Reciprocally, the serpins that show variation in their RCL in Fig. 2.2c are part of the serpin classes that are less distinct in their conservation in the order Brassicales (Fig. 2.2b). One exception is serpin encoded at locus At3g45240, which clusters with clade 3. Remarkably, in the serpin encoded by At3g45220, both P2 and P1 positions are variant. A change in such position would be predicted to be critical to protease specificity, which is discussed below.

Clearly, plant serpins have a rapid pace of evolution when compared to prominent animal lineages. The more rapid evolution has enabled the plant to adapt more specifically to its environmental conditions to maintain selective fitness. Whether the sequence diversity is due to acquiring novel serpin functions is not known and is the subject of future research.

2.6 Functions for Intracellular Serpins

The most conserved group of plant serpins belongs to the LR lineage. At least one gene encoding a serpin containing P2–P1 Leu-Arg appears in all plant species. The *Arabidopsis* AtSerpin1 LR gene is the most highly expressed serpin in *Arabidopsis* and appears in all organs. It is most closely related to vertebrate Clade B serpins. These serpins lack clearly detectable N-terminal signal peptides, reside within nucleo-cytosolic compartments, and are thought to safeguard cells against the lethal effects of uncontrolled proteolytic activity (Silverman et al. 2004; Mangan et al. 2008; Kaiserman and Bird 2005). In white blood cells, SERPINB1 is localized to the cytoplasm that surrounds lytic azurophilic granules. It is thought that SERPINB1 could neutralize the major proteases contained within the granules and prevent damage due to inadvertent release. Similarly, in cytotoxic lymphocytes, SERPINB9 was found to be associated with the cytoplasmic side of granules that contain granzyme B and could dampen activity of released proteases (Hirst et al. 2003). In an example of cytoplasmic-localized serpin expression from the non-vertebrate

Caenorhabditis elegans, serpin SRP-6 was found to function by protecting against stress-induced protease leakage from lysosomes (Luke et al. 2007). In nematodes proteases are released from lytic compartments due to osmotic shock. The release is nonlethal in wild-type nematodes, but nematodes lacking SRP-6 die, presumably due to uncontrolled lytic activity. In the light of these biological concepts, it is of interest to examine the biological role of the cytoplasmic AtSerpin1.

The X-ray crystal structure of AtSerpin1 (pdb: 3LE2) (Lampl et al. 2010) features three β-sheets (sA, sB, and sC), nine canonical α-helices (hA–I), and a protruding RCL. The loop joining s2B and s3B contains a plant-specific motif (YXXGXDXRXF) between Tyr225 and Phe234. This structure confers to the plant serpin "breach" region a more open conformation compared to other native serpin structures. The electrostatic potential map for AtSerpin1 shows a striking positively charged surface. Similar positive potentials under the RCL of mammalian C1 inhibitor (SERPING1) may act in a "charge sandwich" mechanism, whereby negative charged molecules are attracted to the positively charged surface. This in turn would attract positively charged proteases to the serpin (Beinrohr et al. 2007). Whether such cofactors play a role in AtSerpin1 interaction with proteases is unknown.

AtSerpin1 has been shown to form covalent complexes with the papain-like protease RESPONSIVE TO DESICCATION-21 (RD21) (Lampl et al. 2010). Importantly, the tetrahedral intermediate that forms between the protease and serpin reactive-center substrate is based on an active site cysteine. Fractionation of seedling extracts by nonreducing SDS-PAGE followed by immunoblotting with AtSerpin1-specific antibodies revealed the presence of an additional slower-migrating complex that was absent when leaves were treated with the specific cysteine protease inhibitor E-64 (*trans*-epoxysuccinyl-L-leucylamido (4-guanidino)butane). RD21 papain-like cysteine protease (PLCP; C1A family) is the major protease labeled with the E-64 derivative DCG-04 in wild-type extracts but not in extracts of mutant plants constitutively overexpressing AtSerpin1, indicating competition between the serpin and label. Direct immunoblotting with RD21-specific antibody revealed that the protease accumulated both as a free enzyme and in a complex with AtSerpin1. Both RD21 and AtSerpin1 knockout mutants lacked the serpin protease complex. The results establish that the major *Arabidopsis* plant serpin interacts with RD21 in vivo.

Within the larger PLCP family of about 30 members, the RD21 protease belongs to a small multigene family of which four members contain a unique granulin domain in the C-terminus (Beers et al. 2004). This domain is enriched in cysteine residues, creating a highly knotted structure. In animals, granulin domains are synthesized as part of multiple repeats and serve functions in gene regulation (Bateman and Bennett 2009). In plants, the granulin domain is restricted to juxtaposition with RD21 and other PLCPs and is of unknown function. However, as in animals, the granulin domain can be cleaved from the protein during protein processing. RD21 was originally characterized as being induced by dehydration and salt stress in *Arabidopsis* (Koizumi et al. 1993), tomato (Harrak et al. 2001), and senescing leaves of sweet potato (Chen et al. 2006), as well as in postharvest broccoli florets (Coupe et al. 2003). Thus, the juxtaposition of the granulin to a PLCP is also conserved in higher plants.

2.7 Concept of Vacuolar-Mediated Cell Death in Plants

The plant cell vacuole is filled with water and storage metabolites that maintain cellular turgor. Vacuoles contain a host of enzymes, particularly hydrolases, which play a major role in autophagy. The vacuole also serves to maintain cellular homeostasis by stabilizing cytoplasmic pH, making the vacuolar interior more acidic. The integrity of the membrane surrounding the vacuole (the tonoplast) ensures compartmentalization of potentially life-endangering hydrolytic enzymes, such as proteases. The PLCP protease RD21 is localized to the vacuole and is the major source of the vacuolar protease activity in *Arabidopsis* (Gu et al. 2012). A role for vacuolar-induced programmed cell death has been described (Hara-Nishimura and Hatsugai 2011). It is brought about by rapid collapse of tonoplast integrity. For example, the sudden release of vacuolar contents into the cytoplasm caused rapid cell death in the hypersensitive response to tobacco mosaic virus (TMV) in tobacco (Hatsugai et al. 2004) and application of toxins such as oxalic acid and fumonisin produced by necrotrophic pathogens (Wolpert et al. 2002). Similarly, such pathways for cellular clearance are likely to be utilized during developmental programmed cell death (e.g., vascular xylogenesis) as the removal of all cellular content by vacuolar proteases is important for complete xylem maturation.

2.8 Serpins and Vacuolar-Mediated Cell Death

If vacuolar integrity were compromised, the subsequent release of proteases such as RD21 would induce cell death. The common fungal toxin, oxalic acid, and a salicylic acid agonist, benzothiadiazole, are known to compromise the integrity of the tonoplast (Lampl et al. 2013). Changes in vacuole permeability occurred with concomitant release of vacuolar protease RD21, and formation of a covalent serpin-RD21 complex with AtSerpin1 that is present in the cytoplasm. Mutant plants lacking RD21, or plants with cytoplasmic-localized AtSerpin1 overexpression, exhibited significantly less elicitor-stimulated cell death than plants lacking AtSerpin1 (Lampl et al. 2013). Thus, survival and death were dictated by the ratio of protease to serpin (Fig. 2.3). Interestingly, necrotrophic phytopathogenic fungi secrete oxalic acid to kill cells and then flourish on dead tissue. Furthermore, consistent with a pro-death function for RD21 protease, fungal growth was accelerated in plants lacking AtSerpin1, but compromised in plants lacking RD21. Hence, a degree of commonalty exists between animal and plant execution of PCD (Fig. 2.3). Thus, the major cytoplasmic-localized serpin of *Arabidopsis*, AtSerpin1, controls the pro-death functions of compartmentalized proteases by determining a set point for their activity.

Fig. 2.3 Vacuolar-mediated cell death

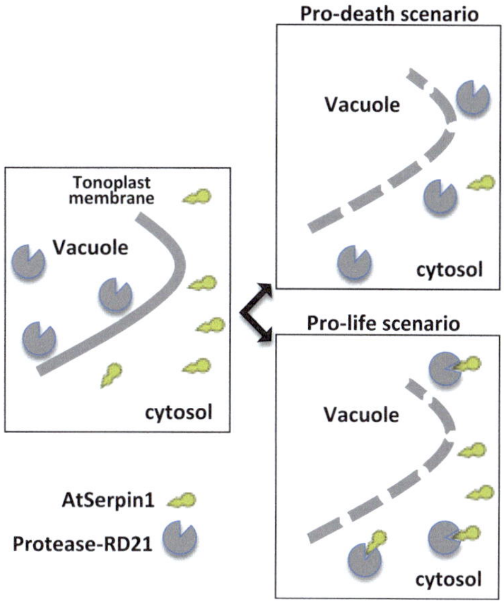

Curiously, AtSerpin1 has also been shown to interact with metacaspase 9 (AtMC9) (Vercammen et al. 2006). Metacaspases are cysteine proteases that are distantly related to animal caspases found in plants, fungi, and protozoa. AtMC9 is an Arg/Lys-specific cysteine-dependent protease (Vercammen et al. 2004). A yeast two-hybrid screen with inactive protease revealed an interaction with AtSerpin1, and this serpin was also a potent inhibitor of AtMC9 activity in vitro. The biological importance of this interaction remains to be elucidated.

The scenario shows the PLCP RD21 in the vacuole in the resting cell (left). During cell death two pathways may be followed. When the ratio of serpin to protease is sufficient, a pro-life scenario unfolds (bottom right), whereas when this ratio is insufficient to deactivate proteases, the cell dies (top, right).

2.9 Other Functions for Plant Serpins

In vitro, plant serpins have been shown to inhibit mammalian proteases of the chymotrypsin family (Vercammen et al. 2006; Dahl et al. 1996; Ostergaard et al. 2000; Yoo et al. 2000). Barley serpin BSZx can inhibit trypsin at the canonical P1 Arg (i.e., using a 17-residue RCL) and chymotrypsin at P2 Leu (Dahl et al. 1996) (using a 16-residue RCL). This ability and the high expression of such serpins in cereal seeds might be important for the in vivo inhibitory specificity of the LR and other serpins. Insects, particularly those from the orders Lepidoptera and Diptera, use enzymes with chymotrypsin-like activity digestive enzymes (Jongsma and Bolter 1997). The presence of these serpins may protect seeds; however, this has yet to be shown. In this respect, it is of interest to note the variability of sequence change among *Arabidopsis* ecotypes (Fig. 2.2c). We suggest that serpins with invariant reactive centers among a large number of accessions are more likely to inhibit endogenous proteases, while those with highly variant reactive centers are more likely to inhibit digestive proteases from pests or pathogens (which would vary between ecosystems).

The direct effect of serpins on the green peach aphid, *Myzus persicae*, was conjectured by the negative correlation of insect growth with the level of the serpin CmPS-1 from pumpkin (*Cucurbita maxima*) (Yoo et al. 2000). However, direct feeding assays were not conclusive. In contrast, AtSerpin1, when directly supplied to *Spodoptera littoralis*, was shown to negatively affect its growth (Alvarez-Alfageme et al. 2011). However, in this case, the normal in vivo levels of AtSerpin1 are far too low to substantiate this attribute as part of its natural function, although the interaction may have an application in pest control. Importantly, considering the functions of AtSerpin1 elaborated above, such use in biotechnology would necessarily impact on general aspects of cell biology, e.g., programmed cell death.

2.10 Novel Serpin Functions in Cell Biology

Plants are subjected to environmental insults that can cause damage to DNA; for example, they produce singlet oxygen in high light due to energy transfer from chlorophyll (Ramel et al. 2012; Rastogi et al. 2014). In *Arabidopsis*, genes encoding AtSRP2 and AtSRP3 were induced 5-fold and 100-fold, respectively, after exposure of seedlings to low concentrations of methyl methanesulfonate (MMS), a model alkylating reagent that causes DNA damage (Ahn et al. 2009). tDNA insertion mutant (single-gene knockout) plants *atsrp2* and *atsrp3* exhibited greater root length, leaf number, and overall size than wild-type plants when exposed to MMS. The results imply that serpins *AtSRP2* and *AtSRP3* are required for normal responses of plants following exposure to genotoxins, but the mechanisms for this response remain unknown.

2.11 Conclusions

Serpins in the plant kingdom evolved during plant speciation and thus might possess an activity adapted to the needs of the species to which they belong. The exceptional and highly conserved LR-type serpin has apparently retained a conserved function and target. Recent research attributed a role in programmed cell death to this serpin. Other serpins may play a role in a range of cellular processes that are related to stress.

Acknowledgments We wish to acknowledge the support of the Lerner Family Plant Science Research Fund and the I-CORE Program of the Planning and Budgeting Committee and the Israel Science Foundation (757/12) to R.F.

References

Ahn JW, Atwell BJ, Roberts TH (2009) Serpin genes *AtSRP2* and *AtSRP3* are required for normal growth sensitivity to a DNA alkylating agent in *Arabidopsis*. BMC Plant Biol. doi:10.1186/1471-2229-9-52

Alvarez-Alfageme F, Maharramov J, Carrillo L, Vandenabeele S, Vercammen D et al (2011) Potential use of a serpin from Arabidopsis for pest control. Plos One. doi:10.1371/journal.pone.0020278

Bateman A, Bennett HP (2009) The granulin gene family: from cancer to dementia. Bioessays 31:1245–1254

Beers EP, Jones AM, Dickerman AW (2004) The S8 serine, C1A cysteine and A1 aspartic protease families in *Arabidopsis*. Phytochemistry 65:43–58

Beinrohr L, Harmat V, Dobo J, Lorincz Z, Gal P et al (2007) C1 inhibitor serpin domain structure reveals the likely mechanism of heparin potentiation and conformational disease. J Biol Chem 282:21100–21109

Cao J, Schneeberger K, Ossowski S, Gunther T, Bender S et al (2011) Whole-genome sequencing of multiple *Arabidopsis thaliana* populations. Nat Genet 43:956–963

Chen HJ, Huang DJ, Hou WC, Liu JS, Lin YH (2006) Molecular cloning and characterization of a granulin-containing cysteine protease SPCP3 from sweet potato (*Ipomoea batatas*) senescent leaves. J Plant Physiol 163:863–876

Coupe SA, Sinclair BK, Watson LM, Heyes JA, Eason JR (2003) Identification of dehydration-responsive cysteine proteases during post-harvest senescence of broccoli florets. J Exp Bot 54:1045–1056

Dahl SW, Rasmussen SK, Hejgaard J (1996) Heterologous expression of three plant serpins with distinct inhibitory specificities. J Biol Chem 271:25083–25088

Francis SE, Ersoy RA, Ahn JW, Atwell BJ, Roberts TH (2012) Serpins in rice: protein sequence analysis, phylogeny and gene expression during development. BMC Genomics. doi:10.1186/1471-2164-13-449

Gettins PGW (2002) Serpin structure, mechanism, and function. Chem Rev 102:4751–4803

Gu C, Shabab M, Strasser R, Wolters PJ, Shindo T et al (2012) Post-translational regulation and trafficking of the granulin-containing protease RD21 of *Arabidopsis thaliana*. Plos One. doi:10.1371/journal.pone.0032422

Hara-Nishimura I, Hatsugai N (2011) The role of vacuole in plant cell death. Cell Death Differ 18:1298–1304

Harrak H, Azelmat S, Baker EN, Tabaeizadeh Z (2001) Isolation and characterization of a gene encoding a drought-induced cysteine protease in tomato (*Lycopersicon esculentum*). Genome 44:368–374

Hatsugai N, Kuroyanagi M, Yamada K, Meshi T, Tsuda S et al (2004) A plant vacuolar protease, VPE, mediates virus-induced hypersensitive cell death. Science 305:855–858

Heit C, Jackson BC, McAndrews M, Wright MW, Thompson DC et al (2013) Update of the human and mouse SERPIN gene superfamily. Hum Genomics 7:22. doi:10.1186/1479-7364-7-22

Herron MD, Hackett JD, Aylward FO, Michod RE (2009) Triassic origin and early radiation of multicellular volvocine algae. Proc Natl Acad Sci USA 106:3254–3258

Hirst CE, Buzza MS, Bird CH, Warren HS, Cameron PU et al (2003) The intracellular granzyme B inhibitor, proteinase inhibitor 9, is up-regulated during accessory cell maturation and effector cell degranulation, and its overexpression enhances CTL potency. J Immunol 170:805–815

Huntington JA, Fan B, Karlsson KE, Deinum J, Lawrence DA et al (1997) Serpin conformational change in ovalbumin. Enhanced reactive center loop insertion through hinge region mutations. Biochemistry 36:5432–5440

Irving JA, Pike RN, Lesk AM, Whisstock JC (2000) Phylogeny of the serpin superfamily. Implications of amino acid conservation for structure and function. Genome Res 10:1845–1864

Jongsma MA, Bolter C (1997) The adaption of insects to plant protease inhibitors. J Insect Physiol 43:885–895
Jordan RE (1983) Antithrombin in vertebrate species – conservation of the heparin-dependent anticoagulant mechanism. Arch Biochem Biophys 227:587–595
Kaiserman D, Bird PI (2005) Analysis of vertebrate genomes suggests a new model for clade B serpin evolution. BMC Genomics. doi:10.1186/1471-2164-6-167
Kaiserman D, Knaggs S, Scarff KL, Gillard A, Mirza G et al (2002) Comparison of human chromosome 6p25 with mouse chromosome 13 reveals a greatly expanded ov-serpin gene repertoire in the mouse. Genomics 79:349–362
Koizumi M, Yamaguchishinozaki K, Tsuji H, Shinozaki K (1993) Structure and expression of 2 genes that encode distinct drought-inducible cysteine proteinases in *Arabidopsis thaliana*. Gene 129:175–182
Lampl N, Budai-Hadrian O, Davydov O, Joss TV, Harrop SJ et al (2010) *Arabidopsis* AtSerpin1: crystal structure and *in vivo* interaction with its target protease responsive to desiccation-21 (RD21). J Biol Chem 285:13550–13560
Lampl N, Alkan N, Davydov O, Fluhr R (2013) Set-point control of RD21 protease activity by AtSerpin1 controls cell death in Arabidopsis. Plant J 74:498–510
Letunic I, Bork P (2011) Interactive Tree Of Life v2: online annotation and display of phylogenetic trees made easy. Nucleic Acids Res. doi:10.1093/nar/gkr201
Luke CJ, Pak SC, Askew YS, Naviglia TL, Askew DJ et al (2007) An intracellular serpin regulates necrosis by inhibiting the induction and sequelae of lysosomal injury. Cell 130:1108–1119
Mangan MS, Kaiserman D, Bird PI (2008) The role of serpins in vertebrate immunity. Tissue Antigens 72:1–10
Ostergaard H, Rasmussen SK, Roberts TH, Hejgaard J (2000) Inhibitory serpins from wheat grain with reactive centers resembling glutamine-rich repeats of prolamin storage proteins. Cloning and characterization of five major molecular forms. J Biol Chem 275:33272–33279
Ramel F, Birtic S, Ginies C, Soubigou-Taconnat L, Triantaphylides C et al (2012) Carotenoid oxidation products are stress signals that mediate gene responses to singlet oxygen in plants. Proc Natl Acad Sci USA 109:5535–5540
Rastogi A, Yadav DK, Szymanska R, Kruk J, Sedlarova M et al (2014) Singlet oxygen scavenging activity of tocopherol and plastochromanol in *Arabidopsis thaliana*: relevance to photo-oxidative stress. Plant Cell Environ 37:392–401
Roberts TH, Hejgaard J (2008) Serpins in plants and green algae. Funct Integr Genomics 8:1–27
Silverman GA, Whisstock JC, Askew DJ, Pak SC, Luke CJ et al (2004) Human clade B serpins (ov-serpins) belong to a cohort of evolutionarily dispersed intracellular proteinase inhibitor clades that protect cells from promiscuous proteolysis. Cell Mol Life Sci 61:301–325
Vercammen D, van de Cotte B, De Jaeger G, Eeckhout D, Casteels P et al (2004) Type II metacaspases Atmc4 and Atmc9 of *Arabidopsis thaliana* cleave substrates after arginine and lysine. J Biol Chem 279:45329–45336
Vercammen D, Belenghi B, van de Cotte B, Beunens T, Gavigan JA et al (2006) Serpin1 of *Arabidopsis thaliana* is a suicide inhibitor for Metacaspase 9. J Mol Biol 364:625–636
Wang DYC, Kumar S, Hedges SB (1999) Divergence time estimates for the early history of animal phyla and the origin of plants, animals and fungi. Proc R Soc B Biol Sci 266:163–171
Wolfe KH, Gouy ML, Yang YW, Sharp PM, Li WH (1989) Date of the monocot dicot divergence estimated from chloroplast DNA-sequence data. Proc Natl Acad Sci USA 86:6201–6205
Wolpert TJ, Dunkle LD, Ciuffetti LM (2002) Host-selective toxins and avirulence determinants: What's in a name? Annu Rev Phytopathol 40:251–285
Yoo BC, Aoki K, Xiang Y, Campbell LR, Hull RJ et al (2000) Characterization of *Cucurbita maxima* phloem serpin-1 (CmPS-1) – a developmentally regulated elastase inhibitor. J Biol Chem 275:35122–35128
Zhou Q, Snipas S, Orth K, Muzio M, Dixit VM et al (1997) Target protease specificity of the viral serpin CrmA. Analysis of five caspases. J Biol Chem 272:7797–7800

Peptides as Modulators of Serpin Action

3

Cathleen Jendrny and Annette G. Beck-Sickinger

Contents

Abstract

Serpins are key players in the regulation of various physiological functions. In addition, their functional loss owing to genetic variations and polymerization causes severe diseases. It is therefore of interest to target serpins for therapeutic interventions. Increasing knowledge of structure, mechanisms of function and dysfunction, as well as natural modulation of human serpins has opened up new possibilities for structure-guided design of modulators. These might provide the possibility of both inhibition and mimicking of serpin function regarding protease inhibition. Also interference with pathological serpin polymerization might be achievable. Peptides are promising molecules for this purpose since they have been shown to be potent and selective inhibitors of protein–protein interactions. This chapter illustrates how serpin action can be modulated by peptides and how these might be useful for therapeutic intervention.

C. Jendrny • A.G. Beck-Sickinger (✉)
Faculty of Biosciences, Pharmacy and Psychology, Institute of Biochemistry, Universität Leipzig, Brüderstraße 34, 04103 Leipzig, Germany
e-mail: beck-sickinger@uni-leipzig.de

M. Geiger et al. (eds.), *The Serpin Family*, DOI 10.1007/978-3-319-22711-5_3

3.1 The Serpin Family: Structural Characteristics and Functional Diversity

3.1.1 Structural Characteristics and Mechanism of Function

The serpin superfamily is a group of large proteins distributed over all areas of life. More than 1500 serpin-related genes have been identified with 36 being expressed in humans (Law et al. 2006). The denotation of serpins is attributed to the finding that the earliest members of this family primarily were serine protease inhibitors (Carrell and Travis 1985). Today, the protein family moreover involves some inhibitors of cysteine proteases as well as cross-class inhibitors that inhibit both serine and cysteine proteases (Izuhara et al. 2008). Additionally, members are known to have lost their capability of protease inhibition and bear other functions instead.

The similarity of all serpins is given in their tertiary structure which shows strong conformity. Since the publication of the first crystal structure in 1984 [cleaved α_1-antitrypsin (Loebermann et al. 1984)], many others followed and revealed a typical fold with 3 β-sheets and 8–9 α-helices. An additional feature is the flexible C-terminal region called reactive center loop (RCL) which is of major importance for the protease inhibitory function.

Inhibition of a serine protease by a serpin is a stepwise process (Fig. 3.1) and goes along with transition from a metastable (RCL exposed) to a hyperstable (RCL inserted) conformation. Starting point is the recognition of the RCL by the protease in a substrate-like manner forming the non-covalent Michaelis complex (Ye et al. 2001; Baglin et al. 2002; Dementiev et al. 2003; Lin et al. 2011). After spatial proximity is achieved, Ser195 of the serine protease's catalytic triad nucleophilically attacks the main chain carbonyl carbon of the scissile bond [between P_1 and P_1' according to nomenclature of Schechter and Berger (1967)] building a tetrahedral intermediate (Liu et al. 2006). Cleavage of the scissile bond yields the acyl ester intermediate (Lawrence et al. 1995; Egelund et al. 1998). Upon this cleavage a conformational rearrangement takes place. Thereby the N-terminal part of the cleaved RCL is inserted as strand 4 into β-sheet A, translocating the still attached protease about 70 Å from the top to the bottom of the serpin (Huntington et al. 2000; Dementiev et al. 2006). In this final serpin–enzyme complex, the active site of the protease is disrupted and catalytic deacylation is prevented (Kaslik et al. 1997; Stratikos and Gettins 1999; Huntington et al. 2000; Dementiev et al. 2006). Therefore, the ability to rapidly insert the RCL into β-sheet A determines between inhibitory and substrate-like character of a serpin.

3.1.2 Serpins in Physiological and Pathological Processes

With their unique mechanism of protease inactivation, serpins are key players in many important physiological functions such as inflammation (Heit et al. 2013), blood coagulation, and fibrinolysis (Rau et al. 2007). Furthermore they participate

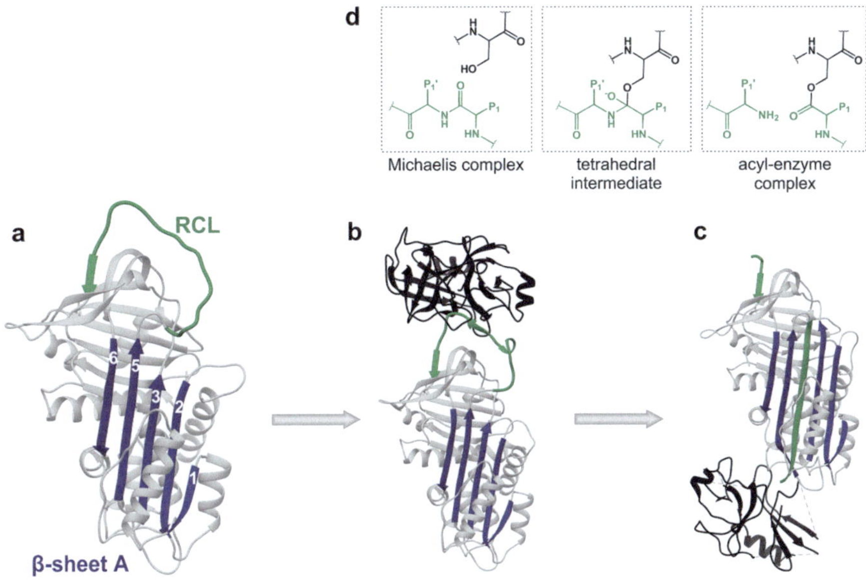

Fig. 3.1 Mechanism of serpin action. Structural changes during inhibition of serine proteases (*black*) by serpins (*gray*; RCL and central β-sheet A colored in *green* and *blue*, respectively); (**a**) crystal structure of an intact serpin (Protein Data Bank (PDB) code 1QLP); (**b**) crystal structure of a non-covalent Michaelis complex between serpin and protease (PDB code 1OPH); (**c**) crystal structure of a final serpin–protease complex (PDB code 1EZX); (**d**) depiction of chemical modifications during complex formation

in the regulation of insulin resistance and defense of microbial infections as shown for the visceral adipose tissue-derived serpin (vaspin) (Heiker et al. 2013) and members of the ov-serpins (Silverman et al. 2001). Serpin expression is altered in several pathological conditions. For example, increased vaspin levels have been linked to obesity and impaired insulin sensitivity in animal models and in humans (Hida et al. 2005; Youn et al. 2008). This has been suggested to be a compensatory mechanism on insulin resistance involving kallikrein 7 as a protease target of vaspin (Heiker et al. 2013).

Some serpins have lost their protease inhibitory properties but bear other functions. Mammary serine protease inhibitor (maspin) and pigment epithelium-derived factor (PEDF) are inhibitors of angiogenesis and involved in tumor suppression (Bodenstine et al. 2012; Filleur et al. 2009). Cortisol- and thyroxine-binding globulins (CBG, TBG) are hormone binders and transporters (Lin et al. 2010; Schussler 2000), heat shock protein 47 (Hsp47) acts as an intracellular chaperone (Sauk et al. 2005), and angiotensinogen is a precursor molecule in the regulation of blood pressure (Morgan et al. 1996).

As serpins are important in the regulation of physiological processes, severe diseases result from serpin dysfunction. The so-called serpinopathies are a group of protein conformational diseases provoked by point mutations that cause serpin polymerization. These diseases are attributed either to serpin accumulation in the cell of origin or to loss of function due to low concentrations that reach the circulation. The Z mutation of α_1-antitrypsin is the best described mutation causing serpinopathy. Its accumulation in the liver is a risk factor for cirrhosis, hepatocellular carcinoma, and neonatal hepatitis, while α_1-antitrypsin loss in the lung is associated with emphysema (Gooptu et al. 2014). Serpinopathies are also reported for neuroserpin, antithrombin, C1 esterase inhibitor, and α_1-antichymotrypsin and manifest in the early-onset dementia "familial encephalopathy with neuroserpin inclusion bodies" (FENIB) (Davis et al. 1999), thrombosis (Bruce et al. 1994; Picard 2003), hereditary angioedema (Aulak et al. 1993; Eldering et al. 1995), and chronic obstructive pulmonary disease (COPD) (Gooptu et al. 2000), respectively.

However, it still remains unclear and is discussed controversially how pathological polymers are formed. For long time it has been accepted that polymers are "loop-sheet" polymers inserting the RCL of one molecule into β-sheet A of another one (Lomas et al. 1992). Since the publication of two crystal structures in 2008 and 2011 (Yamasaki et al. 2008, 2011), the possibility of "domain-swap" polymers characterized by the insertion of larger serpin domains arose. The comparison of both general models with existing data does not reveal an ultimate statement and is controversial, favoring either the "loop-sheet" (Ekeowa et al. 2010; Santangelo et al. 2012) or the "domain-swapping" model (Yamasaki et al. 2010, 2011; Huntington and Whisstock 2010).

3.1.3 Natural Modulation of Serpin Action and Specificity

According to their mechanism of function, the specificity of serpins should be predominantly determined by the P_1 and the adjacent residues. However the often high specificity of serpins toward their target proteases is in contrast to the low variability in their P_1 residue. It turned out that other features like exosites and cofactors are important in defining selectivity. Exosites are additional interaction sites outside the typical recognition region which can enlarge and specify the binding surface between serpin and target protease. For example, exosites are found to play a role in protease inhibition of α_1-antitrypsin, plasminogen activator inhibitor-1 (PAI-1), kallistatin, and α_2-antiplasmin (Gettins and Olson 2009). They can also be used to bind cofactors as modulators to gain specificity or to accelerate activity for a certain protease. In most cases these are glycosaminoglycans like heparin. While sometimes heparin acts as an allosteric modulator of serpin activity (Whisstock et al. 2000; Huntington 2000; Baglin et al. 2002), in most cases it is reported to provide a bridging mechanism, which brings serpin and protease in proximity to each other and thereby accelerates the rate of complex formation (Li et al. 2004; Dementiev et al. 2004; Li and Huntington 2008; Huang et al. 2011). Including vitronectin and protein Z, there are also protein modulators that facilitate serpin modulation. Vitronectin stabilizes the active conformation of PAI-1 and arranges its localization on its site of action (Zhou et al. 2003). The serpin cofactor protein Z has a comparable function for protein Z-dependent protease inhibitor (ZPI) as it anchors the serpin on phospholipid surfaces whereby interactions with membrane-bound factor Xa are promoted (Wei et al. 2009; Huang et al. 2010; Qureshi et al. 2014).

Another way of natural serpin regulation is latency transition which is best described for PAI-1 but also shown for other serpins (Mottonen et al. 1992; Chang 1998; Mushunje et al. 2004; Ricagno et al. 2010). Thereby the serpin spontaneously inserts its non-cleaved RCL into β-sheet A which prevents protease inhibition.

3.2 Peptides as Modulators of Serpin Functions and Dysfunctions

Owing to their versatile functions in humans, serpin modulation gives many opportunities for therapeutic intervention. Based on the knowledge of the serpin mechanism of protease inhibition and its modulation, approaches were taken to manipulate serpin action and polymerization by interactions with small molecules, antibodies, chemical chaperones, proteins, and peptides. Peptidic approaches were mostly performed to gain better understanding of serpin characteristics such as protease inhibition, specificity and selectivity, structural rearrangements, and polymerization. Moreover, peptides are very interesting molecules from the therapeutic point of view. Although having some drawbacks such as low bioavailability, peptides are often able to specifically and selectively target protein–protein interactions. They are much smaller than proteins, more facile to produce and offer the advantage of being easily modified (Mäde et al. 2014).

Peptides derived from the maspin RCL (Ngamkitidechakul et al. 2003; Ravenhill et al. 2010), from the C-terminal part of α_1-antitrypsin (Congote and Temmel 2004; Congote 2006; Congote et al. 2008; Jia et al. 2012), and from PEDF (Mirochnik et al. 2009) are reported to be biologically active with respect to some non-protease-related functions. The following examples however focus on modulation that directly interferes with the protease inhibitory function of serpins or with serpin polymerization.

3.2.1 Peptides Inhibiting Serpin Action

3.2.1.1 Blocking of Reactive Center Loop Insertion into β-Sheet A

25 years ago, it was shown for the first time that peptides mimicking the reactive center loop are capable of inhibiting serpin action. After it has been hypothesized that the N-terminal part of the RCL inserts into β-sheet A as one important step involved in the unique serpin mechanism (Loebermann et al. 1984; Carrell and Travis 1985), it was tested whether the isolated peptide sequences derived from this part of the protein can be incorporated similarly. Incubation of α_1-antitrypsin with a peptide related to its RCL sequence resulted in 1:1 complex formation (Schulze

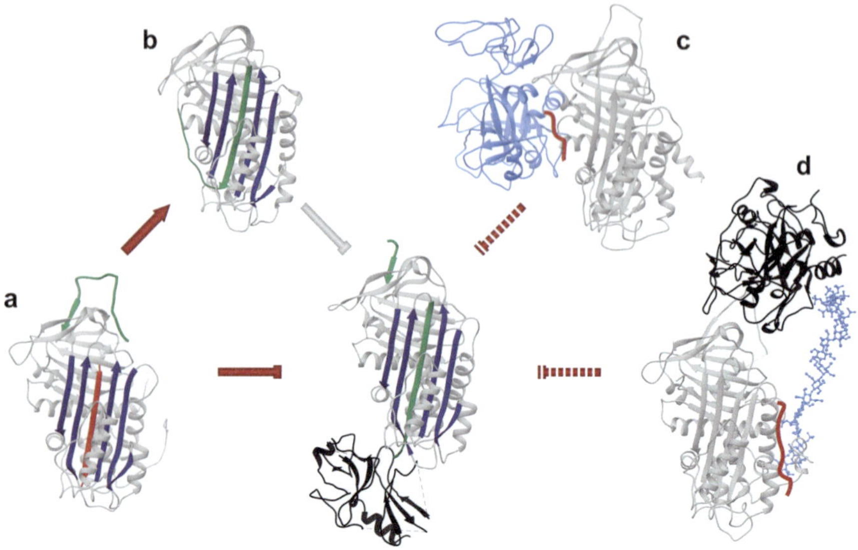

Fig. 3.2 Peptidic inhibitors of serpin function. Peptides and their effects on serpin inactivation are depicted in *red*. (**a**) Peptides inserted into β-sheet A (*dark blue*) (PDB code 1BR8) avoid serpin function of protease (*black*) inhibition either directly or by induction of the latent conformation (**b**; PDB code 1DVN). (**c**, **d**) Peptides blocking the interaction of serpins and their cofactors (*light blue*) slow down serpin activity. (**c**) Interaction of ZPI with protein Z (PDB code 3H5C). (**d**) Heparin, bridging antithrombin and thrombin (PDB code 1TB6)

et al. 1990) (Fig. 3.2a). Concurrently, and in contrast to native α_1-antitrypsin, this one was no longer able to form a stable complex with its target protease trypsin and to provide antitryptic activity. A variety of similar peptides derived from different members of the serpin family were synthesized. They all were able to convert the serpin from an inhibitor into a substrate of the target protease and proved the concept of serpin inhibition by peptides derived from their RCL sequences.

Shortening of the primarily identified α_1-antitrypsin peptide showed that C-terminal truncations are accepted to retain serpin inactivation, while N-terminal truncations are not (Schulze et al. 1992). Studies on short PAI-1 peptides revealed that also the mechanism of serpin inhibition by RCL-derived peptides might be variable depending on peptide length and position of binding. Comparing the effects of two PAI-1-derived peptides representing P_{14}–P_9 and P_8–P_3, respectively, it was demonstrated that both peptides were able to inhibit complex formation between PAI-1 and its protease targets and inhibited PAI-1 activity (D'Amico et al. 2012). While the more N-terminal part (P_{14}–P_9) induced substrate behavior of PAI-1, the more C-terminal part (P_8–P_3) however led to latency transition of PAI-1 (Fig. 3.2b).

It has been proven by crystal structures of antithrombin (Skinner et al. 1998) and PAI-2 (Jankova 2001) complexed with their 12-14-mer RCL peptides that the serpin inhibitory properties of RCL-derived peptides in fact result from peptide

Table 3.1 Serpin-inhibiting peptides

Origin of peptide	Peptide sequence	Target	References
α_1-antitrypsin	...**TEAAGAMFLEAIPM**...		
P_{15}–P_1	GTEAAGAMFLEAIPMY	α_1-antitrypsin α_1-antichymotrypsin antithrombin	Mast et al. (1992)
P_{15}–P_3	Ac-GTEAAGAMFLEAI	α_1-antitrypsin antithrombin	Chang et al. (1996)
P_{14}–P_1	Ac-TEAAGAMFLEAIVM	α_1-antitrypsin	Schulze et al. (1990)
P_{14}–P_4	Ac-TEAAGAMFLEA	α_1-antitrypsin	Schulze et al. (1992)
P_{14}–P_8	Ac-TEAAGAM	α_1-antitrypsin	Schulze et al. (1992)
P_{12}–P_1	Ac-AAGAMFLEAIVM	α_1-antitrypsin	Schulze et al. (1992)
antithrombin	...**EAAASTAVVIAGR**...		
P_{14}–P_1	Ac-SEAAASTAVVIAGR	antithrombin	Björk et al. (1992)
P_{14}–P_2	Ac-SEAAASTAVVIAG	antithrombin	Carrell et al. (1991)
P_{14}–P_3	Ac-SEAAASTAVVIA	antithrombin	Chang et al. (1996), Skinner et al. (1998)
		α_1-antitrypsin	Chang et al. (1996)
P_{14}–P_7	Ac-SEAAAS	antithrombin	Carrell et al. (1991)
PAI-1	...**TVASSSTAVIVSAR**...		
P_{14}–P_1	Ac-TVASSSTAVIVSAR	PAI-1	Eitzman et al. (1995)
P_{14}–P_7	Ac-TVASSSTA	PAI-1	Lawrence (1995)
P_{14}–P_{10}	Ac-TVASS-NH_2	PAI-1	Xue et al. (1998)
P_{14}–P_9	TVASSS	PAI-1	D'Amico et al. (2012)
P_8–P_3	TAVIVS	PAI-1	D'Amico et al. (2012)
PAI-2	...**TEAAAGTGGVMTGR**...		
P_{14}–P_1	Ac-TEAAAGTGGVMTGR	PAI-2	Jankova (2001)
non-serpin derived			
high molecular weight kininogen	GKHKNKGKKNGKHNGWK	PAI-1/vitronectin	Chavakis (2002)

insertion as a fourth strand into β-sheet A. Crystallization studies on a PAI-1 mutant and a short peptide representing residues P_{14}–P_{10} of its reactive center loop showed a complex with two molecules of the peptide (Xue et al. 1998). Both were localized in β-sheet A, but one molecule occupied the same space as P_{14}–P_{10} in cleaved or latent PAI-1, whereas the other occupied the space of P_6–P_2 of the cleaved loop.

The peptide annealing approaches showed that with peptides derived from the RCL (Table 3.1), serpin modulation is possible. Although these peptides might be useful tools for studying serpin structure and function, there are some limitations that impede their use in therapeutic applications. On the one hand, unspecific insertion of peptides into several different serpins (Mast et al. 1992; Chang et al. 1996) might cause undesirable side effects. On the other hand, peptide insertion was experimentally difficult. Solely, binding of PAI-1 peptides could be achieved fast as complete complex formation was observed already after 15–60 min at 37 °C (Eitzman et al. 1995; D'Amico et al. 2012). Insertion of α_1-antitrypsin, antithrombin, and PAI-2 peptides into their corresponding serpins needed high excess of peptide and long incubation times of 12–48 h or temperature highly above 37 °C (Schulze et al. 1990; Mast et al. 1992; Björk et al. 1992; Chang et al. 1996; Skinner et al. 1998; Jankova 2001) and are therefore not applicable for in vivo usage.

3.2.1.2 Inhibitors of Serpin Cofactor Interactions

In cases where cofactors are needed for high rates of protease inhibition by serpins, peptides that competitively block the serpin cofactor interaction are valuable for slowing down protease inhibition (Fig. 3.2c, d). Such modulation has been shown for PAI-1 and its cofactor vitronectin. Based on studies showing that PAI-1 and two-chain high molecular weight kininogen (HKa) compete for adjacent binding sites on vitronectin (Chavakis et al. 2000), it was investigated whether a HKa peptide can interfere with vitronectin-dependent PAI-1 functions. A 17-mer peptide derived from domain 5 of HKa (Table 3.1) was found to inhibit binding of PAI-1 to both isolated and extracellular matrix-associated vitronectin (Chavakis 2002). This effect led to reduction of PAI-1 antifibrinolytic activity and inhibited its high-affinity binding to fibrin.

Inhibition of serpin activity by peptides blocking the accelerating effect of cofactors might be a very interesting approach and has advantages over other strategies. Firstly, those peptides would not lead to complete loss but rather slow down serpin action. This means there is still some active serpin left. Secondly, inhibition is reversible which is easier to handle with respect to therapeutic applications. And finally, the activity of one and the same serpin can be selectively inhibited with regard to one target protease while keeping inhibition of other proteases that do not need the cofactor interaction. Thus, side effects could be diminished.

3.2.2 Peptides Inhibiting Serpin Polymerization

3.2.2.1 Blocking of Reactive Center Loop Insertion into β-Sheet A

As it has been accepted that serpin polymerization occurs by inserting the RCL of one serpin into the β-sheet A of another one (Lomas et al. 1992), it was the next step to test peptides that were shown to insert into this sheet also with respect to their polymerization inhibition.

Interestingly, RCL peptides of α_1-antitrypsin and antithrombin that had already been shown to inhibit serpin activity by insertion into β-sheet A (Carrell et al. 1991; Mast et al. 1992; Chang et al. 1996) also inhibited serpin polymerization (Mast et al. 1992; Lomas et al. 1992; Fitton et al. 1997; Chang et al. 1997) (Fig. 3.3) and were able to degrade preexisting polymers (Lomas et al. 1993).

It was suggested that partial loop insertion into the upper part of β-sheet A opens the lower sheet to be more accessible for polymerization. This was confirmed by observations that peptides inserting into the upper β-sheet A induced

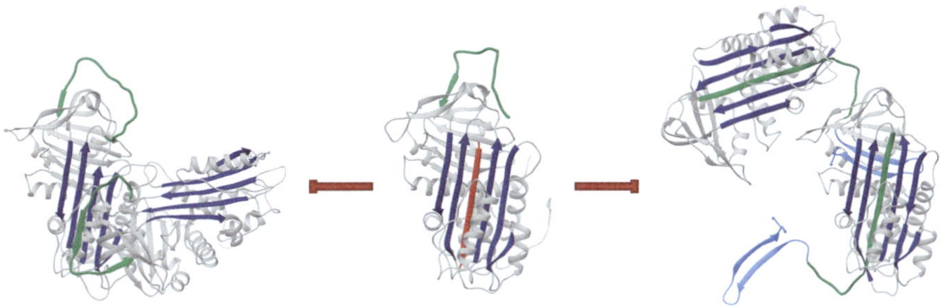

Fig. 3.3 Inhibition of serpin polymerization. Schematic representation of a "loop-sheet" dimer (*left*; monomeric serpin: PDB code 1QLP) and a "domain-swap" dimer (*right*; monomeric serpin: PDB code 3T1P). Larger domain swap includes strands 4 and 5 of β-sheet B (*light blue*). A peptide (*red*) inserted into β-sheet A (*dark blue*) inhibits polymerization by avoiding intra- and intermolecular RCL (*green*) insertion

polymerization (Chang et al. 1997; Fitton et al. 1997). It has therefore been concluded that insertion of long external peptides might be hindered and peptides targeting the lower part of β-sheet A should be more effective inhibitors of polymerization (Mahadeva 2002). To test this hypothesis an α_1-antitrypsin 6-mer peptide (Ac-FLEAIG), targeting only the lower part of sheet A was chosen. This peptide annealed selectively to Z-(mutated) but not to M-(native) α_1-antitrypsin and inhibited Z-α_1-antitrypsin polymerization completely (Mahadeva 2002). Thus, non-pathologic M-α_1-antitrypsin was not limited in its physiological functions. By systematic alanine scanning and shortening of the peptide, other 4-6-mer peptides were identified which bound to Z-α_1-antitrypsin and inhibited its polymerization equally or even better, but selectivity against M-α_1-antitrypsin was widely lost (Chang et al. 2006).

In another study a serpin-unrelated peptide derived from cholecystokinin (WMDF-NH_2) has been found to efficiently block both α_1-antitrypsin and antithrombin polymerization (Zhou et al. 2004). Crystallographic structures of β-sheet-opened antithrombin in complex with this peptide indicated structural requirements for hydrophobic residues at P_4 and P_6 which became even more apparent when comparing other sequences of RCL-derived and exogenous peptides. Using this knowledge a new short peptide (formyl-VVII) was rationally designed that shared the features of the previously identified peptide Ac-FLEAIG (Mahadeva 2002) with respect to binary complex formation with α_1-antitrypsin and selectivity of Z- over M-α_1-antitrypsin binding (Zhou et al. 2004).

Two additional approaches identified further 4-mer peptides interfering with serpin polymerization. In a combinatorial approach, peptide Ac-TTAI-NH_2 was identified that formed a binary complex with Z-α_1-antitrypsin much faster than the previously discovered peptides, inhibited polymer formation, and also promoted α_1-antitrypsin polymer dissociation (Chang et al. 2009). Under the experimental conditions of short-time incubation, binary complex formation occurred specifically with Z- but not with M-α_1-antitrypsin. The computationally designed peptide

Table 3.2 Inhibitors of serpin polymerization

Origin of peptide	Peptide sequence	Target	References
antithrombin	...**SEAAASTAVVIAGR**...		
P_{14}–P_2	Ac-SEAAASTAVVIAG	α_1-antitrypsin	Lomas et al. (1992, 1993)
P_{14}–P_3	Ac-SEAAASTAVVIA	α_1-antitrypsin antithrombin	Fitton et al. (1997), Chang et al. (1997), Zhou et al. (2004)
P_8–P_3	Ac-TAVVIA	α_1-antitrypsin	Zhou et al. (2004)
P_7–P_3	Ac-AVVIA	antithrombin	Zhou et al. (2004)
α_1-antitrypsin	...**TEAAGAMFLEAIPM**...		
P_{15}–P_1	GTEAAGAMFLEAIPMY	α_1-antitrypsin	Mast et al. (1992)
P_7–P_2	Ac-FLEAIG	α_1-antitrypsin	Mahadeva (2002)
P_7–P_2	Ac-FLEAAG	α_1-antitrypsin	Chang et al. (2006)
P_7–P_3	Ac-FLEAA-NH_2	α_1-antitrypsin	Chang et al. (2006)
P_7–P_2	Ac-FLAAIG	α_1-antitrypsin	Chang et al. (2006)
P_7–P_4	Ac-FLAA-NH_2	α_1-antitrypsin	Chang et al. (2006)
P_6–P_2	Ac-LAAIG-NH_2	α_1-antitrypsin	Chang et al. (2006)
non-serpin derived			
cholecystokinin	WMDF-NH_2	α_1-antitrypsin antithrombin	Zhou et al. (2004)
	formyl-VVII	α_1-antitrypsin	Zhou et al. (2004)
	Ac-TTAI-NH_2	α_1-antitrypsin	Chang et al. (2009)
	VIKF	α_1-antitrypsin	Chowdhury et al. (2007)

VIKF was capable of depolymerizing α_1-antitrypsin polymers as was investigated by fluorescence correlation spectroscopy (Chowdhury et al. 2007).

Although the mechanism of polymerization is still controversial, all models are in agreement with an inserted reactive center loop. It is therefore not surprising that the peptides for inhibition of polymerization shown here (Table 3.2) are based on the same principle like the abovementioned serpin inhibitors, namely, peptide insertion into β-sheet A. Hence, the same limitations of their usage have to be considered. It is important to keep in mind that peptides based on this mechanism will not only inhibit polymerization but also serpin function itself. Supplement therapies are necessary to compensate for loss-of-function diseases. The aim must therefore be to find peptides that selectively inhibit polymerization (and function) of mutated, but not of native, serpins as was indicated for peptide Ac-FLEAIG (Mahadeva 2002).

3.2.2.2 Allosteric Blocking of Polymerization

Another starting point for a peptidic α_1-antitrypsin polymerization inhibitor arose from studies on structural changes during the transition from the meta- to the hyperstable serpin conformation. The studies identified a large hydrophobic cavity present in monomeric α_1-antitrypsin that is diminished during polymerization (Elliott et al. 2000) and therefore is concluded to be a possible target (Parfrey et al. 2003; Patschull et al. 2011). Small molecules were identified that target this cavity and inhibit polymerization but unexpectedly also abolish serpin inhibitory function (Mallya et al. 2007). As no peptide approach has emerged so far, it is discussable if targeting this cavity by peptides is possible at all (Patschull et al. 2011).

3.2.3 Peptides Mimicking Serpin Action

A peptide derived from the RCL sequence of α_1-antitrypsin (Table 3.3) was found to be an inhibitor of porcine pancreatic elastase (PPE) in a mass spectrometric approach. Kinetic analyses indicated that the peptide acts as an uncompetitive inhibitor, but inhibition was weak (Wright et al. 2000).

A different approach of protease inhibition by a serpin-derived peptide was undertaken for furin inhibition. The bioengineered serpin α_1-antitrypsin Portland which carries RIPR instead of AIPM in its P_4–P_1 positions was found to inhibit the subtilisin-like proprotein convertase furin (Anderson et al. 1993). Furin does not belong to the trypsin-like serine proteases that are typically inhibited by α_1-antitrypsin. Nevertheless it is believed that the mechanism of furin inhibition by α_1-antitrypsin Portland is serpin-like (Jean et al. 1998; Dufour et al. 1998). Based on the sequence of the serpin, acyclic and cyclic RCL peptides (Table 3.3) were synthesized and both variants were found to inhibit furin in vitro (Basak and Lotfipour 2005). The mechanism of inhibition is not completely understood but involves slow peptide binding and, in contrast to α_1-antitrypsin Portland, no cleavage of the inhibitor.

These examples show that peptides derived from the RCL of serpins might have inhibitory properties. Of course, lacking the serpin mechanism and exosites for cofactor binding, it will be challenging to achieve acceptable inhibition rates. On the other hand, serine proteases have been successfully inhibited by peptides. These were substrate-like, but could be turned into inhibitors by chemical modification of their scissile bond. The underlying mechanisms involve trapping the protease–substrate complex in the tetrahedral or in the acyl-enzyme intermediate (Walker and Lynas 2001; Hedstrom 2002). Additionally, natural occurring peptides such as the sunflower trypsin inhibitor-1 (SFTI-1) have been shown to inhibit serine proteases and can serve as natural scaffolds for inhibitor design (Luckett et al. 1999). Combining knowledge about RCL sequence preferences of different serine proteases with such chemical modifications or scaffolds might lead to the development of potent and selective protease inhibitors in future.

Table 3.3 Peptides mimicking serpin action

Origin of peptide	Peptide sequence	Target	References
α_1-antitrypsin Portland	CKGTEAAGAMFLERLPRSIPPEVK FNKPFC-NH_2 (acyclic/cyclic)	furin	Basak and Lotfipour (2005)
α_1-antitrypsin	MFLEAIPM	PPE	Wright et al. (2000)

3.3 Conclusion

Various approaches demonstrated the usage of peptides as tools to gain knowledge about serpin structure and function. Nevertheless, it is discussed controversially whether they evolve therapeutic potential. Many of the here presented peptides show one or more limitations such as unfavorable peptide length, low potency, or poor selectivity. Nevertheless, some of the short and selective peptides developed for inhibition of serpin action or polymerization might be promising. To better evaluate the benefits of the peptides, in vivo data and data concerning bioavailability and proteolytic stability have to be obtained. In summary, even if there are some good approaches, the design of peptidic serpin modulators for therapeutic intervention remains challenging. However, peptides are important tools to study the mode of action of many serpins.

Acknowledgments We gratefully acknowledge the graduate school "Leipzig School of Natural Sciences—Building with Molecules and Nano-objects" (BuildMoNa) and the "Deutsche Forschungsgemeinschaft" (DFG; SFB1052/C2) for funding.

References

Anderson ED, Thomas L, Hayflick JS, Thomas G (1993) Inhibition of HIV-1 gp160-dependent membrane fusion by a furin-directed α_1-antitrypsin variant. J Biol Chem 268 (33):24887–24891

Aulak KS, Eldering E, Hack CE, Lubbers YP, Harrison RA, Mast A, Cicardi M, Davis AE (1993) A hinge region mutation in C1-inhibitor ($Ala^{436} \rightarrow$ Thr) results in nonsubstrate-like behavior and in polymerization of the molecule. J Biol Chem 268(24):18088–18094

Baglin TP, Carrell RW, Church FC, Esmon CT, Huntington JA (2002) Crystal structures of native and thrombin-complexed heparin cofactor II reveal a multistep allosteric mechanism. Proc Natl Acad Sci USA 99(17):11079–11084. doi:10.1073/pnas.162232399

Basak A, Lotfipour F (2005) Modulating furin activity with designed mini-PDX peptides: synthesis and in vitro kinetic evaluation. FEBS Lett 579(21):4813–4821. doi:10.1016/j.febslet.2005.07.062

Björk I, Ylinenjärvi K, Olson ST, Bock PE (1992) Conversion of antithrombin from an inhibitor of thrombin to a substrate with reduced heparin affinity and enhanced conformational stability by binding of a tetradecapeptide corresponding to the P_1 to P_{14} region of the putative reactive bond loop of the inhibitor. J Biol Chem 267(3):1976–1982

Bodenstine TM, Seftor REB, Khalkhali-Ellis Z, Seftor EA, Pemberton PA, Hendrix MJC (2012) Maspin: molecular mechanisms and therapeutic implications. Cancer Metastasis Rev 31 (3-4):529–551. doi:10.1007/s10555-012-9361-0

Bruce D, Perry DJ, Borg J-Y, Carrell RW, Wardell MR (1994) Thromboembolic disease due to thermolabile conformational changes of antithrombin Rouen-VI (187 Asn $\rightarrow$ Asp). J Clin Invest 94(6):2265–2274. doi:10.1172/JCI117589

Carrell R, Travis J (1985) α_1-Antitrypsin and the serpins: variation and countervariation. Trends Biochem Sci 10(1):20–24. doi:10.1016/0968-0004(85)90011-8

Carrell RW, Evans DL, Stein PE (1991) Mobile reactive centre of serpins and the control of thrombosis. Nature 353(6344):576–578. doi:10.1038/353576a0

Chang W-SW (1998) Latent α_1-antichymotrypsin: a molecular explanation for the inactivation of α_1-antichymotrypsin in chronic bronchitis and emphysema. J Biol Chem 273(6):3695–3701. doi:10.1074/jbc.273.6.3695

Chang WS, Wardell MR, Lomas DA, Carrell RW (1996) Probing serpin reactive-loop conformations by proteolytic cleavage. Biochem J 314(Pt 2):647–653

Chang W-SW, Whisstock J, Hopkins PC, Lesk AM, Carrell RW, Wardell MR (1997) Importance of the release of strand 1C to the polymerization mechanism of inhibitory serpins. Protein Sci 6 (1):89–98. doi:10.1002/pro.5560060110

Chang Y-P, Mahadeva R, Chang W-SW, Shukla A, Dafforn TR, Chu Y-H (2006) Identification of a 4-mer peptide inhibitor that effectively blocks the polymerization of pathogenic Z α_1-antitrypsin. Am J Respir Cell Mol Biol 35(5):540–548. doi:10.1165/rcmb.2005-0207OC

Chang Y-P, Mahadeva R, Chang W-SW, Lin S-C, Chu Y-H (2009) Small-molecule peptides inhibit Z α_1-antitrypsin polymerization. J Cell Mol Med 13(8b):2304–2316. doi:10.1111/j.1582-4934.2008.00608.x

Chavakis T (2002) A novel antithrombotic role for high molecular weight kininogen as inhibitor of plasminogen activator inhibitor-1 function. J Biol Chem 277(36):32677–32682. doi:10.1074/jbc.M204010200

Chavakis T, Kanse SM, Lupu F, Hammes H-P, Müller-Esterl W, Pixley RA, Colman RW, Preissner KT (2000) Different mechanisms define the antiadhesive function of high molecular weight kininogen in integrin- and urokinase receptor–dependent interactions. Blood 96 (2):514–522

Chowdhury P, Wang W, Lavender S, Bunagan MR, Klemke JW, Tang J, Saven JG, Cooperman BS, Gai F (2007) Fluorescence correlation spectroscopic study of serpin depolymerization by computationally designed peptides. J Mol Biol 369(2):462–473. doi:10.1016/j.jmb.2007.03.042

Congote LF (2006) The C-terminal 26-residue peptide of serpin A1 is an inhibitor of HIV-1. Biochem Biophys Res Commun 343(2):617–622. doi:10.1016/j.bbrc.2006.02.190

Congote LF, Temmel N (2004) The C-terminal 26-residue peptide of serpin A1 stimulates proliferation of breast and liver cancer cells: role of protein kinase C and CD47. FEBS Lett 576(3):343–347. doi:10.1016/j.febslet.2004.09.035

Congote LF, Temmel N, Sadvakassova G, Dobocan MC (2008) Comparison of the effects of serpin A1, a recombinant serpin A1-IGF chimera and serpin A1 C-terminal peptide on wound healing. Peptides 29(1):39–46. doi:10.1016/j.peptides.2007.10.011

D'Amico S, Martial JA, Struman I (2012) A peptide mimicking the C-terminal part of the reactive center loop induces the transition to the latent form of plasminogen activator inhibitor type-1. FEBS Lett 586(6):686–692. doi:10.1016/j.febslet.2012.02.013

Davis RL, Shrimpton AE, Holohan PD, Bradshaw C, Feiglin D, Collins GH, Sonderegger P, Kinter J, Becker LM, Lacbawan F, Krasnewich D, Muenke M, Lawrence DA, Yerby MS, Shaw CM, Gooptu B, Elliott PR, Finch JT, Carrell RW, Lomas DA (1999) Familial dementia caused by polymerization of mutant neuroserpin. Nature 401(6751):376–379. doi:10.1038/43894

Dementiev A, Simonovic M, Volz K, Gettins PGW (2003) Canonical inhibitor-like interactions explain reactivity of α_1-Proteinase inhibitor Pittsburgh and antithrombin with proteinases. J Biol Chem 278(39):37881–37887. doi:10.1074/jbc.M305195200

Dementiev A, Petitou M, Herbert J-M, Gettins PGW (2004) The ternary complex of antithrombin-anhydrothrombin-heparin reveals the basis of inhibitor specificity. Nat Struct Mol Biol 11 (9):863–867. doi:10.1038/nsmb810

Dementiev A, Dobo J, Gettins PGW (2006) Active site distortion is sufficient for proteinase inhibition by serpins: structure of the covalent complex of α1-proteinase inhibitor with porcine pancreatic elastase. J Biol Chem 281(6):3452–3457. doi:10.1074/jbc.M510564200

Dufour EK, Denault J-B, Hopkins PC, Leduc R (1998) Serpin-like properties of α1-antitrypsin Portland towards furin convertase. FEBS Lett 426(1):41–46. doi:10.1016/S0014-5793(98)00307-X

Egelund R, Rodenburg KW, Andreasen PA, Rasmussen MS, Guldberg RE, Petersen TE (1998) An ester bond linking a fragment of a serine proteinase to its serpin inhibitor. Biochemistry 37 (18):6375–6379. doi:10.1021/bi973043+

Eitzman DT, Fay WP, Lawrence DA, Francis-Chmura AM, Shore JD, Olson ST, Ginsburg D (1995) Peptide-mediated inactivation of recombinant and platelet plasminogen activator inhibitor-1 in vitro. J Clin Invest 95(5):2416–2420. doi:10.1172/JCI117937

Ekeowa UI, Freeke J, Miranda E, Gooptu B, Bush MF, Perez J, Teckman J, Robinson CV, Lomas DA (2010) Defining the mechanism of polymerization in the serpinopathies. Proc Natl Acad Sci USA 107(10):17146–17151. doi:10.1073/pnas.1004785107

Eldering E, Verpy E, Roem D, Meo T, Tosi M (1995) COOH-terminal substitutions in the serpin C1 inhibitor that cause loop overinsertion and subsequent multimerization. J Biol Chem 270 (6):2579–2587. doi:10.1074/jbc.270.6.2579

Elliott PR, Pei XY, Dafforn TR, Lomas DA (2000) Topography of a 2.0 Å structure of α_1-antitrypsin reveals targets for rational drug design to prevent conformational disease. Protein Sci 9(7):1274–1281. doi:10.1110/ps.9.7.1274

Filleur S, Nelius T, de Riese W, Kennedy RC (2009) Characterization of PEDF: a multi-functional serpin family protein. J Cell Biochem 106(5):769–775. doi:10.1002/jcb.22072

Fitton HL, Pike RN, Carrell RW, Chang WS (1997) Mechanisms of antithrombin polymerisation and heparin activation probed by the insertion of synthetic reactive loop peptides. Biol Chem 378(9):1059–1063. doi:10.1515/bchm.1997.378.9.1059

Gettins PGW, Olson ST (2009) Exosite determinants of serpin specificity. J Biol Chem 284 (31):20441–20445. doi:10.1074/jbc.R800064200

Gooptu B, Hazes B, Chang W-SW, Dafforn TR, Carrell RW, Read RJ, Lomas DA (2000) Inactive conformation of the serpin α_1-antichymotrypsin indicates two-stage insertion of the reactive loop: Implications for inhibitory function and conformational disease. Proc Natl Acad Sci USA 97(1):67–72. doi:10.1073/pnas.97.1.67

Gooptu B, Dickens JA, Lomas DA (2014) The molecular and cellular pathology of α_1-antitrypsin deficiency. Trends Mol Med 20(2):116–127. doi:10.1016/j.molmed.2013.10.007

Hedstrom L (2002) Serine protease mechanism and specificity. Chem Rev 102(12):4501–4524. doi:10.1021/cr000033x

Heiker JT, Klöting N, Kovacs P, Kuettner EB, Sträter N, Schultz S, Kern M, Stumvoll M, Blüher M, Beck-Sickinger AG (2013) Vaspin inhibits kallikrein 7 by serpin mechanism. Cell Mol Life Sci 70(14):2569–2583. doi:10.1007/s00018-013-1258-8

Heit C, Jackson BC, McAndrews M, Wright MW, Thompson DC, Silverman GA, Nebert DW, Vasiliou V (2013) Update of the human and mouse SERPIN gene superfamily. Hum Genomics 7(1):22. doi:10.1186/1479-7364-7-22

Hida K, Wada J, Eguchi J, Zhang H, Baba M, Seida A, Hashimoto L, Okada T, Yasuhara A, Nakatsuka A, Shikata K, Hourai S, Futami J, Watanabe E, Matsuki Y, Hiramatsu R, Akagi S, Makino H, Kanwar YS (2005) Visceral adipose tissue-derived serine protease inhibitor: a unique insulin-sensitizing adipocytokine in obesity. Proc Natl Acad Sci USA 102 (30):10610–10615. doi:10.1073/pnas.0504703102

Huang X, Dementiev A, Olson ST, Gettins PGW (2010) Basis for the specificity and activation of the serpin protein Z-dependent proteinase inhibitor (ZPI) as an inhibitor of membrane-associated factor Xa. J Biol Chem 285(26):20399–20409. doi:10.1074/jbc.M110.112748

Huang X, Rezaie AR, Broze GJ, Olson ST (2011) Heparin is a major activator of the anticoagulant serpin, protein Z-dependent protease inhibitor. J Biol Chem 286(11):8740–8751. doi:10.1074/jbc.M110.188375

Huntington JA (2000) The conformational activation of antithrombin: a 2.85-Å structure of a fluorescein derivative reveals an electrostatic link between the hinge and heparin binding regions. J Biol Chem 275(20):15377–15383. doi:10.1074/jbc.275.20.15377

Huntington JA, Whisstock JC (2010) Molecular contortionism – on the physical limits of serpin 'loop-sheet' polymers. Biol Chem 391(8):973–982. doi:10.1515/bc.2010.085

Huntington JA, Read RJ, Carrell RW (2000) Structure of a serpin-protease complex shows inhibition by deformation. Nature 407(6806):923–926. doi:10.1038/35038119

Izuhara K, Ohta S, Kanaji S, Shiraishi H, Arima K (2008) Recent progress in understanding the diversity of the human ov-serpin/clade B serpin family. Cell Mol Life Sci 65(16):2541–2553. doi:10.1007/s00018-008-8049-7

Jankova L (2001) Crystal structure of the complex of plasminogen activator inhibitor 2 with a peptide mimicking the reactive center loop. J Biol Chem 276(46):43374–43382. doi:10.1074/jbc.M103021200

Jean F, Stella K, Thomas L, Liu G, Xiang Y, Reason AJ, Thomas G (1998) α_1-Antitrypsin Portland, a bioengineered serpin highly selective for furin: application as an antipathogenic agent. Proc Natl Acad Sci USA 95(13):7293–7298. doi:10.1073/pnas.95.13.7293

Jia Q, Jiang X, Yu F, Qiu J, Kang X, Cai L, Li L, Shi W, Liu S, Jiang S, Liu K (2012) Short cyclic peptides derived from the C-terminal sequence of α1-antitrypsin exhibit significant anti-HIV-1 activity. Bioorg Med Chem Lett 22(7):2393–2395. doi:10.1016/j.bmcl.2012.02.037

Kaslik G, Kardos J, Szabó E, Szilágyi L, Závodszky P, Westler WM, Markley JL, Gráf L (1997) Effects of serpin binding on the target proteinase: global stabilization, localized increased structural flexibility, and conserved hydrogen bonding at the active site. Biochemistry 36 (18):5455–5464. doi:10.1021/bi962931m

Law RH, Zhang Q, McGowan S, Buckle AM, Silverman GA, Wong W, Rosado CJ, Langendorf CG, Pike RN, Bird PI et al (2006) An overview of the serpin superfamily. Genome Biol 7 (5):216. doi:10.1186/gb-2006-7-5-216

Lawrence DA (1995) The acid stabilization of plasminogen activator inhibitor-1 depends on protonation of a single group that affects loop insertion into β-sheet A. J Biol Chem 270 (46):27942–27947. doi:10.1074/jbc.270.46.27942

Lawrence DA, Ginsburg D, Day DE, Berkenpas MB, Verhamme IM, Kvassman J-O, Shore JD (1995) Serpin-protease complexes are trapped as stable acyl-enzyme intermediates. J Biol Chem 270(43):25309–25312. doi:10.1074/jbc.270.43.25309

Li W, Huntington JA (2008) The heparin binding site of protein C inhibitor is protease-dependent. J Biol Chem 283(51):36039–36045. doi:10.1074/jbc.M805974200

Li W, Johnson DJD, Esmon CT, Huntington JA (2004) Structure of the antithrombin-thrombin-heparin ternary complex reveals the antithrombotic mechanism of heparin. Nat Struct Mol Biol 11(9):857–862. doi:10.1038/nsmb811

Lin H-Y, Muller YA, Hammond GL (2010) Molecular and structural basis of steroid hormone binding and release from corticosteroid-binding globulin. Mol Cell Endocrinol 316(1):3–12. doi:10.1016/j.mce.2009.06.015

Lin Z, Jiang L, Yuan C, Jensen JK, Zhang X, Luo Z, Furie BC, Furie B, Andreasen PA, Huang M (2011) Structural basis for recognition of urokinase-type plasminogen activator by plasminogen activator inhibitor-1. J Biol Chem 286(9):7027–7032. doi:10.1074/jbc.M110.204537

Liu B, Schofield CJ, Wilmouth RC (2006) Structural analyses on intermediates in serine protease catalysis. J Biol Chem 281(33):24024–24035. doi:10.1074/jbc.M600495200

Loebermann H, Tokuoka R, Deisenhofer J, Huber R (1984) Human α_1-proteinase inhibitor. Crystal structure analysis of two crystal modifications, molecular model and preliminary analysis of the implications for function. J Mol Biol 177(3):531–557. doi:10.1016/0022-2836(84)90298-5

Lomas DA, Evans DL, Finch JT, Carrell RW (1992) The mechanism of Z α_1-antitrypsin accumulation in the liver. Nature 357(6379):605–607. doi:10.1038/357605a0

Lomas DA, Evans DL, Stone SR, Chang WSW, Carrell RW (1993) Effect of the Z mutation on the physical and inhibitory properties of α_1-antitrypsin. Biochemistry 32(2):500–508. doi:10.1021/bi00053a014

Luckett S, Garcia RS, Barker JJ, Konarev AV, Shewry PR, Clarke AR, Brady RL (1999) High-resolution structure of a potent, cyclic proteinase inhibitor from sunflower seeds. J Mol Biol 290(2):525–533. doi:10.1006/jmbi.1999.2891

Mäde V, Els-Heindl S, Beck-Sickinger AG (2014) Automated solid-phase peptide synthesis to obtain therapeutic peptides. Beilstein J Org Chem 10:1197–1212. doi:10.3762/bjoc.10.118

Mahadeva R (2002) 6-mer peptide selectively anneals to a pathogenic serpin conformation and blocks polymerization: implications for the prevention of Z α1-antitrypsin-related cirrhosis. J Biol Chem 277(9):6771–6774. doi:10.1074/jbc.C100722200

Mallya M, Phillips RL, Saldanha SA, Gooptu B, Leigh Brown SC, Termine DJ, Shirvani AM, Wu Y, Sifers RN, Abagyan R, Lomas DA (2007) Small molecules block the polymerization of Z α_1-antitrypsin and increase the clearance of intracellular aggregates. J Med Chem 50 (22):5357–5363. doi:10.1021/jm070687z

Mast AE, Enghild JJ, Salvesen G (1992) Conformation of the reactive site loop of α_1-proteinase inhibitor probed by limited proteolysis. Biochemistry 31(10):2720–2728. doi:10.1021/bi00125a012

Mirochnik Y, Aurora A, Schulze-Hoepfner FT, Deabes A, Shifrin V, Beckmann R, Polsky C, Volpert OV (2009) Short pigment epithelial-derived factor-derived peptide inhibits angiogenesis and tumor growth. Clin Cancer Res 15(5):1655–1663. doi:10.1158/1078-0432.CCR-08-2113

Morgan L, Broughton Pipkin F, Kalsheker N (1996) Angiotensinogen: molecular biology, biochemistry and physiology. Int J Biochem Cell Biol 28(11):1211–1222. doi:10.1016/S1357-2725(96)00086-6

Mottonen J, Strand A, Symersky J, Sweet RM, Danley DE, Geoghegan KF, Gerard RD, Goldsmith EJ (1992) Structural basis of latency in plasminogen activator inhibitor-1. Nature 355 (6357):270–273. doi:10.1038/355270a0

Mushunje A, Evans G, Brennan SO, Carrell RW, Zhou A (2004) Latent antithrombin and its detection, formation and turnover in the circulation. J Thromb Haemost 2(12):2170–2177. doi:10.1111/j.1538-7836.2004.01047.x

Ngamkitidechakul C, Warejcka DJ, Burke JM, O'Brien WJ, Twining SS (2003) Sufficiency of the reactive site loop of maspin for induction of cell-matrix adhesion and inhibition of cell invasion: conversion of ovalbumin to a maspin-like molecule. J Biol Chem 278 (34):31796–31806. doi:10.1074/jbc.M302408200

Parfrey H, Mahadeva R, Ravenhill NA, Zhou A, Dafforn TR, Foreman RC, Lomas DA (2003) Targeting a surface cavity of α_1-antitrypsin to prevent conformational disease. J Biol Chem 278(35):33060–33066. doi:10.1074/jbc.M302646200

Patschull AOM, Segu L, Nyon MP, Lomas DA, Nobeli I, Barrett TE, Gooptu B (2011) Therapeutic target-site variability in α_1-antitrypsin characterized at high resolution. Acta Crystallogr Sect F: Struct Biol Cryst Commun 67(12):1492–1497. doi:10.1107/S1744309111040267

Picard V (2003) Antithrombin Phe229Leu: a new homozygous variant leading to spontaneous antithrombin polymerization in vivo associated with severe childhood thrombosis. Blood 102 (3):919–925. doi:10.1182/blood-2002-11-3391

Qureshi SH, Lu Q, Manithody C, Yang L, Rezaie AR (2014) Characterization of the protein Z-dependent protease inhibitor interactive-sites of protein Z. BBA-Proteins Proteom 1844 (9):1631–1637. doi:10.1016/j.bbapap.2014.06.011

Rau JC, Beaulieu LM, Huntington JA, Church FC (2007) Serpins in thrombosis, hemostasis and fibrinolysis. J Thromb Haemost 5(Suppl 1):102–115. doi:10.1111/j.1538-7836.2007.02516.x

Ravenhill L, Wagstaff L, Edwards DR, Ellis V, Bass R (2010) G-helix of maspin mediates effects on cell migration and adhesion. J Biol Chem 285(47):36285–36292. doi:10.1074/jbc.M110.177253

Ricagno S, Pezzullo M, Barbiroli A, Manno M, Levantino M, Santangelo MG, Bonomi F, Bolognesi M (2010) Two latent and two hyperstable polymeric forms of human neuroserpin. Biophys J 99(10):3402–3411. doi:10.1016/j.bpj.2010.09.021

Santangelo MG, Noto R, Levantino M, Cupane A, Ricagno S, Pezzullo M, Bolognesi M, Mangione MR, Martorana V, Manno M (2012) On the molecular structure of human neuroserpin polymers. Proteins 80(1):8–13. doi:10.1002/prot.23197

Sauk JJ, Nikitakis N, Siavash H et al (2005) Hsp47 a novel collagen binding serpin chaperone, autoantigen and therapeutic target. Front Biosci 10(2):107–118. doi:10.2741/1513

Schechter I, Berger A (1967) On the size of the active site in proteases. I. Papain. Biochem Biophys Res Commun 27(2):157–162. doi:10.1016/S0006-291X(67)80055-X

Schulze AJ, Baumann U, Knof S, Jaeger E, Huber R, Laurell C-B (1990) Structural transition of α_1-antitrypsin by a peptide sequentially similar to β-strand s4A. Eur J Biochem 194(1):51–56. doi:10.1111/j.1432-1033.1990.tb19425.x

Schulze AJ, Frohnert PW, Engh RA, Huber R (1992) Evidence for the extent of insertion of the active site loop of intact α_1 proteinase inhibitor in β-sheet A. Biochemistry 31(33):7560–7565. doi:10.1021/bi00148a017

Schussler GC (2000) The thyroxine-binding proteins. Thyroid 10(2):141–149. doi:10.1089/thy.2000.10.141

Silverman GA, Bird PI, Carrell RW, Church FC, Coughlin PB, Gettins PGW, Irving JA, Lomas DA, Luke CJ, Moyer RW, Pemberton PA, Remold-O'Donnell E, Salvesen GS, Travis J, Whisstock JC (2001) The serpins are an expanding superfamily of structurally similar but functionally diverse proteins: evolution, mechanism of inhibition, novel functions, and a revised nomenclature. J Biol Chem 276(36):33293–33296. doi:10.1074/jbc.R100016200

Skinner R, Chang W-SW, Jin L, Pei X, Huntington JA, Abrahams J-P, Carrell RW, Lomas DA (1998) Implications for function and therapy of a 2.9 Å structure of binary-complexed antithrombin. J Mol Biol 283(1):9–14. doi:10.1006/jmbi.1998.2083

Stratikos E, Gettins PG (1999) Formation of the covalent serpin-proteinase complex involves translocation of the proteinase by more than 70 Å and full insertion of the reactive center loop into β-sheet A. Proc Natl Acad Sci USA 96(9):4808–4813. doi:10.1073/pnas.96.9.4808

Walker B, Lynas JF (2001) Strategies for the inhibition of serine proteases. Cell Mol Life Sci 58 (4):596–624. doi:10.1007/PL00000884

Wei Z, Yan Y, Carrell RW, Zhou A (2009) Crystal structure of protein Z-dependent inhibitor complex shows how protein Z functions as a cofactor in the membrane inhibition of factor X. Blood 114(17):3662–3667. doi:10.1182/blood-2009-04-210021

Whisstock JC, Pike RN, Jin L, Skinner R, Pei XY, Carrell RW, Lesk AM (2000) Conformational changes in serpins: II. The mechanism of activation of antithrombin by heparin. J Mol Biol 301 (5):1287–1305. doi:10.1006/jmbi.2000.3982

Wright PA, Rostom AA, Robinson CV, Schofield CJ (2000) Mass spectrometry reveals elastase inhibitors from the reactive centre loop of α_1-antitrypsin. Bioorg Med Chem Lett 10 (11):1219–1221. doi:10.1016/S0960-894X(00)00194-3

Xue Y, Björquist P, Inghardt T, Linschoten M, Musil D, Sjölin L, Deinum J (1998) Interfering with the inhibitory mechanism of serpins: crystal structure of a complex formed between cleaved plasminogen activator inhibitor type 1 and a reactive-centre loop peptide. Structure 6 (5):627–636. doi:10.1016/S0969-2126(98)00064-1

Yamasaki M, Li W, Johnson DJD, Huntington JA (2008) Crystal structure of a stable dimer reveals the molecular basis of serpin polymerization. Nature 455(7217):1255–1258. doi:10.1038/nature07394

Yamasaki M, Sendall TJ, Harris LE, Lewis GMW, Huntington JA (2010) Loop-sheet mechanism of serpin polymerization tested by reactive center loop mutations. J Biol Chem 285 (40):30752–30758. doi:10.1074/jbc.M110.156042

Yamasaki M, Sendall TJ, Pearce MC, Whisstock JC, Huntington JA (2011) Molecular basis of α_1-antitrypsin deficiency revealed by the structure of a domain-swapped trimer. EMBO Rep 12 (10):1011–1017. doi:10.1038/embor.2011.171

Ye S, Cech AL, Belmares R, Bergstrom RC, Tong Y, Corey DR, Kanost MR, Goldsmith EJ (2001) The structure of a Michaelis serpin-protease complex. Nat Struct Mol Biol 8(11):979–983. doi:10.1038/nsb1101-979

Youn B-S, Kloting N, Kratzsch J, Lee N, Park JW, Song E-S, Ruschke K, Oberbach A, Fasshauer M, Stumvoll M, Bluher M (2008) Serum vaspin concentrations in human obesity and type 2 diabetes. Diabetes 57(2):372–377. doi:10.2337/db07-1045

Zhou A, Huntington JA, Pannu NS, Carrell RW, Read RJ (2003) How vitronectin binds PAI-1 to modulate fibrinolysis and cell migration. Nat Struct Biol 10(7):541–544. doi:10.1038/nsb943

Zhou A, Stein PE, Huntington JA, Sivasothy P, Lomas DA, Carrell RW (2004) How small peptides block and reverse serpin polymerisation. J Mol Biol 342(3):931–941. doi:10.1016/j.jmb.2004.07.078

Genomic and Functional Studies of Serpins from Basal Vertebrates Provide Insight into the Evolution of Haemostatic Control and Blood Pressure Regulation

4

Hermann Ragg and Yunjie Wang

Contents

Abstract

Lampreys and hagfish are the only extant lineages of jawless fish that diverged from jawed vertebrates more than half a billion years ago. Lampreys, in particular, are regarded as a model system for the study of ancestral vertebrate-specific traits, though they also have developed lineage-specific features. Genetics, genomics and biochemistry of serpins from basal vertebrates may provide important insights into the emergence and regulation of the endothelium-lined haemostatic system and the pump-driven pressurised circulation that are characteristic for these metazoans. Recent studies indicate that heparin cofactor II (HCII), SpnV4_1 and angiotensinogen are early appearing regulators of coagulation and blood pressure modulation, respectively, in vertebrates. Cartilaginous fish possess almost all serpins characteristic for the advanced haemostatic control system of tetrapods, thus dating emergence of a nearly fully developed inhibitor ensemble to the base of gnathostomes. The conservation of exon-intron organisation of serpin genes from lampreys to mammals provides strong support for the gene structure-based classification of vertebrate serpins.

H. Ragg (✉) • Y. Wang
Faculty of Technology, Bielefeld University, 33615 Bielefeld, Germany
e-mail: hermann.ragg@uni-bielefeld.de

M. Geiger et al. (eds.), *The Serpin Family*, DOI 10.1007/978-3-319-22711-5_4

4.1 Lamprey Genomics Helps to Decipher the Evolution of Serpins Involved in Haemostasis

Database screening and literature searches reveal about 11 different serpin genes represented in the genomes of at least one of three lamprey species, including the sea lamprey *Petromyzon marinus* (*P. marinus*), the European river lamprey *Lampetra fluviatilis* (*L. fluviatilis*), and the Japanese lamprey *Lethenteron japonicum* (also known as *Lethenteron camtschaticum*) (Ragg et al. 2009; Kimura et al. 2009; Smith et al. 2013; Mehta et al. 2013). Due to the high G + C content of lamprey genomes, some of these serpin genes, however, are still incomplete, and some even may have escaped detection. With respect to exon-intron organisation, all lamprey serpin genes identified to date perfectly fit into the gene structure-based classification scheme that categorises vertebrate serpins into six groups, V1 to V6 (Ragg et al. 2001). Group V1 that exclusively consists of ovalbumin-type intracellular serpins encompasses three members in lampreys. Group V2 comprises two prominent members that code for angiotensinogen and HCII, respectively. No α_1-antitrypsin-like genes that, together with angiotensinogen, account for the clade A serpins (Silverman et al. 2001) are known from lampreys. Group V3 consists of a single representative, while two group V4 paralogues have been identified. Antithrombin, the only known associate of group V5, appears to be absent in lampreys. Group V6 that may include up to three HSP47-like genes in some fish lineages (Ragg et al. 2009) includes a single member in lampreys and in most gnathostomes.

Lamprey serpins and the genes encoding them provide useful insights into the evolution of the gene family across vertebrates and beyond. Serpin genes from vertebrates display distinctly divergent exon-intron patterns that differ strongly from those of other metazoans. It appears that a set of five groups of serpin genes (V1-V4,V6), each characterised by a particular gene structure, arose with or before the emergence of the vertebrate lineage, as none of these gene structures can be traced further back in metazoan history. The serpin genes of lancelets (cephalochordates), for instance, share a single conserved intron position with vertebrates, where, in total, 24 different intron positions map to the conserved serpin core (Ragg et al. 2009). Leaving gene structures aside, there is little hard data that could help to disentangle the roots of vertebrate serpins. Profound genomic alterations affecting both organisation and sequences of serpin genes may have been responsible for this conspicuous discontinuity. The only exception is neuroserpin (serpinI1) that has orthologues in lancelets and sea urchins (Kumar and Ragg 2008).

Once established, the distinct types of serpin gene organisation were maintained essentially unchanged across all vertebrates. The only modifications occur in some fish lineages, where a few serpin genes apparently acquired extra introns (Ragg et al. 2009). Some of these gene groups expanded to a dozen or even more members as observed with groups V1 and V2 in mammals. The only representatives of group V2 in lampreys are HCII and angiotensinogen, suggesting that one of these or their common ancestor may have served as template for generation of clade A serpins that, without exception, display group V2-specific gene structures. The relative scarcity of clade A serpins also in sharks (Sect. 4.4) argues in favour of the first alternative, though it is conceivable that lampreys and sharks may have lost these serpins. The origin of group V5 is currently obscure. The absence of its single member, antithrombin, in lampreys may either reflect lineage-specific loss or gain of the gene by an ancestral jawed vertebrate.

The division of vertebrate serpins into six categories with distinct gene structures that are maintained nearly invariably across all species thus provides a useful alternative to classify serpins, especially in cases where standard methods of phylogenetic analysis are problematic.

4.2 Functions of Lamprey Serpins Provide Insight into the Onsets of Coagulation Control and Blood Pressure Regulation

The recognised function of angiotensinogen in gnathostomes is that of a precursor for a family of hormones, the angiotensins that exert multiple functions, including blood pressure modulation and osmoregulation. Little is known about the role of the serpin scaffold of angiotensinogen from jawed vertebrates except that some single nucleotide polymorphisms are associated with increased blood pressure and other cardiovascular diseases (Jeunemaitre et al. 1997). There is some confusion about the existence of an endogenous angiotensinogen and angiotensin in lampreys. However, several lines of evidence strongly indicate the presence of a proprietary angiotensinogen in these basal vertebrates. The lamprey protein in question (GenBank accession: CAV16871.1), a serpin, contains a decapeptide sequence with 60 % identity (80 % similarity) to human angiotensin I. All residues of human angiotensin important for angiotensin receptor binding and activation are invariably conserved in the lamprey homologue (Wang and Ragg 2011; Wang et al. 2014). Moreover, the angiotensin sequence of lampreys and all other vertebrates resides at the very end of the N-terminal extension protruding from the serpin scaffold of all angiotensinogen orthologues known. Angiotensinogen genes from lampreys to man share the exon-intron pattern of group V2, thus corroborating affiliation of the lamprey gene and its protein product with the angiotensinogen family (Ragg et al. 2009).

Lamprey angiotensinogen, however, also shows features that are not associated with any other orthologue. The angiotensin sequence of this agnathan fish includes a five-residue N-terminal cluster of negatively charged amino acids of unknown functionality. Lamprey angiotensinogen also depicts potent protease-inhibitory activity, a feature that is unparalleled. In the presence of appropriate glycosaminoglycans, lamprey angiotensinogen and HCII from vertebrates (including lampreys) are efficient thrombin inhibitors that exhibit the standard serpin inhibitory mechanism. During the reaction, large-scale conformational changes are known to affect several structural elements of the serpin scaffold (Huntington et al. 2000). Since angiotensinogens of gnathostomes are non-inhibitory, sequence constraints imposed by the serpin reaction mechanism may have been relieved, thus fostering acquisition and conservation of mutations in the serpin scaffold. Ancestral sequence concordances, in contrast, may have been maintained in HCII and angiotensinogen of lampreys due to functional needs. This unique attribute may be responsible, in part, for the results of some phylogenetic classification methods that deem lamprey angiotensinogen as an HCII orthologue.

Several other features suggest indeed that HCII and angiotensinogen are offspring of a previous duplication event. As discussed above, the genes coding for angiotensinogen and HCII possess the characteristic exon-intron pattern (including conserved intron phases) associated with group V2 serpins. Angiotensinogen and HCII proteins share long N-terminal extensions (though of low similarity), about 80–100 residues in length. The reactive site loop (RSL) sequences of lamprey angiotensinogen and HCII are very similar (Wang and Ragg 2011), and both proteins inhibit thrombin. Though sequence similarity may ensue from convergent evolution, the formation of antiproteolytically active angiotensinogen from an ancestor unable to perform the serpin inhibitory reaction mechanism is not an appealing hypothesis. Gene structures, sequence data, and biochemical properties rather support a gene duplication scenario for the emergence of HCII and angiotensinogen. The actual lamprey HCII gene is easily identifiable through its phylogenetically conserved position within an intron of the *pi4ka* gene (Wang et al. 2014). Lamprey HCII also shares additional characteristic features with its orthologues, including the nearly invariable helix D region (Ragg et al. 2009) that is crucially involved in dermatan sulphate and heparin binding (Ragg et al. 1990). It is well known that standard phylogenetic approaches for evolutionary classification of proteins with short, highly conserved hormone sequences embedded in an otherwise divergent environment may pose problems (Mirabeau and Joly 2013).

There are reports on the presence of a teleost-type angiotensin II (NRVY[V/I]HPF) in lampreys that, however, is mostly not found in fasting animals (Wong et al. 2012). There is currently no evidence, in the *P. marinus* genome, for a gene encoding such a sequence; however, it is possible that the high G+C content of lamprey genomes impedes the identification of genes. Alternatively, the ectoparasitic lifestyle may be responsible for the presence of teleost-type angiotensin peptides in some lamprey tissues.

Mammalian angiotensinogen is proteolytically cleaved primarily via renin that liberates, from the N-terminus, the decapeptide angiotensin I that is further processed through angiotensin-converting enzyme 1 (ACE1) to the octapeptide angiotensin II. Renin is a very specific enzyme that often does not accept heterologous angiotensinogens (Hatae et al. 1994). Alternative pathways of angiotensin II formation and production of further biologically active cleavage products are known, involving participation of chymase and/or other proteases (Benigni et al. 2010). Angiotensin II, the best characterised of these peptides, exerts its numerous effects, including pressure modulation and osmoregulation through binding and activation of G protein-coupled receptors (GPCRs), angiotensin type

1 receptor (AGTR1) and angiotensin type 2 receptor (AGTR2). Little is known about the processing path(s) of angiotensinogen and mode of action(s) of its potential cleavage products in lampreys. Database searches in the *P. marinus* genome suggest the existence of an ACE1-like protein in lampreys (ENSEMBL, release 71: ENSPMAG00000007309); however, evidence for a renin orthologue is lacking. Initial cleavage of lamprey angiotensinogen might involve two proline residues adjacent to the angiotensin sequence (EEDYDERPYMQPFHLI**PPS**). Data mining also uncovers the presence of a truncated GPCR sequence in the sea lamprey genome (ENSEMBL accession: ENSPMAG00000005736) that clusters, in phylogenetic analyses, with AGTR1 from humans and other gnathostomes, though with modest bootstrap support. We recently isolated an apparently intact orthologue of this receptor from *L. fluviatilis* that encodes a protein of 352 amino acids. Several residues important for angiotensin/AGTR1 interaction or receptor activation in mammals are perfectly conserved in the lamprey protein (Wang et al. 2014). It thus appears that lampreys contain an intact angiotensin/AGTR1 signalling axis. To date, lampreys represent the most ancestral lineage with such a signalling system, though some other metazoans are known to possess individual constituents of the signalling pathway (Salzet et al. 2001). The intracellular effects potentially triggered by activation of lamprey AGTR1 remain to be analysed. Presumably, they affect blood pressure modulation and osmoregulation, since injection of lamprey angiotensin II (100–1000 pmol/kg) in cannulated conscious lampreys provokes either a transient vasodepressor effect or a biphasic cardiovascular response, depending on the dose used (Wong and Takei 2011).

Lamprey angiotensinogen serves as source of hormones and acts as protease inhibitor. In the presence of heparin, human thrombin is rapidly and effectively inhibited (Wang and Ragg 2011). Heparin accelerates the enzyme/inhibitor reaction in a bell-shaped dose-response curve, implying ternary complex formation as the underlying mechanism, similar to heparin-mediated thrombin/antithrombin interaction. Our recent investigations also showed that, from a number of proteases investigated, thrombin is the only enzyme reacting with lamprey angiotensinogen, thus indicating the high target selectivity of the inhibitor. In the light of the close relationship to HCII, this is not necessarily surprising, since it is not uncommon that gene duplication results in copies with specialised tasks (subfunctionalisation). This also appears to be the case with HCII and angiotensinogen following their emergence more than half a billion years ago. The two descendants of the ancestral thrombin inhibitor prefer different cofactors. Dermatan sulphate and to a lesser extent heparin are the best activators of HCII both from humans and from lampreys. The stimulating effect of dermatan sulphate on lamprey angiotensinogen/thrombin interaction, in contrast, is modest in comparison with heparin or heparan sulphate. Together, angiotensinogen and HCII might initially have regulated different aspects of thrombin's manifold roles that, in gnathostomes, are mainly controlled by antithrombin.

The coagulation cascade of ancestral vertebrates, as inferred from the lamprey model, is a basic version of the modern blood clotting system. Essential constituents of the primordial coagulation system include factor Xa (FXa), in addition to factors FVII, thrombin, fibrinogen and others (Doolittle 2009, 2013). However, since genes coding for antithrombin or protein Z-dependent inhibitor (PZPI), which is also known as serpinA10, are not detected in the lamprey genome, it remained an unresolved issue whether there is a serpin-type FXa inhibitor in these animals. Biochemical characterisation of a recently identified serpin from *L. fluviatilis*, Lfl_SpnV4_1 (Ragg et al. 2009), revealed that the bacterially expressed and refolded protein shows potent inhibitory activity towards human FXa (Ragg and Wang 2013; Wang et al. 2014). Similar to antithrombin, Lfl_SpnV4_1 exhibits arginine at the presumed P1 position. The protein, a serpin belonging to group V4, is neither closely related to PZPI (group V2) nor to antithrombin (group V5). It thus appears that lampreys may exert anti-FXa activity through serpins, though different family members execute this job in jawless and jawed vertebrates. Interestingly, the genes coding for SpnV4_1 and HCII are situated in immediate proximity and in parallel orientation within an intron of the sea lamprey *pi4ka* gene. The appearance of paralogous serpin genes at the same chromosomal locus thus provides additional evidence for the proposed kinship between groups V2 and V4. Lampreys presumably have kept an ancestral genomic state ensuing from an ancient gene duplication that involved the progenitor of serpin groups V2 and V4. Gnathostomes, in contrast, eventually lost the SpnV4_1-like FXa inhibitor, as no orthologue appears to present in this lineage.

Current evidence thus suggests that lampreys contain at least three serpins that interfere with two different procoagulant key enzymes. The maintenance of angiotensinogen and HCII as potent and selective thrombin inhibitors for more than 500 million years identifies these proteins as primordial thrombin regulators in the basal coagulation cascade. The synteny of HCII and SpnV4_1 in the lamprey genome marks the latter serpin as ancestral antagonist of FXa activity. Composition and properties of the anticoagulant serpin spectrum of lampreys provide plausible explanations for some events presumably associated with the advent of the complex coagulation cascade of gnathostomes. A completely new aspect of early vertebrate physiology, still maintained in present-day lampreys, is the intimate interlacing of blood pressure modulation, osmoregulation and coagulation control mediated through one and the same molecule. Though the exact physiological context of this linkage is not clear, it should be kept in mind that, in gnathostomes, cross talk between coagulation and pressure regulation also exists (Borissoff et al. 2009). The study of lamprey serpins thus has yielded important and unexpected insight into the origin of complex traits associated with the emergence of vertebrates.

4.3 Novel Serpins Emerged with the Increasingly Complex Coagulation Network of Gnathostomes

With the advent of the increasingly elaborate coagulation network in gnathostomes (Doolittle 2009), novel serpins interfering with procoagulant activity and clot dissolution appeared. The elephant fish *Callorhinchus milii* (*C. milii*), an archaic cartilaginous fish, and sharks contain an antithrombin gene, arguing for the emergence of the most important circulating thrombin inhibitor of gnathostomes soon after their split from cyclostomes. However, it is also possible that lampreys have lost the gene. The antithrombin genes of both *C. milii* (GenBank accession number: NW_006890154.1) (Venkatesh et al. 2014) and from the brown-banded bamboo shark *Chiloscyllium punctatum* (KP128093.1) possess the standard gene organisation known from their orthologues in tetrapods and most fish lineages. The antithrombin genes of *Danio rerio* (*D. rerio*) and some other fish lineages have an additional intron at position 262c [numbering refers to the sequence of mature human α_1-antitrypsin (Ragg et al. 2001)], suggesting intron gain. The antithrombin proteins of cartilaginous fishes share with their orthologues strongly conserved RSL sequences (brown-banded bamboo shark: NEEGSEAAGASTVIVMGR↓SLRP) and nearly invariable helix D regions, the most important structural features determining specificity and reactivity of the inhibitor towards thrombin, factor Xa and other clotting proteases. The appearance of the broad-spectrum inhibitor antithrombin at the base of gnathostomes may have influenced the fate of the more ancestral inhibitors, resulting in loss of SpnV4_1 and inactivation of angiotensinogen's antiproteolytic activity.

PZPI (serpinA10) of mammals is an anticoagulant serpin with a tyrosine residue at the P1 position. In the presence of protein Z, lipid, calcium and heparin, PZPI inactivates coagulation factors FXa and FXIa. Several charged surface loops contribute to the specificity of PZPI towards the target enzymes and cofactors (Wei et al. 2009). Gene organisation reveals PZPI as a member of group V2 serpins that, in lampreys, solely include angiotensinogen and HCII. In *C. milii*, group V2 comprises at least six members (GenBank accession numbers in brackets): angiotensinogen (NP_001279673.1), HCII (XP_007900003.1), HCII-L, a HCII paralogue (XP_007904387.1) and three additional serpins (XP_007902080.1, which includes two serpins erroneously merged to a single protein, and NP_001279794.1). Though, with the exception of HCII and angiotensinogen, the family relationships of group V2 serpins are difficult to unravel, several features point out to the presence of PZPI in cartilaginous fishes. *C. milii* contains a serpin (part of XP_007902080.1) with an RSL sequence showing a tyrosine residue at P1 (GTEAAAVTAVTATAY↓SLPKIL). The protein also displays the IYEE signature close to the C-terminal end that, with little variation, is also present in most PZPI orthologues. Conservation of genomic synteny corroborates that PZPI is an early constituent of the coagulation control system. From *C. milii* to man, the *serpina10* and *ppp4r4* genes are immediate genomic neighbours with antiparallel orientation (Fig. 4.1).

The contribution of protease nexin-1 (PN-1, serpinE2) to the regulation of coagulation and fibrinolysis has only recently been recognised. PN-1 is a powerful thrombin inhibitor in the presence of heparin that also inhibits several other proteases although with significantly lower reaction rates (Bouton et al. 2012). PN-1 is barely detectable in plasma, but it is found in many organs, where it is thought to act in the vessel walls. In humans and other gnathostomes, the *PN-1* gene is part of a gene triad [*serpine2* (→)—*wdfy1* (→)—*ap1s3* (→)], (relative

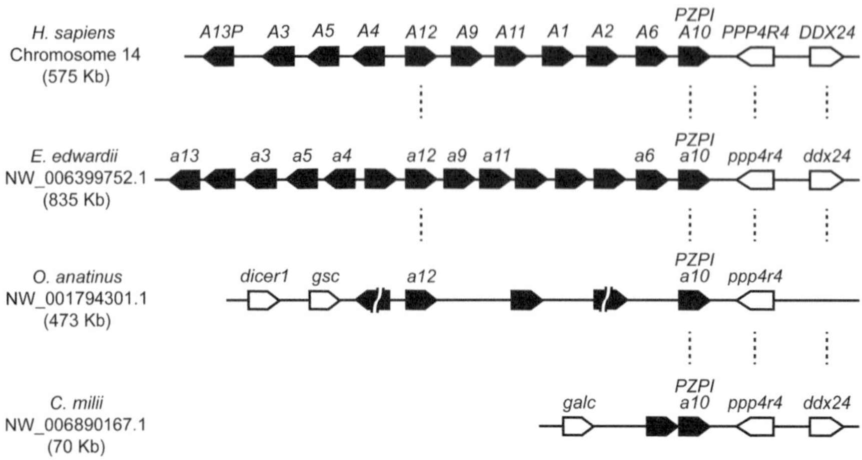

Fig. 4.1 Comparative genomics enables reconstruction of the evolutionary history of a group V2 serpin gene cluster. Serpin genes are highlighted in *black*, and neighbouring genes are represented by *white boxes*. In humans, the cluster encompasses 11 genes coding for clade A serpins, including four genes with antiparallel orientation. The group V2 serpin gene cluster of the Cape elephant shrew, *Elephantulus edwardii*, an insectivorous mammal, contains 14 serpin genes, including *serpina5*. In platypus (*Ornithorhynchus anatinus*), an egg-laying mammal, two of five identified V2 genes are still incomplete (marked by *wavy lines*). At the homologous genomic locus of the elephant fish *C. milii*, only two serpin genes, including *pzpi*, are present. Genes and gene distances are not drawn to scale

orientation of genes in bracketed arrows) that is also present in *C. milii* (GenBank accession: XP_007886566.1). Though the conserved microsynteny is a strong indicator for the presence of *PN-1* in cartilaginous fishes, it is not clear whether the gene product fulfils a role similar to that in mammals. The basic helix D of human PN-1 plays an important role for heparin-activated thrombin inhibition. The homologous sequence of the elephant fish, however, contains significantly less positively charged residues than the human protein. There are also some differences in the RSL sequences that might differently affect the inhibitory activity of PN-1 from elephant fish (IIMLK↓SIG) and mammals (ILIAR↓SSP), respectively.

Human protein C inhibitor (PCI, SERPINA5) can inhibit activated protein C, thrombin, some other procoagulant enzymes and the thrombin/thrombomodulin complex that activates protein C to an anticoagulant enzyme. PCI thus can interfere with pro- and anticoagulant reaction steps, but the role of PCI in haemostasis is still unclear (Geiger 2007). The absence of PCI in plasma of mice also underscores the ambiguous significance of PCI during the formation of the haemostatic network. Comparative genomics reveals that PCI is probably an innovation that arose at the base of mammalian radiation. In the Cape elephant shrew *Elephantulus edwardii*, an insectivorous mammal, the PCI gene is located within a conserved microsyntenic arrangement that includes three group V2 serpins [*serpina4* (→)—*serpina5* (→)—*serpina3* (→)]. This block of serpin genes appears to be present only in mammals but not in other vertebrates (Fig. 4.1). The emergence of this gene triad may antedate the split of monotremes, but genomic data are still insufficient to draw firm conclusions. The PCI protein of basal mammals shares with its orthologues a patch of basic residues along the helix H region that is involved in heparin binding and the sequence FR↓SA[W/Y/R] at positions P2 to P3′ of the RSL.

4.4 Serpins Affecting Fibrinolysis

Plasminogen activator inhibitor-1 (PAI-1, serpinE1) and α_2-antiplasmin (serpinF2) are serpins that contribute to the regulation of fibrinolysis. PAI-1 is a fast-acting inhibitor of tissue-type plasminogen activator and urokinase that convert plasminogen into the active, fibrin-degrading form. PAI-1 also plays a role in various other processes, including cell migration, adhesion and proliferation (Dupont et al. 2009). PAI-1 is a group V3 serpin and thus related to neuroserpin and nexin-1. The presence of PAI-1 in *C. milii* (NP_001279494.1) within a conserved gene triad including *serpine1* (→), *ap1s1* (→) and *vgf* (←) indicates that the gene emerged during the formation of gnathostomes at the latest. The biochemical functions of PAI-1 are maintained without changes as suggested by the little varying RSL core sequences [Y/S][S/A]R↓M[A/S].

α_2-Antiplasmin (serpinF2) inactivates plasmin, an enzyme participating in clot dissolution. Apart from *D. rerio* and some other fish lineages, vertebrates from *C. milii* to humans exhibit a syntenic block of four genes [*wdr81* (→)—*serpinf2* (→)—*serpinf1* (→)—*myo1c* (←)] thus documenting that α_2-antiplasmin originated at the base of gnathostomes at the latest. All α_2-antiplasmin orthologues contain C-terminal extensions of varying sizes and sequences that end with the characteristic, though functionally dispensable PK dipeptide motif. With the emergence of mammals, the RSL of α_2-antiplasmin underwent a three-nucleotide insertion (Fig. 4.2) resulting in a novel, alternative RSL cleavage site (Potempa et al. 1988). Unravelling the roots of serpin-mediated antifibrinolytic activity requires further investigation, since it not known how clot lysis in basal vertebrates such as lampreys (Ponczek et al. 2012) is regulated.

```
Elephant fish (XP_007894683.1)   KEEGVEAAAATVVAMSR-SFSMFSINRPF
Spotted gar (XP_006641161.1)     NEQGAEAAAATSVVISR-SSFFFSLNRPF
Zebrafish (NP_001073479.1)       NEEGAEAVAATSVVISR-SNPSFTVNQPF
Coelacanth (XP_006006155.1)      QEDGAEAAAATSITMSR-SFVGFHVNSPF
Painted turtle (XP_005298341.1)  AEDGVEASAATSVMMSR-SLSTFSLNQPF
Chicken (XP_003642447.1)         KEDGVEASAATGVMISR-SLSAFSIDRPF
Platypus (XP_007668099.1)        NEEGVEASAATSISMSRMSAASFRVNRPF
Elephant shrew (XP_006891086.1)  SEAGVEAAAATGTAMSRMSLSSFSVNRPF
Cattle (XP_005220124.1)          SEAGVQAAAATSTAMSRMSLSSFIVNRPF
Human (XP_005256760.1)           SEVGVEAAAATSIAMSRMSLSSFSVNRPF
Mouse (NP_032904.1)              SEAGVEAAAATSVAMNRMSLSSFTVNRPF
                                  * *  * ***     * *   *    **
```

Fig. 4.2 Alignment of α_2-antiplasmin RSL sequences from various vertebrates. An insertion of three nucleotides in the mammalian lineage resulted in a novel, additional RSL cleavage site. GenBank accession numbers are given in *brackets*

Acknowledgement We thank Ingrid Verhamme for constructive comments and helping in manuscript proofreading.

References

Benigni A, Cassis P, Remuzzi G (2010) Angiotensin II revisited: new roles in inflammation, immunology and aging. EMBO Mol Med 2:247–257

Borissoff JI, Spronk HM, Heeneman S et al (2009) Is thrombin a key player in the 'coagulation-atherogenesis' maze? Cardiovasc Res 82:392–403

Bouton MC, Boulaftali Y, Richard B et al (2012) Emerging role of serpinE2/protease nexin-1 in hemostasis and vascular biology. Blood 119:2452–2457

Doolittle RF (2009) Step-by-step evolution of vertebrate blood coagulation. Cold Spring Harb Symp Quant Biol 74:35–40

Doolittle RF (2013) The evolution of vertebrate blood clotting. University Science Books, Wilsted & Taylor Publishing Services

Dupont DM, Madsen JB, Kristensen T et al (2009) Biochemical properties of plasminogen activator inhibitor-1. Front Biosci 14:1337–1361

Geiger M (2007) Protein C inhibitor, a serpin with functions in- and outside vascular biology. Thromb Haemost 97:343–347

Hatae T, Takimoto E, Murakami K et al (1994) Comparative studies on species-specific reactivity between renin and angiotensinogen. Mol Cell Biochem 131:43–47

Huntington JA, Read RJ, Carrell RW (2000) Structure of a serpin-protease complex shows inhibition by deformation. Nature 407:923–926

Jeunemaitre X, Inoue I, Williams C et al (1997) Haplotypes of angiotensinogen in essential hypertension. Am J Hum Genet 60:1448–1460

Kimura A, Ikeo K, Nonaka M (2009) Evolutionary origin of the vertebrate blood complement and coagulation systems inferred from liver EST analysis of lamprey. Dev Comp Immunol 33:77–87

Kumar A, Ragg H (2008) Ancestry and evolution of a secretory pathway serpin. BMC Evol Biol 8:250

Mehta TK, Ravi V, Yamasaki S et al (2013) Evidence for at least six Hox clusters in the Japanese lamprey (*Lethenteron japonicum*). Proc Natl Acad Sci USA 110:16044–16049

Mirabeau O, Joly JS (2013) Molecular evolution of peptidergic signaling systems in bilaterians. Proc Natl Acad Sci USA 110:E2028–E2037

Ponczek MB, Bijak MZ, Nowak PZ (2012) Evolution of thrombin and other hemostatic proteases by survey of protochordate, hemichordate, and echinoderm genomes. J Mol Evol 74:319–331

Potempa J, Shieh BH, Travis J (1988) Alpha-2-antiplasmin: a serpin with two separate but overlapping reactive sites. Science 241:699–700

Ragg H, Ulshöfer T, Gerewitz J (1990) Glycosaminoglycan-mediated leuserpin-2/thrombin interaction. Structure-function relationships. J Biol Chem 265:22386–22391

Ragg H, Lokot T, Kamp PB et al (2001) Vertebrate serpins: construction of a conflict-free phylogeny by combining exon-intron and diagnostic site analyses. Mol Biol Evol 18:577–584

Ragg H, Kumar A, Köster K et al (2009) Multiple gains of spliceosomal introns in a superfamily of vertebrate protease inhibitor genes. BMC Evol Biol 9:208

Ragg H, Wang Y (2013) Serpins interfering with procoagulant activity and pressure regulation in a basal vertebrate. J Thromb Haemost 11(Suppl 2):853

Salzet M, Deloffre L, Breton C et al (2001) The angiotensin system elements in invertebrates. Brain Res Brain Res Rev 36:35–45

Silverman GA, Bird PI, Carrell RW et al (2001) The serpins are an expanding superfamily of structurally similar but functionally diverse proteins. Evolution, mechanism of inhibition, novel functions, and a revised nomenclature. J Biol Chem 276:33293–33296

Smith JJ, Kuraku S, Holt C et al (2013) Sequencing of the sea lamprey (*Petromyzon marinus*) genome provides insights into vertebrate evolution. Nat Genet 45:415–421
Venkatesh B, Lee AP, Ravi V et al (2014) Elephant shark genome provides unique insights into gnathostome evolution. Nature 505:174–179
Wang Y, Ragg H (2011) An unexpected link between angiotensinogen and thrombin. FEBS Lett 585:2395–2399
Wang Y, Köster K, Lummer M et al (2014) Origin of serpin-mediated regulation of coagulation and blood pressure. PLoS One 9:e97879
Wei Z, Yan Y, Carrell RW et al (2009) Crystal structure of protein Z-dependent inhibitor complex shows how protein Z functions as a cofactor in the membrane inhibition of factor X. Blood 114:3662–3667
Wong MK, Takei Y (2011) Characterization of a native angiotensin from an anciently diverged serine protease inhibitor in lamprey. J Endocrinol 209:127–137
Wong MK, Sower SA, Takei Y (2012) The presence of teleost-type angiotensin components in lamprey buccal gland suggests a role in endocrine mimicry. Biochimie 94:637–648

Regulation of Neutrophil Serine Proteases by Intracellular Serpins

5

Charaf Benarafa

Contents

Abstract

Neutrophil granules contain serine proteases that are central components of the antimicrobial weapons of the innate immune system. Neutrophil proteases also contribute to the amplification and resolution of inflammatory responses through defined proteolytic cleavage of mediators, cell surface receptors, and extracellular matrix proteins. In the blood and at mucosal surfaces, neutrophil serine

C. Benarafa, D.V.M., Ph.D. (✉)
Theodor Kocher Institute, University of Bern, Freiestrasse 1, 3012 Bern, Switzerland
e-mail: charaf.benarafa@tki.unibe.ch

M. Geiger et al. (eds.), *The Serpin Family*, DOI 10.1007/978-3-319-22711-5_5

proteases are regulated by serpins found in plasma and by non-serpin secreted inhibitors. Distinct mechanisms leading to neutrophil cell death have been described for the granule serine proteases, neutrophil elastase, cathepsin G, and proteinase-3. Granule leakage in neutrophils triggers death pathways mediated by cathepsin G and proteinase-3, and both proteases are tightly regulated by their inhibitor SERPINB1 in a cell intrinsic manner. Although stored in the same types of granules, neutrophil elastase does not significantly contribute to cell death following intracellular release from granules into the cytoplasm. However, heterozygous mutations in *ELANE*, the gene encoding elastase, are the cause of severe congenital neutropenia, a life-threatening condition characterized by the death of neutrophils at an early precursor stage in the bone marrow. This chapter focuses on recent work exploring the biology of clade B intracellular serpins that inhibit neutrophil serine proteases and their functions in neutrophil homeostasis and serine protease control at sites of inflammation.

5.1 Neutrophil Granules and Their Proteases

5.1.1 Neutrophil Biology and Neutrophil Serine Proteases

Neutrophils are white blood cells with a prominent function in the control of microbes at mucosal surfaces and in tissues. The containment and elimination of bacteria and fungi can be severely affected if too few neutrophils are available or if there is a defect in their ability to reach the infection site. Neutrophils carry a vast repertoire of antimicrobial molecules and enzymes that directly contribute to killing pathogens. While neutrophils are essential to fight bacterial and fungal infections, the presence and death of neutrophils at inflammatory sites has been associated with delayed healing, tissue damage, and pathogenesis of chronic diseases. For these reasons, neutrophils and their potentially harmful granule cargo have been considered a double-edged sword in inflammatory diseases. Neutrophils are now emerging as a central orchestrator of inflammatory responses and their resolution through continuous interactions with components of the innate and adaptive responses (Nathan 2006; Kruger et al. 2015). In this chapter, the function of the neutrophil granule serine proteases in neutrophil life cycle and their function in host defense and in inflammatory diseases will be reviewed with a special focus on their regulation by intracellular serpins.

Neutrophils develop in the bone marrow from hematopoietic stem cells and differentiate progressively through progenitors with more restricted lineage potential known as the common myeloid progenitors and the granulocyte-macrophage progenitors (GMPs). The transcription factor PU.1 is essential for the generation of GMPs. At the GMP stage, the expression of the transcription factors, CCAAT/enhancer-binding protein-α (C/EBPα) and growth factor independent-1 (GFI-1), appears essential for commitment toward the neutrophil lineage, whereas PU.1

promotes monocyte differentiation (Cheng et al. 1996; Zhang et al. 1997; Dahl et al. 2003). C/EBP-α and GFI-1 promote proliferation of neutrophil precursors from myeloblasts, promyelocytes, and myelocytes, known as the mitotic pool of neutrophils in the bone marrow. These transcription factors are also key for the expression of neutrophil serine proteases and other primary granule proteins such as myeloperoxidase and defensins at the promyelocyte stage. As C/EPB-α and GFI-1 levels diminish, terminal granulocytic differentiation to mature neutrophils is under the transcriptional regulation of C/EBP-ε, which controls the expression of secondary and tertiary granule proteins. These stages are also known as the postmitotic pool of neutrophils. Mature neutrophils accumulate in the bone marrow as a large reserve pool that can be rapidly released to the circulation. During the last phase of differentiation, neutrophils start expressing receptors to transduce signals for mobilization to the blood and for effector functions. Once in the blood circulation, neutrophils have a relatively short life span of a few hours. They can sense tissue injury through cytokines, peptides, and lipid mediators secreted and/or presented by endothelial cells. Once firmly arrested on the vascular endothelium, neutrophils migrate into the tissue in a highly directional manner toward the injury site. Recruited neutrophils are highly efficient phagocytes and kill microbes using a broad range of antimicrobial peptides and enzymes. In the absence of inflammation, the neutrophil life cycle ends by efferocytosis, where neutrophils undergo apoptosis and are phagocytosed. Efferocytosis of neutrophils occurs in all tissues but, in steady-state conditions, it occurs principally in the spleen, the liver, and the bone marrow. This form of cell death is important for several reasons (Vandivier et al. 2006). First, it dampens the production of inflammatory mediators through the production of TGF-β, for example. Second, and in contrast to necrotic neutrophils, apoptotic neutrophils do not release their highly reactive contents, thus preventing damage to neighboring cells and extracellular matrix proteins. Third, impaired removal of apoptotic cells is associated with autoimmune disease. Last but not least, phagocytosis of apoptotic neutrophils reduces the production of IL-23 by the macrophages. IL-23 induces the production of IL-17 by αβ and γδ T cells. IL-17, in turn, upregulates G-CSF production by stromal cells leading to increased granulopoiesis (Stark et al. 2005). Therefore, the removal of apoptotic cells in the periphery sends a negative feedback loop that ultimately regulates neutrophil production in the bone marrow.

5.1.2 Granule Subsets and Protease Packaging

Neutrophil serine proteases (NSPs) contribute to many aspects of neutrophil biology in host defense and diseases associated with neutrophilic inflammation (Pham 2006). Human neutrophils contain four NSPs with an active catalytic triad: neutrophil elastase (NE), proteinase-3 (PR3), cathepsin G (CG), and the recently characterized neutrophil serine protease-4 (NSP4). NE and PR3 are elastinolytic proteases, and CG has chymotrypsin-like activity, whereas NSP4 is a trypsin-like protease. Azurocidin is encoded by a phylogenetically related serine protease gene, but its catalytic domain is mutated rendering the protease inactive. NSPs evolved from a common ancestor and are phylogenetically related to other serine proteases found in innate immune cells including granzymes and mast cell proteases (Ahmad et al. 2014). NSPs are conserved in mice, and in contrast to granzymes and mast cell proteases, there is a single mouse homolog for each of the human NSP genes.

Neutrophils contain three types of granules (primary, secondary, and tertiary) and secretory vesicles. NSPs are principally stored in primary, also termed azurophil, granules. However, PR3 was also reported in other types of granules and secretory vesicles. Together with NSPs, primary granules carry myeloperoxidase, lysozyme, and antimicrobial peptides such as α-defensins and cathelicidin. How each cargo protein ends up in the correct granule appears largely controlled by transcriptional timing (Cowland and Borregaard 1999; Theilgaard-Mönch et al. 2005). As described above, different transcription factors dominate at different phases of neutrophil differentiation and mRNA expression of NSPs is highest in promyelocytes, which is the time where primary granules are formed.

NSPs are synthesized as pre-pro-enzymes and require posttranslational cleavage at the N-terminus and C-terminus to gain their fully mature active form. First, the signal peptide is cleaved by a signal peptidase. The resulting pro-enzymes are then further processed by dipeptidyl peptidase I (DPPI), also known as cathepsin C, which cleaves the N-terminal dipeptide in the endoplasmic reticulum (ER) and Golgi apparatus before packaging into granules. Cleavage of the pro-dipeptide by DPPI is essential for the enzymatic activity of the NSPs. Neutrophils of *dppi*$^{-/-}$ mice have no detectable NSP activity, and CG protein is not detectable in neutrophil lysates (Adkison et al. 2002). The C-terminus of the pro-enzymes is also cleaved before packaging into granules, but the protease responsible for this process has not yet been identified. Mutants of NE and CG lacking the C-terminal peptide and expressed in RBL cell line are correctly routed to granules and are enzymatically active indicating that the C-terminal domain is not required for these functions (Gullberg et al. 1995).

The mechanisms of targeting and retention of NSPs in primary granules are not fully elucidated. Cleavage of the C-terminus of NE appears to promote the routing of NE to granules by two mechanisms: first, the C-terminus contains a putative transmembrane domain that leads to membrane anchoring and leads to targeting to the plasma membrane; second, the cleavage of the C-terminus leads to the uncovering of a binding domain for adaptor protein complex 3 (AP3), which may favor protein distribution to granules (Benson et al. 2003). The proteoglycan serglycin contributes to the retention of proteases and inflammatory mediators in leukocytes and endothelial cells (Kolset and Tveit 2008). In neutrophils of serglycin-deficient mice, only NE appeared to require serglycin for proper localization into granules, whereas sorting of CG and PR3 was not affected (Niemann et al. 2007). While early studies showed that PR3 localized principally in primary granules (Borregaard and Cowland 1997), PR3 has also been reported in secretory vesicles and other types of granules, which may explain the more rapid mobilization of PR3 to the plasma membrane (Witko-Sarsat et al. 1999; Loison et al. 2014).

5.2 NSP Inhibitors

5.2.1 Secreted Clade A Serpins

Several inhibitors of the clade A and the clade B serpins, as well as non-serpin inhibitors of the macroglobulin and chelonianin families, contribute to the regulation of the activity of NSPs in different compartments of the cell, tissues, and the whole organism.

The clade A serpins α1-antitrypsin (AAT, SERPINA1) and α1-antichymotrypsin (ACT, SERPINA3) are the two principal serpins found in plasma that inhibit NSPs. AAT inhibits NE, CG, and PR3 by the classical suicide-substrate mechanism of serpins (Huntington et al. 2000). The three NSPs cleave the reactive center loop (RCL) of AAT at the same P1 site Met-358. ACT, in contrast, inhibits only CG. AAT and ACT are produced by the liver and respiratory airway epithelial cells and function by regulating NSPs in blood and extracellular compartments when NSPs are released from neutrophils following degranulation or necrosis.

AAT is also expressed in neutrophils during the late stages of granulopoiesis in the bone marrow and is found at relatively higher levels in mature differentiated cells (Missen et al. 2006). AAT protein is found within neutrophil granules and can be secreted upon stimulation (Mason et al. 1991; Clemmensen et al. 2011). In circulating neutrophils, AAT levels in neutrophils are increased as the secretory vesicles are formed by engulfing plasma proteins that can later be released in response to inflammatory stimuli (Borregaard et al. 1992). The relative importance and function of AAT within the different granules of neutrophils in homeostatic and inflammatory conditions remain to be defined. There is also growing evidence that AAT functions as an anti-inflammatory mediator independently of NSP inhibition (Jonigk et al. 2013). One mechanism may be through interference with TNF-α signaling pathways (Bergin et al. 2014).

In mice and rats, AAT and ACT genes have been duplicated multiple times. The number of AAT genes and plasma concentrations appear to vary between different laboratory mouse strains. Modeling AAT deficiency by deleting the mouse genes has proved to be even more challenging because of embryonic lethality (Wang et al. 2011).

5.2.2 Intracellular Clade B Serpins

Clade B serpins regroup phylogenetically related proteins with nuclear and/or cytoplasmic localization. Vertebrate clade B serpins evolved from a single serpinB1-like gene that has remained conserved in fish, birds, and mammals (Kaiserman and Bird 2005; Benarafa and Remold-O'Donnell 2005). In man, there are 13 clade B serpins found on two loci. *SERPINB1*, *SERPINB6*, and *SERPINB9* genes are found on chromosome 6p25, and the ten other clade B serpin genes are clustered on chromosome 18q21. Serpins encoded in the 6p25 cluster inhibit granule serine proteases of leukocytes, and high expression of these serpins is found in cells that also carry target proteases.

SERPINB1 was the first cytoplasmic inhibitor of elastase to be identified in neutrophils and monocytes/macrophages (Remold-O'Donnell 1985, 1989; Potempa et al. 1988). It was thus named monocyte/neutrophil elastase inhibitor (MNEI) and leukocyte elastase inhibitor (LEI). SERPINB1 also inhibits CG and PR3 through the classical serpin complex inhibition mechanism. NE and PR3 target the RCL of SERPINB1 at Cys-344 at the classical P1 position, corresponding to Met-358 of AAT. In contrast, CG and other chymotrypsin-like proteases cleave the reactive center loop (RCL) of SERPINB1 at Phe-343 (Cooley et al. 2001). SERPINB1 is expressed in most tissues with higher levels in lymphoid organs such as the bone marrow and the spleen as well as in the pancreas and the lungs. In hematopoietic cells, SERPINB1 is expressed in stem cells, all leukocyte lineages, and platelets. SERPINB1 levels are the highest in the neutrophil lineage, where highest mRNA expression is found in the early stages of granulopoiesis and protein levels remain high in bone marrow and blood neutrophils (Benarafa et al. 2002; Missen et al. 2006). Its role in neutrophil homeostasis is further described below.

SERPINB6, previously described as proteinase inhibitor-6 (PI-6), inhibits CG but not NE and PR3 (Scott et al. 1999). SERPINB6 is also broadly expressed in various tissues and cells with high levels in all myeloid cells with relatively higher levels in monocytic/dendritic cells (Scarff et al. 2003; Missen et al. 2006).

First identified as squamous cell carcinoma antigen (SCCA), SERPINB4 (SCCA2) is also an inhibitor of CG but with slower association rate constants than SERPINB6 and SERPINB1 (Schick et al. 1997). Its closest homolog in mice, *serpinb3a*, also inhibits CG but is not expressed in hematopoietic cells (Al-Khunaizi et al. 2002; Askew et al. 2004).

5.2.3 Non-serpin Inhibitors of NSPs

α2-Macroglobulin (α2M) is a very large (725 kDa) protease inhibitor found in plasma. It inhibits NSPs as well as many other serine proteases by a cleavage-induced conformational change that traps the proteases into a cavity within α2M (Barrett and Starkey 1973). In contrast to serpin-protease complexes, proteases trapped by α2M can still cleave small molecular weight substrates that can reach the proteolytic domain of the protease. Because of its large size, the diffusion of α2M into tissues during inflammation was considered to be limited and its realm of activity is thus likely limited to the blood circulation, where it regulates fibrinolysis, coagulation, complement, and NSPs (de Boer et al. 1993). However, α2M is found in the lung epithelial lining fluid in adult respiratory distress syndrome (ARDS), where it traps elastase and prevents cleavage of large substrates (Wewers et al. 1988).

Secretory leukocyte protease inhibitor (SLPI) and elafin are two chelonianins that are secreted by epithelial cells and contribute to inhibition of extracellular NSPs at mucosal surfaces and in tissues (Sallenave 2010). They use a reversible keyhole type of inhibition, where the small approximately 10 kDa inhibitors bind the catalytic pocket of the protease with high affinity. SLPI inhibits both NE and CG, while elafin inhibits NE and PR3.

The relative importance of each inhibitor depends on expression levels, compartmentalization, and posttranslational modifications. In addition, because NSPs are usually released together, inhibitors that are inactivated by proteolysis by one NSP may be disarmed before inhibiting their target. While redundancy and compensatory mechanisms are expected and observed between the different inhibitors, essential physiological functions have also been attributed to individual NSP inhibitors.

5.3 Neutrophil Death in Steady-State Conditions

SERPINB1 has emerged as an important regulator of neutrophil survival. A cytoprotective function for intracellular serpins was postulated based on the presence of these inhibitors in cells carrying cell death-inducing granule proteases, such as granzymes (Bird 1999). In the case of SERPINB1, expression profiling during granulopoiesis supported this hypothesis since mRNA and protein levels of SERPINB1 peaked in promyelocyte and myelocyte stages, which coincided with the high but transient transcription of NSPs and other primary granule proteins (Theilgaard-Mönch et al. 2005; Benarafa et al. 2011). The evidence that SERPINB1 is cytoprotective for neutrophil has been elucidated in a series of studies of mice with a targeted deletion of *serpinb1a* (Benarafa et al. 2007), the mouse ortholog of the human gene (Benarafa et al. 2002). In these mice, neutrophil numbers in the bone marrow are approximately 50 % of the levels of wild-type mice. The phenotype is highly penetrant as it is observed in 129S6 and C57BL/6J backgrounds. Circulating blood neutrophil numbers in *serpinb1a*$^{-/-}$ mice are only marginally lower than in wild-type mice in normal conditions, but a faster decline in circulating neutrophils is observed after myeloablation with cyclophosphamide (Benarafa et al. 2011). In mice, reduction of neutrophil numbers in the bone marrow is physiologically relevant as it holds over 90 % of mature neutrophils. Indeed, laboratory strains of mice have on average 1×10^6 neutrophils/ml of blood with a total volume of 2 ml, and a mouse femur represents approximately 6 % of the total bone marrow and contains 4–6 $\times$ 10^6 neutrophils depending on the mouse strain.

Neutropenia in *serpinb1a*$^{-/-}$ mice is due to a cell intrinsic survival defect affecting differentiated neutrophils, while the development of mitotic neutrophil precursors is normal (Benarafa et al. 2011). The mechanism of cell death in the absence of *serpinb1a* is dependent on CG in steady state in vivo. Indeed, neutropenia in *serpinb1a*$^{-/-}$ mice was fully rescued by deficiency in CG (*Ctsg*$^{-/-}$). In contrast, NE deficiency had no effect on neutrophil survival in *serpinb1a*$^{-/-}$ mice (Baumann et al. 2013). Competitive stem cell transplantation of irradiated mice showed a survival advantage of wild-type neutrophils over *serpinb1a*$^{-/-}$ neutrophils but not over *Ctsg*$^{-/-}$.*serpinb1a*$^{-/-}$ double-deficient neutrophils. These mixed bone marrow chimera experiments demonstrated that neutrophil death was cell intrinsic in vivo and support intracellular leakage of granule proteins in mature neutrophils as a mechanism of cell death in the absence of SERPINB1. Importantly, mice lacking one or all three NSPs, as in *dppi*$^{-/-}$ mice, do not present increased neutrophil numbers in bone marrow and blood. These findings indicate that in homeostatic conditions, neutrophil homeostasis is not critically regulated by NSPs unless the intracellular anti-protease shield is deficient.

SERPINB6 may also contribute to neutrophil homeostasis in steady state as it is expressed in the cytoplasm of neutrophils and inhibits CG. However, mice deficient in *serpinb6a*, the ortholog of the human gene (Kaiserman et al. 2002), do not present an obvious defect in neutrophil numbers in vivo (Scarff et al. 2004). This somewhat surprising result may be explained by a compensatory increase in serpinb1a expression in bone marrow cells of *serpinb6a*$^{-/-}$ mice. The characterization of mice deficient in the two serpins may reveal complementary functions.

5.4 Spontaneous Neutrophil Death In Vitro

Mechanisms regulating neutrophil cell death are often investigated in vitro, where isolated neutrophils rapidly die by caspase-dependent apoptosis. This process is often referred to as spontaneous apoptosis. The survival of isolated neutrophils can be extended by addition of synthetic caspase inhibitors or by acting on upstream survival pathways. *Serpinb1a*$^{-/-}$ neutrophils also demonstrate reduced survival in vitro that is only partly inhibited by caspase inhibitors. In contrast, the survival defect is fully rescued if neutrophils are concomitantly deficient in CG, indicating that in the absence of serpinb1, CG induce neutrophil death through both caspase-dependent and caspase-independent pathways. While CG has been shown to activate caspase-7 (Zhou and Salvesen 1997), the targets of CG leading to caspase-independent cell death are not yet know.

Interestingly, isolated neutrophils of mice deficient in PR3 (*prtn3*$^{-/-}$) show improved survival in vitro compared to WT neutrophils and PR3 is found in the cytoplasm of aging neutrophils in vitro. Inhibitory complexes between PR3 and serpinb1 are found in the cytoplasm of aging neutrophils, supporting the notion that granule leakage promotes cell death. In contrast to CG or NE, PR3 directly activates pro-caspase-3 leading to apoptosis (Loison et al. 2014). Therefore, serpinb1 appears to act at the intersection of death pathways mediated by CG and PR3. It is conceivable that the two proteases function in a cascade because CG deletion alone rescues the cell intrinsic death pathway in *serpinb1a*$^{-/-}$ mice in vitro and in steady-state conditions in vivo. This hypothesis could be tested in *prtn3*$^{-/-}$.*serpinb1a*$^{-/-}$ double knockout mice. Alternatively, CG and PR3 could induce neutrophil death through parallel pathways that would enhance each other by depletion of the serpin shield.

5.5 Neutrophil Elastase and Severe Congenital Neutropenia

5.5.1 Genetic Causes of Severe Congenital Neutropenia

Severe congenital neutropenia (SCN) is a rare inherited disease characterized by persistently low neutrophil counts (<500/μl) in blood and a maturation arrest of neutrophil precursors in the bone marrow. Infants with SCN consequently develop mouth ulcers and pneumonia that can be fatal. Mutations in the neutrophil elastase gene (*ELANE* or *ELA2*) are the cause of most cases of SCN and of all cases of cyclic neutropenia (CyN) (Dale et al. 2000; Dale and Link 2009). CyN is a milder form of SCN where blood neutrophil counts oscillate between normal and very low counts with a cyclic period of 21 days (Horwitz et al. 1999).

SCN was first described in the 1950s by Dr. Rolf Kostmann in a family in northern Sweden. The genetic cause of the original cases of Kostmann's syndrome has recently been attributed to a homozygous mutation in the *HAX1* (HCLS1-associated protein X-1) gene (Klein et al. 2007). Other infrequent causes of SCN include mutations in adenylate kinase 2 (*AK2*), the transcriptional repressor (*GFI1*), and the glucose-6-phosphatase *G6PC3* genes (Klein 2011). X-linked neutropenia has also been described in a subset of patients with Wiskott-Aldrich syndrome (WAS) carrying activating mutations in the *WAS* gene (Albert et al. 2011). Finally, a significant proportion of SCN cases remain genetically undefined, and with the advances in sequencing technology, new genetic defects leading to SCN will certainly be discovered (Boztug and Klein 2011). Because it is genetically heterogeneous, several molecular pathways are likely leading to SCN.

5.5.2 ELANE Mutations in SCN and CyN

Close to two hundred distinct mutations in *ELANE* have been identified in patients with SCN and CyN (Germeshausen et al. 2013; Makaryan et al. 2015). How mutant elastase leads to neutropenia is not completely elucidated, but the current working paradigm is that the mutated NE protein induces apoptosis in developing neutrophils in the bone marrow via the initiation of the unfolded protein response (UPR) (Köllner et al. 2006; Grenda et al. 2007). NE mutants are predicted to have defective folding as the protein is synthesized in the ER. Misfolded proteins are detected by ER sensors, which trigger the UPR that can ultimately lead to apoptosis. As very large amounts of mutant misfolded NE are produced at the promyelocyte stage, the UPR triggers apoptosis resulting in the absence of differentiated neutrophils in the bone marrow and blood. Genotype-phenotype analysis studies suggest that the pattern of mutations in *ELANE* is different in patients that develop SCN or CyN (Makaryan et al. 2015). However, some mutations can lead to both SCN and CyN suggesting that additional disease modifiers may affect the severity of the phenotype (Germeshausen et al. 2013). Furthermore, about 25 % of mutations leading to SCN, but none to CyN, are associated with the development of myelodysplasia (MDS) and acute myeloid leukemia (AML). However, only a fraction of the patients with these mutations also develop MDS/AML, indicating again that other genetic, epigenetic, and environmental factors contribute to disease initiation and progression.

5.5.3 A Role for Serpins in SCN?

Whether mutations in intracellular serpin genes contribute to the severity of SCN or to the subsequent development of MDS/AML has yet to be demonstrated in humans. Studies presented above suggest that CG and PR3 can contribute to neutrophil death if the serpin shield is deficient as in *serpinb1a*$^{-/-}$ mice. In addition, *serpinb1a*$^{-/-}$ mice are more susceptible and succumb to lung infections with *Pseudomonas aeruginosa* (Benarafa et al. 2007). However, the level of neutropenia observed in *serpinb1a*$^{-/-}$ mice is milder than that observed in SCN and no maturation arrest is observed in the bone marrow (Benarafa et al. 2011).

Several studies have investigated NE activity in the context of *ELANE* mutations, and the combined evidence does not support a role for altered NE activity as the direct cause of neutrophil precursor death in the bone marrow. In transfection studies, recombinant NE mutants showed variable activity on peptide substrates ranging from the absence of activity, reduction in activity, to a higher activity than wild-type NE. In addition, AAT inhibited all active mutants in vitro (Li and Horwitz 2001). Because SCN patients have heterozygous mutations in *ELANE*, it was hypothesized that mutant NE might interfere with wild-type NE. The overall NE activity was tested in blood neutrophils of SCN patients with or without *ELANE* mutations. In single cell assays by flow cytometry, which are associated with technical caveats such as dye loading standardization, average NE activity was similar in neutrophils of SCN patients and those of normal subjects, but the variation between individuals was greater in the SCN patient group (Germeshausen et al. 2013). In the same study, NE activity in lysates of blood neutrophils of a small subset of SCN patients with mutated NE was significantly lower than the activity of neutrophil lysates of SCN patients without *ELANE* mutation and than that of normal donors. Therefore, *ELANE* mutations do not appear to be linked with consistently lower or higher NE activity. In addition, *elane*$^{-/-}$ mice have normal granulopoiesis and neutrophil counts (Belaaouaj et al. 1998), further indicating that reduced or absent NE activity is not required for neutrophil survival and differentiation. Measurements of CG and PR3 activity and levels of SERPINB1 and SERPINB6 in neutrophils of SCN patients with or without *ELANE* mutations have not been investigated and may provide additional mechanistic clues on pathways leading to neutropenia.

5.5.4 Therapy and Mouse Models for SCN

G-CSF therapy is the standard treatment for almost all types of neutropenia, whether congenital or induced by chemotherapy. Long-term G-CSF therapy has considerably improved the quality of life and reduced mortality from infections in SCN patients (Rosenberg et al. 2008). Yet, as in other severe inherited diseases with defects in hematopoiesis, there is a high long-term risk of developing malignancy. Therefore, better understanding of disease mechanisms may help design better targeted therapies for SCN. In *serpinb1a*$^{-/-}$ mice, G-CSF treatment increased precursor proliferation and accumulation of mature neutrophils in the bone marrow to levels comparable to those of untreated wild-type mice. However, G-CSF treatment failed to accumulate neutrophils in blood of *serpinb1a*$^{-/-}$ mice indicating that G-CSF therapy does not fully block NSP-mediated death in circulating cells (Baumann et al. 2013).

Attempts at modeling SCN in mice expressing *ELANE* mutants found in SCN patients unfortunately did not live up to expectations. Knock-in mice expressing the human V72M mutant of NE instead of mouse NE did not develop SCN and had normal granulopoiesis in steady-state and in stress conditions (Grenda et al. 2002). Another knock-in mouse expressing the G193X mutation that leads to the truncation of the carboxy-terminal 27 amino acids of the mature NE protein was also generated but similarly showed normal granulopoiesis in steady-state conditions (Nanua et al. 2011). However, neutrophil precursors of mice carrying the G193X mutation showed reduced proliferation and survival after tunicamycin and bortezomib treatment, which block N-linked glycosylation and proteasome activity, respectively. These effects were associated with endoplasmic reticulum stress and markers of unfolded protein responses suggesting that modulating the unfolded protein response may be of therapeutic benefit. Finally, the overall lack of effect of these mutants may be due to the significantly lower expression levels of NE in mice compared to human neutrophils (Nanua et al. 2011). Expressing *ELANE* mutants under the control of a stronger promoter may provide a model.

5.6 NSPs and Intracellular Serpins Beyond Neutrophils

5.6.1 Inflammatory Lung Disease

AAT, SLPI, and elafin have been long established as part of the antiprotease shield against NSPs in the lungs (Greene and McElvaney 2009). Most prominently, patients with AAT deficiency develop lung emphysema at a young age (Laurell and Eriksson 1963). This finding is one of the pillars of protease-antiprotease paradigm of chronic lung disease, where NSPs destroy the lung elastin fibers and other matrix proteins leading to emphysema. Single nucleotide polymorphism that alters the protein sequence of ACT has also been associated with chronic obstructive pulmonary disease (Lomas and Silverman 2001), suggesting that CG may contribute to pathogenesis.

SERPINB1 is also found in airway fluids during lung inflammatory disease (Cooley et al. 2011; Davies et al. 2010; Yasumatsu et al. 2006). Indeed, SERPINB1 can be secreted by an alternative mechanism used by other leaderless cytoplasmic proteins such as IL-1 family members (Keller et al. 2008). However, the relative importance of this secreting pathway and the release of SERPINB1 in the extracellular milieu after cell death remain to be established. Because AAT and SERPINB1 are both fast inhibitors of all three NSPs, a role for SERPINB1 in preventing the development and the severity of pulmonary emphysema in aging mice and following cigarette smoke exposure was tested in *serpinb1a*$^{-/-}$ mice. Serpinb1a was expressed in the lungs of control mice and expression was higher after cigarette smoke exposure. However, stereological analysis of the lungs and lung function tests revealed that *serpinb1a*$^{-/-}$ mice did not develop early onset emphysema as they aged. In addition, they developed cigarette smoke-induced emphysema to a comparable extent as wild-type mice after 6-month exposure (Cremona et al. 2013). These findings suggest distinct functions for the intracellular and extracellular serpins in emphysema development.

5.6.2 Lung Infection Models

Excessive inflammatory host response increases morbidity and mortality associated with seasonal respiratory influenza, and highly pathogenic virus strains are characterized by massive infiltration of leukocytes that produce a storm of injurious cytokines. Following up on studies showing increased production of inflammatory cytokines in *P. aeruginosa* infection of *serpinb1a*$^{-/-}$ mice (Benarafa et al. 2007), a role for SERPINB1 in influenza A virus infection was investigated. After challenge with a high-dose influenza A/Philadelphia/82 (H3N2), the survival of *serpinb1a*$^{-/-}$ mice was significantly reduced. Sublethally infected animals suffered increased morbidity, delayed resolution of lung injury, and increased immune cell death (Gong et al. 2011). Importantly, early virus-induced cytokine and chemokine burst and influx of PMNs and monocytes were also normal, and these responses were associated with normal viral clearance in *serpinb1a*$^{-/-}$ mice compared to wild-type. Whereas initial cytokines and chemokines rapidly decreased in WT mice, TNF-α, IL-6, KC/CXCL1, G-CSF, IL-17A, and MCP-1/CCL2 remained elevated in *serpinb1a*$^{-/-}$ mice. Monocyte-derived cells were the dominant immune cells in influenza-infected lungs, and those from *serpinb1a*$^{-/-}$ mice produced more IL-6 and TNF-α when tested ex vivo (Gong et al. 2011). Because viral clearance was unimpaired, the study highlights the critical role of serpinB1 in mitigating tissue injury, restricting inflammatory cytokine production, and reducing morbidity.

In lung infection models, *serpinb1a*$^{-/-}$ deficiency is characterized by defective microbial clearance and increased production of inflammatory cytokines (Benarafa et al. 2007; Gong et al. 2011). Neutrophil extracellular traps (NETs) are web of nuclear DNA, histones, and antimicrobial molecules released by neutrophils following stimulation by endogenous and pathogen-associated inflammatory mediators. The generation of NETs, or NETosis, was initially described as the ultimate effort by neutrophils to contain microbes and prevent their dissemination by executing this form of programmed cell death (Brinkmann et al. 2004; Fuchs et al. 2007). However, some agonists, such as *Staphylococcus aureus*, may not require neutrophil death for NET release (Pilsczek et al. 2010); and GM-CSF-primed neutrophils release NETs composed of mitochondrial DNA following stimulation with complement C5a (Yousefi et al. 2009). In vitro, *serpinb1a*$^{-/-}$ neutrophils release more NETs than wild-type neutrophils in response to agonists that induce NETs via reactive oxygen species-dependent and species-independent routes (Farley et al. 2012). These findings suggest that SERPINB1 regulates a conserved portion of the NET release pathway. SERPINB1 was also shown to regulate DNA release from activated neutrophils in vivo following *Pseudomonas* infection. Yet, despite producing more NETs, neutropenic *serpinb1a*$^{-/-}$ mice fail to control *Pseudomonas* infection in the lung and cannot prevent systemic bacterial spreading (Benarafa et al. 2007; Farley et al. 2012). On the contrary, increased NET generation in these mice may contribute to increased inflammation and tissue injury as observed during influenza infection (Gong et al. 2011).

Several questions thus remain on the potential molecular partners of SERPINB1 in DNA release and associated inflammation. Potential mechanistic pathways include inhibition of NE activity, which is required for NET generation (Papayannopoulos et al. 2010). Moreover, *prtn3*$^{-/-}$ mice show reduced inflammation and increased survival in a peritonitis model (Loison et al. 2014). Because SERPINB1 translocates to the nucleus during NET generation and is associated with chromatin, it was also hypothesized that SERPINB1 is involved in chromatin decondensation (Popova et al. 2006; Farley et al. 2012).

5.6.3 SERPINB6 and Deafness

A homozygous truncating mutation in *SERPINB6* is associated with non-syndromic sensorineural hearing loss in humans (Sirmaci et al. 2010). Most strikingly, progressive age-related hearing loss is also observed in *serpinb6a*$^{-/-}$ mice (Tan et al. 2013). Serpinb6a is expressed in cells of the cochlea and its absence leads to degeneration of the organ of Corti, which is composed of hair cells required for transmitting auditory signals. Whether inhibition of proteases, and CG in particular, is involved in this process remains to be determined.

Acknowledgments Charaf Benarafa is funded by the Swiss National Science Foundation (Grant number 310030_149790).

References

Adkison AM, Raptis SZ, Kelley DG, Pham CTN (2002) Dipeptidyl peptidase I activates neutrophil-derived serine proteases and regulates the development of acute experimental arthritis. J Clin Invest 109:363–371. doi:10.1172/JCI13462

Ahmad J, Bird PI, Kaiserman D (2014) Analysis of the evolution of granule associated serine proteases of immune defence (GASPIDs) suggests a revised nomenclature. Biol Chem 395:1253–1262. doi:10.1515/hsz-2014-0174

Albert MH, Notarangelo LD, Ochs HD (2011) Clinical spectrum, pathophysiology and treatment of the Wiskott-Aldrich syndrome. Curr Opin Hematol 18:42–48. doi:10.1097/MOH.0b013e32834114bc
Al-Khunaizi M, Luke CJ, Askew YS et al (2002) The serpin SQN-5 is a dual mechanistic-class inhibitor of serine and cysteine proteinases. Biochemistry 41:3189–3199
Askew DJ, Askew YS, Kato Y et al (2004) Comparative genomic analysis of the clade B serpin cluster at human chromosome 18q21: amplification within the mouse squamous cell carcinoma antigen gene locus. Genomics 84:176–184. doi:10.1016/j.ygeno.2004.01.015
Barrett AJ, Starkey PM (1973) The interaction of alpha 2-macroglobulin with proteinases. Characteristics and specificity of the reaction, and a hypothesis concerning its molecular mechanism. Biochem J 133:709–724
Baumann M, Pham CTN, Benarafa C (2013) SerpinB1 is critical for neutrophil survival through cell-autonomous inhibition of cathepsin G. Blood 121:3900–3907. doi:10.1182/blood-2012-09-455022
Belaaouaj AA, McCarthy R, Baumann M et al (1998) Mice lacking neutrophil elastase reveal impaired host defense against gram negative bacterial sepsis. Nat Med 4:615–618
Benarafa C, Remold-O'Donnell E (2005) The ovalbumin serpins revisited: perspective from the chicken genome of clade B serpin evolution in vertebrates. Proc Natl Acad Sci USA 102:11367–11372. doi:10.1073/pnas.0502934102
Benarafa C, Cooley J, Zeng W et al (2002) Characterization of four murine homologs of the human ov-serpin monocyte neutrophil elastase inhibitor MNEI (SERPINB1). J Biol Chem 277:42028–42033. doi:10.1074/jbc.M207080200
Benarafa C, Priebe GP, Remold-O'Donnell E (2007) The neutrophil serine protease inhibitor serpinb1 preserves lung defense functions in Pseudomonas aeruginosa infection. J Exp Med 204:1901–1909. doi:10.1084/jem.20070494
Benarafa C, LeCuyer TE, Baumann M et al (2011) SerpinB1 protects the mature neutrophil reserve in the bone marrow. J Leukoc Biol 90:21–29. doi:10.1189/jlb.0810461
Benson KF, Li F-Q, Person RE et al (2003) Mutations associated with neutropenia in dogs and humans disrupt intracellular transport of neutrophil elastase. Nat Genet 35:90–96. doi:10.1038/ng1224
Bergin DA, Reeves EP, Hurley K et al (2014) The circulating proteinase inhibitor α-1 antitrypsin regulates neutrophil degranulation and autoimmunity. Sci Transl Med 6:217ra1. doi:10.1126/scitranslmed.3007116
Bird PI (1999) Regulation of pro-apoptotic leucocyte granule serine proteinases by intracellular serpins. Immunol Cell Biol 77:47–57. doi:10.1046/j.1440-1711.1999.00787.x
Borregaard N, Cowland JB (1997) Granules of the human neutrophilic polymorphonuclear leukocyte. Blood 89:3503–3521
Borregaard N, Kjeldsen L, Rygaard K et al (1992) Stimulus-dependent secretion of plasma proteins from human neutrophils. J Clin Invest 90:86–96. doi:10.1172/JCI115860
Boztug K, Klein C (2011) Genetic etiologies of severe congenital neutropenia. Curr Opin Pediatr 23:21–26. doi:10.1097/MOP.0b013e32834262f8
Brinkmann V, Reichard U, Goosmann C et al (2004) Neutrophil extracellular traps kill bacteria. Science 303:1532–1535. doi:10.1126/science.1092385
Cheng T, Shen H, Giokas D et al (1996) Temporal mapping of gene expression levels during the differentiation of individual primary hematopoietic cells. Proc Natl Acad Sci USA 93:13158–13163
Clemmensen SN, Jacobsen LC, Rørvig S et al (2011) Alpha-1-antitrypsin is produced by human neutrophil granulocytes and their precursors and liberated during granule exocytosis. Eur J Haematol 86:517–530. doi:10.1111/j.1600-0609.2011.01601.x
Cooley J, Takayama TK, Shapiro SD et al (2001) The serpin MNEI inhibits elastase-like and chymotrypsin-like serine proteases through efficient reactions at two active sites. Biochemistry 40:15762–15770

Cooley J, Sontag MK, Accurso FJ, Remold-O'Donnell E (2011) SerpinB1 in cystic fibrosis airway fluids: quantity, molecular form and mechanism of elastase inhibition. Eur Respir J 37:1083–1090. doi:10.1183/09031936.00073710

Cowland JB, Borregaard N (1999) The individual regulation of granule protein mRNA levels during neutrophil maturation explains the heterogeneity of neutrophil granules. J Leukoc Biol 66:989–995

Cremona TP, Tschanz SA, von Garnier C, Benarafa C (2013) SerpinB1 deficiency is not associated with increased susceptibility to pulmonary emphysema in mice. Am J Physiol Lung Cell Mol Physiol 305:L981–9. doi:10.1152/ajplung.00181.2013

Dahl R, Walsh JC, Lancki D et al (2003) Regulation of macrophage and neutrophil cell fates by the PU.1:C/EBPalpha ratio and granulocyte colony-stimulating factor. Nat Immunol 4:1029–1036. doi:10.1038/ni973

Dale DC, Link DC (2009) The many causes of severe congenital neutropenia. N Engl J Med 360:3–5. doi:10.1056/NEJMp0806821

Dale DC, Person RE, Bolyard AA et al (2000) Mutations in the gene encoding neutrophil elastase in congenital and cyclic neutropenia. Blood 96:2317–2322

Davies PL, Spiller OB, Beeton ML et al (2010) Relationship of proteinases and proteinase inhibitors with microbial presence in chronic lung disease of prematurity. Thorax 65:246–251. doi:10.1136/thx.2009.116061

de Boer JP, Creasey AA, Chang A et al (1993) Alpha-2-macroglobulin functions as an inhibitor of fibrinolytic, clotting, and neutrophilic proteinases in sepsis: studies using a baboon model. Infect Immun 61:5035–5043

Farley K, Stolley JM, Zhao P et al (2012) A SerpinB1 regulatory mechanism is essential for restricting neutrophil extracellular trap generation. J Immunol 189:4574–4581. doi:10.4049/jimmunol.1201167

Fuchs TA, Abed U, Goosmann C et al (2007) Novel cell death program leads to neutrophil extracellular traps. J Cell Biol 176:231–241. doi:10.1083/jcb.200606027

Germeshausen M, Deerberg S, Peter Y et al (2013) The spectrum of ELANE mutations and their implications in severe congenital and cyclic neutropenia. Hum Mutat 34:905–914. doi:10.1002/humu.22308

Gong D, Farley K, White M et al (2011) Critical role of serpinB1 in regulating inflammatory responses in pulmonary influenza infection. J Infect Dis 204:592–600. doi:10.1093/infdis/jir352

Greene CM, McElvaney NG (2009) Proteases and antiproteases in chronic neutrophilic lung disease – relevance to drug discovery. Br J Pharmacol 158:1048–1058. doi:10.1111/j.1476-5381.2009.00448.x

Grenda DS, Johnson SE, Mayer JR et al (2002) Mice expressing a neutrophil elastase mutation derived from patients with severe congenital neutropenia have normal granulopoiesis. Blood 100:3221–3228. doi:10.1182/blood-2002-05-1372

Grenda DS, Murakami M, Ghatak J et al (2007) Mutations of the ELA2 gene found in patients with severe congenital neutropenia induce the unfolded protein response and cellular apoptosis. Blood 110:4179–4187. doi:10.1182/blood-2006-11-057299

Gullberg U, Lindmark A, Lindgren G et al (1995) Carboxyl-terminal prodomain-deleted human leukocyte elastase and cathepsin G are efficiently targeted to granules and enzymatically activated in the rat basophilic/mast cell line RBL. J Biol Chem 270:12912–12918

Horwitz M, Benson KF, Person RE et al (1999) Mutations in ELA2, encoding neutrophil elastase, define a 21-day biological clock in cyclic haematopoiesis. Nat Genet 23:433–436. doi:10.1038/70544

Huntington JA, Read RJ, Carrell RW (2000) Structure of a serpin-protease complex shows inhibition by deformation. Nature 407:923–926. doi:10.1038/35038119

Jonigk D, Al-Omari M, Maegel L et al (2013) Anti-inflammatory and immunomodulatory properties of α1-antitrypsin without inhibition of elastase. Proc Natl Acad Sci USA 110:15007–15012. doi:10.1073/pnas.1309648110

Kaiserman D, Bird PI (2005) Analysis of vertebrate genomes suggests a new model for clade B serpin evolution. BMC Genomics 6:167. doi:10.1186/1471-2164-6-167
Kaiserman D, Knaggs S, Scarff KL et al (2002) Comparison of human chromosome 6p25 with mouse chromosome 13 reveals a greatly expanded ov-serpin gene repertoire in the mouse. Genomics 79:349–362. doi:10.1006/geno.2002.6716
Keller M, Rüegg A, Werner S, Beer H-D (2008) Active caspase-1 is a regulator of unconventional protein secretion. Cell 132:818–831. doi:10.1016/j.cell.2007.12.040
Klein C (2011) Genetic defects in severe congenital neutropenia: emerging insights into life and death of human neutrophil granulocytes. Annu Rev Immunol 29:399–413. doi:10.1146/annurev-immunol-030409-101259
Klein C, Grudzien M, Appaswamy G et al (2007) HAX1 deficiency causes autosomal recessive severe congenital neutropenia (Kostmann disease). Nat Genet 39:86–92. doi:10.1038/ng1940
Köllner I, Sodeik B, Schreek S et al (2006) Mutations in neutrophil elastase causing congenital neutropenia lead to cytoplasmic protein accumulation and induction of the unfolded protein response. Blood 108:493–500. doi:10.1182/blood-2005-11-4689
Kolset SO, Tveit H (2008) Serglycin--structure and biology. Cell Mol Life Sci 65:1073–1085. doi:10.1007/s00018-007-7455-6
Kruger P, Saffarzadeh M, Weber ANR et al (2015) Neutrophils: between host defence, immune modulation, and tissue injury. PLoS Pathog 11, e1004651. doi:10.1371/journal.ppat.1004651
Laurell CB, Eriksson S (1963) The electrophoretic α; 1-globulin pattern of serum in α; 1-antitrypsin deficiency. Scand J Clin Lab Invest 15:132–140. doi:10.1080/00365516309051324
Li FQ, Horwitz M (2001) Characterization of mutant neutrophil elastase in severe congenital neutropenia. J Biol Chem 276:14230–14241. doi:10.1074/jbc.M010279200
Loison F, Zhu H, Karatepe K et al (2014) Proteinase 3-dependant caspase-3 cleavage modulates neutrophil death and inflammation. J Clin Invest. doi:10.1172/JCI76246
Lomas DA, Silverman EK (2001) The genetics of chronic obstructive pulmonary disease. Respir Res 2:20–26
Makaryan V, Zeidler C, Bolyard AA et al (2015) The diversity of mutations and clinical outcomes for ELANE-associated neutropenia. Curr Opin Hematol 22:3–11. doi:10.1097/MOH.0000000000000105
Mason DY, Cramer EM, Massé JM et al (1991) Alpha 1-antitrypsin is present within the primary granules of human polymorphonuclear leukocytes. Am J Pathol 139:623–628
Missen MA, Haylock D, Whitty G et al (2006) Stage specific gene expression of serpins and their cognate proteases during myeloid differentiation. Br J Haematol 135:715–724. doi:10.1111/j.1365-2141.2006.06360.x
Nanua S, Murakami M, Xia J et al (2011) Activation of the unfolded protein response is associated with impaired granulopoiesis in transgenic mice expressing mutant Elane. Blood 117:3539–3547. doi:10.1182/blood-2010-10-311704
Nathan C (2006) Neutrophils and immunity: challenges and opportunities. Nat Rev Immunol 6:173–182. doi:10.1038/nri1785
Niemann CU, Åbrink M, Pejler G et al (2007) Neutrophil elastase depends on serglycin proteoglycan for localization in granules. Blood 109:4478–4486. doi:10.1182/blood-2006-02-001719
Papayannopoulos V, Metzler KD, Hakkim A, Zychlinsky A (2010) Neutrophil elastase and myeloperoxidase regulate the formation of neutrophil extracellular traps. J Cell Biol 191:677–691. doi:10.1083/jcb.201006052
Pham CTN (2006) Neutrophil serine proteases: specific regulators of inflammation. Nat Rev Immunol 6:541–550. doi:10.1038/nri1841
Pilsczek FH, Salina D, Poon KKH et al (2010) A novel mechanism of rapid nuclear neutrophil extracellular trap formation in response to Staphylococcus aureus. J Immunol 185:7413–7425. doi:10.4049/jimmunol.1000675

Popova EY, Claxton DF, Lukasova E et al (2006) Epigenetic heterochromatin markers distinguish terminally differentiated leukocytes from incompletely differentiated leukemia cells in human blood. Exp Hematol 34:453–462. doi:10.1016/j.exphem.2006.01.003
Potempa J, Dubin A, Watorek W, Travis J (1988) An elastase inhibitor from equine leukocyte cytosol belongs to the serpin superfamily. Further characterization and amino acid sequence of the reactive center. J Biol Chem 263:7364–7369
Remold-O'Donnell E (1985) A fast-acting elastase inhibitor in human monocytes. J Exp Med 162:2142–2155
Remold-O'Donnell E, Nixon JC, Rose RM (1989) Elastase inhibitor. Characterization of the human elastase inhibitor molecule associated with monocytes, macrophages, and neutrophils. J Exp Med 169:1071–1086
Rosenberg PS, Alter BP, Link DC et al (2008) Neutrophil elastase mutations and risk of leukaemia in severe congenital neutropenia. Br J Haematol 140:210–213. doi:10.1111/j.1365-2141.2007.06897.x
Sallenave J-M (2010) Secretory leukocyte protease inhibitor and elafin/trappin-2: versatile mucosal antimicrobials and regulators of immunity. Am J Respir Cell Mol Biol 42:635–643. doi:10.1165/rcmb.2010-0095RT
Scarff KL, Ung KS, Sun J, Bird PI (2003) A retained selection cassette increases reporter gene expression without affecting tissue distribution in SPI3 knockout/GFP knock-in mice. Genesis 36:149–157. doi:10.1002/gene.10210
Scarff KL, Ung KS, Nandurkar H et al (2004) Targeted disruption of SPI3/Serpinb6 does not result in developmental or growth defects, leukocyte dysfunction, or susceptibility to stroke. Mol Cell Biol 24:4075–4082
Schick C, Kamachi Y, Bartuski AJ et al (1997) Squamous cell carcinoma antigen 2 is a novel serpin that inhibits the chymotrypsin-like proteinases cathepsin G and mast cell chymase. J Biol Chem 272:1849–1855
Scott FL, Hirst CE, Sun J et al (1999) The intracellular serpin proteinase inhibitor 6 is expressed in monocytes and granulocytes and is a potent inhibitor of the azurophilic granule protease, cathepsin G. Blood 93:2089–2097
Sirmaci A, Erbek S, Price J et al (2010) A truncating mutation in SERPINB6 is associated with autosomal-recessive nonsyndromic sensorineural hearing loss. Am J Hum Genet 86:797–804. doi:10.1016/j.ajhg.2010.04.004
Stark MA, Huo Y, Burcin TL et al (2005) Phagocytosis of apoptotic neutrophils regulates granulopoiesis via IL-23 and IL-17. Immunity 22:285–294. doi:10.1016/j.immuni.2005.01.011
Tan J, Prakash MD, Kaiserman D, Bird PI (2013) Absence of SERPINB6A causes sensorineural hearing loss with multiple histopathologies in the mouse inner ear. Am J Pathol 183:49–59. doi:10.1016/j.ajpath.2013.03.009
Theilgaard-Mönch K, Jacobsen LC, Borup R et al (2005) The transcriptional program of terminal granulocytic differentiation. Blood 105:1785–1796. doi:10.1182/blood-2004-08-3346
Vandivier RW, Henson PM, Douglas IS (2006) Burying the dead: the impact of failed apoptotic cell removal (efferocytosis) on chronic inflammatory lung disease. Chest 129:1673–1682. doi:10.1378/chest.129.6.1673
Wang D, Wang W, Dawkins P et al (2011) Deletion of Serpina1a, a murine α1-antitrypsin ortholog, results in embryonic lethality. Exp Lung Res 37:291–300. doi:10.3109/01902148.2011.554599
Wewers MD, Herzyk DJ, Gadek JE (1988) Alveolar fluid neutrophil elastase activity in the adult respiratory distress syndrome is complexed to alpha-2-macroglobulin. J Clin Invest 82:1260–1267. doi:10.1172/JCI113724
Witko-Sarsat V, Cramer EM, Hieblot C et al (1999) Presence of proteinase 3 in secretory vesicles: evidence of a novel, highly mobilizable intracellular pool distinct from azurophil granules. Blood 94:2487–2496
Yasumatsu R, Altiok O, Benarafa C et al (2006) SERPINB1 upregulation is associated with in vivo complex formation with neutrophil elastase and cathepsin G in a baboon model of

bronchopulmonary dysplasia. Am J Physiol Lung Cell Mol Physiol 291:L619–L627. doi:10.1152/ajplung.00507.2005
Yousefi S, Mihalache C, Kozlowski E et al (2009) Viable neutrophils release mitochondrial DNA to form neutrophil extracellular traps. Cell Death Differ 16:1438–1444. doi:10.1038/cdd.2009.96
Zhang DE, Zhang P, Wang ND et al (1997) Absence of granulocyte colony-stimulating factor signaling and neutrophil development in CCAAT enhancer binding protein alpha-deficient mice. Proc Natl Acad Sci USA 94:569–574
Zhou Q, Salvesen GS (1997) Activation of pro-caspase-7 by serine proteases includes a non-canonical specificity. Biochem J 324(Pt 2):361–364

SerpinB1: A Global Regulator of Neutrophil Homeostasis

6

Lifei Hou, Kalamo Farley, and Eileen Remold-O'Donnell

Contents

Abstract

SerpinB1 is an inhibitory serpin, which is broadly expressed at highly variable levels. It is encoded by an ancient gene—possibly the extant homologue of the serpin gene at the root of the clade B tree. Studies of gene-deleted mice in models of lung injury, infection, and inflammation identified SerpinB1 as a

L. Hou • K. Farley
Program in Cellular and Molecular Medicine, Harvard Medical School, Boston, MA 02115, USA

Department of Pediatrics, Harvard Medical School, Boston, MA 02115, USA
e-mail: Lifei.Hou@childrens.harvard.edu; Kalamo.Farley@gmail.com

E. Remold-O'Donnell (✉)
Program in Cellular and Molecular Medicine, Harvard Medical School, Boston, MA 02115, USA

Division of Hematology/Oncology, Boston Children's Hospital, Harvard Medical School, Boston, MA 02115, USA

Department of Pediatrics, Harvard Medical School, Boston, MA 02115, USA
e-mail: Eileen.Remold-Odonnell@childrens.harvard.edu

M. Geiger et al. (eds.), *The Serpin Family*, DOI 10.1007/978-3-319-22711-5_6

protective anti-inflammatory immunomodulator. Other studies document its disparate cell-specific functions. Here we review these studies and present a unifying hypothesis, which is that the multiplicity and diversity of known functions each reflects a different aspect of SerpinB1's overruling function as a global regulator of neutrophil homeostasis.

6.1 Localization

SerpinB1, previously known as MNEI (monocyte/neutrophil elastase inhibitor) and LEI (leukocyte elastase inhibitor), is a 42 kDa non-glycosylated clade B serpin (Dubin et al. 1992; Remold-O'Donnell et al. 1992). Whereas most mammalian serpins are secreted via the classical pathway, those of clade B (previously called ovalbumin-related serpins) are distinguished by the absence of a hydrophobic cleavable signal sequence (Remold-O'Donnell 1993) and have a nucleocytoplasmic location (Bird et al. 2001). Nonetheless, SerpinB1 is also found extracellularly—in media of cultured macrophages (Remold-O'Donnell and Lewandrowski 1983) and in human lung fluids, particularly bronchoalveolar lavage fluid of cystic fibrosis patients (Cooley et al. 2011). Whether this extracellular presence is due to passive release from lysed cells or active secretion is unclear; secretion by an unconventional caspase-1-mediated mechanism has been proposed (Keller et al. 2008).

6.2 Inhibition of Proteases

Serine proteases inhibited by SerpinB1 include the elastinolytic enzymes, neutrophil elastase, proteinase 3, and pancreatic elastase, and the chymotrypsin specificity enzymes, cathepsin G, chymotrypsin, chymase, granzyme H, and kallikrein-3 (prostate-specific antigen, PSA), but not trypsin-specific proteases (Cooley et al. 2001; Wang et al. 2013). The stoichiometry of inhibition (SI) values is close to 1 for all these inhibitions, and the rate constants are remarkably rapid: $>10^7\ M^{-1}\ s^{-1}$ for the elastinolytic enzymes, $>10^6\ M^{-1}\ s^{-1}$ for cathepsin G and chymotrypsin, $>10^5\ M^{-1}\ s^{-1}$ for chymase and granzyme H, and $\sim 10^4\ M^{-1}\ s^{-1}$ for kallikrein-3. The dual specificity of inhibition has its molecular correlate in the use of two reactive sites: the canonical P_1 Cys^{344} for inhibition of elastinolytic proteases and Phe^{343} (canonical P_2) for inhibition of chymotrypsin-specific proteases. The chymotrypsin-specific protease kallikrein-3 is an exception because Cys^{344} is P1, possibly due to steric features (Cooley et al. 2001). Of note, the list of inhibited proteases is not yet complete in that preliminary evidence was recently reported for cross-class inhibition of cysteine cathepsins by SerpinB1 (Zhao et al. 2014).

6.3 Evolutionary Perspective

Completion of the chicken genome in 2004 (Hillier et al. 2004) made possible phylogenetic analyses of the clade B serpins, which in combination with cross-species comparisons of gene and protein structures, provided perspective on the evolutionary pathway, allowing identification of conserved serpins and "newer" serpins such as ovalbumin (*serpinb14*) that may have contributed to vertebrate adaptation (Benarafa and Remold-O'Donnell 2005; Kaiserman and Bird 2005). The comparative genomic analyses placed SerpinB1 at the root of the clade B serpins (Fig. 6.1), suggesting that its highly efficient protease inhibitory activity may have been conserved since the origin of vertebrates. Consistent with an ancient origin, no biochemical activity other than protease inhibition has been identified for SerpinB1 and, thus far, no modifying interactions. The lack of accessory function is consistent with the seven-exon gene structure that SerpinB1 shares with SerpinB6, B8, and B9 (Benarafa and Remold-O'Donnell 2005; Kaiserman and Bird 2005) (Fig. 6.1). These genes, unlike the eight-exon clade B genes, lack the variable length loop between helices C and D. The C–D loop, which does not disrupt the serpin fold (Huber and Carrell 1989), has evolved to support additional binding and localization motifs in individual serpins, e.g., transglutaminase acceptor sites that

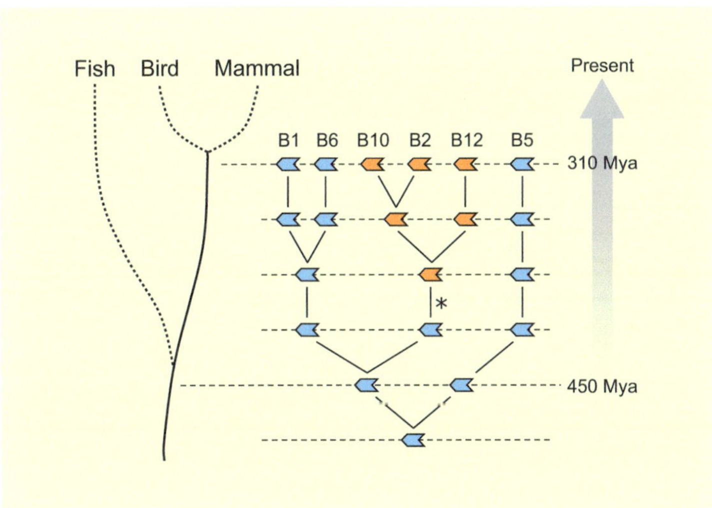

Fig. 6.1 Early evolution of the clade B serpins. The model shows serpin genes as *blue* (seven-exon) or *orange* (eight-exon) *block arrows*. Shown, starting at the *bottom*, is the ancestral locus with a single seven-exon gene, followed by subsequent stages to generate the ancestor locus of birds and mammals shown here at the *top*. Within the tetrapod lineage, the *asterisk* represents an event in which additional sequence, including a CD loop, was added within a seven-exon gene, generating an eight-exon gene. Subsequent duplications within the mammalian serpins generated additional seven-exon (B8 and B9) and eight-exon (B3, B4, B7, B11, and B13) genes (not shown). Reprinted with permission from Benarafa and Remold-O'Donnell, Proc Natl Acad Sci USA, 102: 11367, 2005

support cross-linking of SerpinB2 (PAI-2) (Jensen et al. 1994), and a nuclear localization signal of SerpinB10 in combination with an “AT hook” motif that supports chromatin condensation of MENT (avian *serpinb10b*) (Grigoryev et al. 1999). It is also consistent with their ancient origin that SerpinB1 and SerpinB6 are broadly expressed (Kaiserman et al. 2002). SerpinB1 expression levels are however highly variable (Benarafa et al. 2002) consistent with the diverse cell-specific functions described below.

6.4 Neutrophils, Pathogens, and Inflammation

Early studies showed high levels of SerpinB1 in myeloid cells especially neutrophils, the short-lived cells recruited early to infection sites to destroy pathogenic bacteria and fungi. Three related SerpinB1-inhibitable neutrophil serine proteases (NSPs), elastase, proteinase-3, and cathepsin G, are laid down in azurophil granules of developing neutrophils (promyelocytes) in the bone marrow. These NSPs constitute a major weapon in the neutrophil’s antimicrobial arsenal. However, if present in excess, as occurs in chronic inflammatory diseases such as cystic fibrosis, chronic obstructive pulmonary disease (COPD), and emphysema, the NSPs injure host tissue and induce inflammation (Janoff 1985). Thus, tight regulation is crucial. Early studies showed that the delivery of recombinant human SerpinB1 protects the lungs of rats in a model of lung injury (hemorrhage) induced by instilling human neutrophil elastase (Rees et al. 1999). Recombinant SerpinB1 also protected rat lungs in a bacterial infection model (Woods et al. 2005). The role of endogenous SerpinB1 in protecting lungs from NSP-mediated injury was first demonstrated in a baboon model of bronchopulmonary dysplasia (chronic lung disease of premature infants) (Yasumatsu et al. 2006).

Whereas human SerpinB1 is encoded by a single gene (*SERPINB1*), the locus in mice includes four homologues, *serpinb1a*, *serpinb1b*, *serpinb1c*, and *serpinb1d*, a pseudogene (Benarafa et al. 2002). Deletion of *serpinb1a*, the murine ortholog of *SERPINB1*, produced mice that have normal phenotype in the absence of challenge, but showed increased mortality on intranasal infection with *Pseudomonas aeruginosa* (Benarafa et al. 2007). The inoculated *serpinb1a*-deficient mice (hereafter *serpinb1*$^{-/-}$ mice) mount early responses to infection including production of cytokines and chemokines, recruitment of neutrophils, and even restriction of bacteria that are not different from the responses of wild-type (WT) mice. However, the *serpinb1*$^{-/-}$ mice develop late-onset failure to clear bacteria in association with increased free neutrophil protease activity and prolonged production of inflammatory cytokines. Ex vivo culture of lung neutrophils of the infected mice revealed a defect of survival of the *serpinb1*$^{-/-}$ neutrophils (Benarafa et al. 2007). Because of the greater number of bacteria in the *serpinb1*$^{-/-}$ mice compared with WT, it was not possible to determine whether the pathological changes were caused by bacteria or by the host response. Overall, SerpinB1 protects neutrophil survival, bacteria clearing capacity, and inflammation restricting capacity of infected mice.

Serpinb1$^{-/-}$ mice have also been studied in influenza virus respiratory infection. This model, high-dose inoculation with the strain Philadelphia/82 (H3N2), causes rapid lung infiltration of neutrophils and monocytes/macrophages that produce a storm of injurious inflammatory cytokines and severe lung histopathology—featuring characteristic of human infection with highly pathogenic influenza strains such as the 1918 H1N1 and the avian H5N1 strains. Mortality and morbidity in the form of extreme weight loss were increased for infected *serpinb1*$^{-/-}$ mice compared with WT mice (Gong et al. 2011). Tissue injury and immune cell death were also increased. Compared to WT mice, the virus-infected *serpinb1*$^{-/-}$ mice displayed increases of TNF-α- and IL-6-producing monocyte/macrophages (the "cytokine storm") and, surprisingly, increases of IL-17 and IL-17-producing T cells (Gong et al. 2011) (detailed below). Importantly, virus clearance was not different between *serpinb1*$^{-/-}$ and WT mice, indicating that the excess pathology is due to an overactive host response. Together, the studies of bacterial and viral infection demonstrate that prevention of excessive and injurious immune cell responses at infection sites is an important physiological function of SerpinB1. The studies indicate that SerpinB1 is a protective anti-inflammatory immunomodulator.

6.5 Protecting the Bone Marrow Reserve of Neutrophils

Quantitation of hematopoietic cells in the bone marrow revealed a specific deficit of neutrophils (neutrophilic granulocytes) in *serpinb1*$^{-/-}$ mice at steady state (Benarafa et al. 2011). In researching the cause, we started with hematopoietic stem cells that differentiate via granulocyte/macrophage progenitors to committed myeloblasts, promyelocytes, and myelocytes, each of which actively proliferates (Akashi et al. 2000). Mitosis ceases with myelocytes, which then differentiate to metamyelocytes, to band cells, and finally to mature neutrophils with characteristic segmented polymorphonuclear structure. We noted that SerpinB1 levels change continually during granulopoiesis, peaking in the transition from myeloblasts to promyelocytes (Benarafa et al. 2011), coincident with synthesis of the NSPs and their storage as sequestered enzymes in the azurophil granules (Theilgaard-Monch et al. 2005). At steady state, the bone marrow holds considerably more mature neutrophils than the periphery (Chervenick et al. 1968; Boxio et al. 2004) and continually releases neutrophils to the circulating blood. The released neutrophils survive for a few hours in the circulation unless recruited to a site of infection or sterile inflammation (Chervenick et al. 1968).

Colony-forming assays of *serpinb1*$^{-/-}$ bone marrow cells produced normal numbers and types of mature colonies, indicating that the neutrophil deficit is not due to defective early progenitor cells. A deficiency of granulocyte-colony stimulating factor (G-CSF) was considered; however, G-CSF levels are fourfold increased in *serpinb1*$^{-/-}$ mice at steady state. Maturational arrest was excluded by the normal expression levels of late differentiation markers, and a proliferation defect was excluded by the normal number of Ki67-positive proliferating neutrophils. On overnight culture of purified bone marrow neutrophils, both apoptosis and necrosis were increased for *serpinb1*$^{-/-}$ cells compared with WT. Collectively, the findings indicate that the deficit of *serpinb1*$^{-/-}$ bone marrow cells is largely restricted to mature post-mitotic neutrophils and is due to increased spontaneous cell death (Benarafa et al. 2011). The findings demonstrate that preserving the bone marrow reserve of mature neutrophils by restricting spontaneous cell death is an important physiological function of SerpinB1.

6.6 Restricting Neutrophil Death Pathways

To identify the protease(s) putatively responsible for the increased spontaneous death of *serpinb1*$^{-/-}$ neutrophils, Bauman et al. studied mice genetically depleted of *serpinb1* in combination with one or more of the NSP genes. Quantitation and phenotyping of bone marrow cells revealed that genetic deletion of *Ctsg* (cathepsin G) or *Ctsg* and *Elane* (elastase) in combination with *serpinb1* rescued the deficit of mature bone marrow neutrophils, but deletion of *Elane* did not (Baumann et al. 2013). The findings indicate that regulation of endogenous cathepsin G by serpinB1 contributes to normal maintenance of the bone marrow pool of mature neutrophils. Exogenously added cathepsin G did not induce the death process. Mechanistic studies established that cathepsin G-dependent death of *serpinb1*$^{-/-}$ bone marrow neutrophils is caspase independent and requires permeabilization of the granule membrane to release endogenous cathepsin G to the cytoplasm (Baumann et al. 2013).

An independent study by Loison et al. that focused on spontaneous death of cultured bone marrow and blood neutrophils identified proteinase-3 as the key enzyme mediating caspase-dependent death of "aging" (i.e., cultured) neutrophils. Caspase-3-dependent neutrophil death also required permeabilization of granule membranes, releasing proteinase-3. This death pathway, which is unique to neutrophils, has many hallmarks of classical apoptotic death but is independent of the activator caspases, caspase-8 and caspase-9. Rather active effector caspase-3 is generated by proteinase-3. Western blots of cytosolic fractions revealed a decrease of free cytosolic SerpinB1 during neutrophil aging in culture and formation of a proteinase-3-SerpinB1 complex, strongly suggesting that the caspase-dependent neutrophil death pathway is regulated by SerpinB1 inhibition of proteinase-3 (Loison et al. 2014). Together, the two studies establish that SerpinB1 carries out the physiological function of protecting mature neutrophils in bone marrow and blood by inhibiting endogenous cathepsin G and proteinase-3.

6.7 Restricting NETosis

SerpinB1 also protects neutrophils from a third death pathway distinct from apoptosis and necrosis. NETosis is a programmed pathway initiated when neutrophils contact pathogens or inflammatory mediators or PMA in vitro. These diverse agents interact with different surface receptors and induce different cytoplasmic signaling pathways that converge at shared downstream events that include transformation of the segmented nucleus to a spherical shape, decondensation of chromatin, and expansion of the nucleus. The process culminates in extrusion of linearized DNA decorated with proteases (elastase, cathepsin G, and proteinase-3) and antimicrobial proteins myeloperoxidase and histones (Brinkmann et al. 2004; Fuchs et al. 2007; Hakkim et al. 2011). At the time of discovery, NETosis was considered protective, a final act by dying neutrophils to prevent or delay systemic infection. However, mounting evidence indicates that NETs, or more specifically excess NETs, inflicts serious local inflammatory damage, particularly endothelial cell death, in a range of diseases, including cystic fibrosis, thrombosis, and fatal murine influenza (Massberg et al. 2010; Narasaraju et al. 2011; Papayannopoulos et al. 2011). NETs also exacerbate the pathology of autoimmune diseases, e.g., systemic lupus erythematosus, in part, because they consist of DNA and RNA decorated with self proteins, which can bind anti-self antibodies and be endocytosed as immune complexes by plasmacytoid dendritic cells (pDCs) to signal via TLR7/9 and induce type I interferon (IFN) responses (Garcia-Romo et al. 2011; Lande et al. 2011).

Evidence that SerpinB1 restricts NETosis includes the increase of NETs in vivo in lungs of *P. aeruginosa*-infected *serpinb1*$^{-/-}$ mice compared with WT mice (Farley et al. 2012). In vitro, isolated *serpinb1*$^{-/-}$ neutrophils generated substantially more NETs than WT neutrophils in response to diverse stimulating agents. These include PMA, which induces reactive oxygen species (ROS)-dependent NETosis, and PAF, MIP-2, and LPS, which induce ROS-independent NETosis (Fig. 6.2) (Farley et al. 2012). In normal neutrophils, SerpinB1 translocates into the nucleus early in NETosis, suggesting that the nucleus is the site of its regulatory action (Farley et al. 2012). The key critical event of NETosis is chromatin decondensation catalyzed by the nuclear enzyme peptidylarginine deiminase-4 (PAD4), which is expressed at high levels in mature neutrophils (Wang et al. 2009; Li et al. 2010). PAD4 deiminates arginine residues on histone tails,

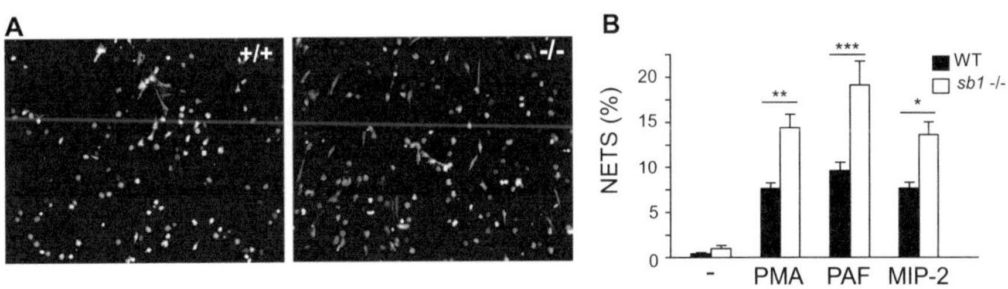

Fig. 6.2 Increased NETosis of *serpinb1*$^{-/-}$ neutrophils. WT and *serpinb1*$^{-/-}$ (*sb1*$^{-/-}$) neutrophils were treated with the indicated agents and stained with Sytox for DNA. (**a**) Micrographs of PAF-treated WT (+/+) and *serpinb1*$^{-/-}$ (−/−) neutrophils. (**b**) Quantitation of NETs. Neutrophils were treated with PMA, PAF, and MIP-2/cxcl2, and NETs were quantified in Sytox-stained micrographs. (**a**) Representative micrograph. (**b**) Means ± SEM of 6 mice per genotype. *$P < 0.05$; **$P < 0.01$; ***$P < 0.001$. Reprinted with permission from Farley et al., J Immunol, 189: 4574, 2012

converting them to citrulline. The resulting loss of positive charges facilitates decondensation of chromatin and expansion of the nucleus.

The regulatory effect of SerpinB1 on NETosis suggests the involvement of one or more proteases. However, there is presently no consistent evidence identifying the responsible protease. A study by Papayannopoulos et al. found that elastase-inhibitor-treated neutrophils failed to generate NETs in response to PMA or *Candida albicans*, and elastase-deficient mice failed to generate NETs on infection with *Klebsiella pneumonia* (Papayannopoulos et al. 2010). However, a study by Chen et al. found no defect of NET generation by elastase-deficient mice in response to soluble immune complexes (Chen et al. 2012). Also, our studies of human neutrophils treated with inhibitors of NSPs and mouse neutrophils genetically depleted for NSPs failed to identify a NETosis regulatory protease (unpublished findings). In a recent study, NETosis was absent in neutrophils from a patient genetically depleted of the cysteine protease cathepsin C (Sorensen et al. 2014). Because cathepsin C (also known as dipeptidyl peptidase I) is required for N-terminal trimming during biosynthesis of the NSPs (Pham et al. 2004), the patient's neutrophils also lack the NSPs (Sorensen et al. 2014) and possibly other proteases.

6.8 IL-17 in Neutrophil Homeostasis

Neutrophil numbers in the circulation are tightly regulated by homeostatic regulatory mechanisms that match neutrophil production (granulopoiesis) to neutrophil clearance, both in steady state and in infection and inflammation (stress/emergency granulopoiesis). G-CSF, a product of bone marrow stromal cells, is the principle cytokine that regulates granulopoiesis (Schwarzenberger et al. 2000) by inducing the commitment of multipotential progenitor cell to myeloid lineages and stimulating proliferation of granulocyte precursors [reviewed in Christopher and Link (2007)].

The major link of G-CSF to neutrophil clearance in the periphery is IL-17, which is produced by two types of T cells (Forlow et al. 2001; Stark et al. 2005). IL-17 is rapidly produced by innate γδ T cells on encountering microbial products or in response to the inflammatory cytokines IL-23 and IL-1β (Martin et al. 2009; Sutton et al. 2009). The delayed IL-17-producing cells are Th17 cells, a specialized adaptive CD4 cell lineage (Harrington et al. 2005; Park et al. 2005). IL-17 activates granulopoiesis by inducing bone marrow stromal cells to produce G-CSF. IL-17 also induces local epithelial cells and fibroblasts to produce chemokines and cytokines that recruit and activate neutrophils (Miyamoto et al. 2003; Fogli et al. 2013). By thus serving as the dominant driver of neutrophil number and neutrophil function at peripheral sites, IL-17 protects the host against infections with fungi and extracellular bacteria (Shibata et al. 2007) but is also a major driver of progressive tissue damage in autoimmune diseases including multiple sclerosis, systemic lupus erythematosus, rheumatoid arthritis, and inflammatory bowel disease.

6.9 SerpinB1 and IL-17

The first hint that SerpinB1 controls IL-17 was the finding of increased IL-17 and IL-17-producing T cells in lungs of influenza virus-infected *serpinb1*$^{-/-}$ mice (Fig. 6.3a) (Gong et al. 2011). The responsible cells in these mice were primarily γδ T cells but also Th17 cells. Subsequent study of noninfected mice showed that IL-17^{+} γδ T cells are already expanded at steady state in *serpinb1*$^{-/-}$ mice compared with WT mice (Zhao et al. 2014). Increased numbers of IL-17^{+} γδ T cells were found in the lungs and spleen but not in the thymus of *serpinb1*$^{-/-}$ mice, indicating that the increase is due to peripheral expansion rather than altered thymic development. The effect of SerpinB1 is specific in that no differences were found between WT and *serpinb1*$^{-/-}$ mice in the number of IFNγ^{+} γδ T cells. Because subsets of γδ T cells that share function are defined by the expression of particular γ- and/or δ-V genes, we could further test the specificity of the SerpinB1 effect by co-staining the cells with V-region-specific TCR-γδ antibodies. We found that the Vγ4^{+} and Vγ6/Vδ1^{+} subsets, which are IL-17 producers, were increased in *serpinb1*$^{-/-}$ mice, but that the Vγ1^{+} γδ cells, which are IFNγ producers, were not different between *serpinb1*$^{-/-}$ and WT mice (Fig. 6.3b). Further studies showed that the percentages of Vγ4^{+} and Vγ6/Vδ1^{+} subsets producing IL-17 were also increased for *serpinb1*$^{-/-}$ mice compared with WT mice (Fig. 6.3c) as well as the percentages expressing Ki-67, a marker for proliferating cells (not shown) (Zhao et al. 2014). Of note, microarray studies by the Immunological Genome Consortium project found that *serpinb1a* is highly and preferentially expressed in mature Vγ4^{+} γδ T cells, but not in immature Vγ4^{+} γδ T cells and not in Vγ1^{+} γδ T cells (Narayan et al. 2012).

We subsequently showed that increased numbers of adaptive Th17 cells are generated in *serpinb1*$^{-/-}$ mice compared with WT, both in vivo after immunization and in vitro on differentiation of naive CD4 cells in the presence of polarizing cytokines (Hou et al. 2014). The effect of SerpinB1 was specific in that no differences were found between *serpinb1*$^{-/-}$ and WT for in vitro generated Th0 cells, Th1 cells, or T-regulatory (Treg) cells. Of note, microarray studies found that *serpinb1a* mRNA was highly induced in Th17 cells compared with other CD4 helper cell subsets (Brustle et al. 2012; Yosef et al. 2013; Heinemann et al. 2014). Efforts are underway to identify the putative proteases that antagonize the effect of serpinB1 in generating IL-17^{+} γδ cells and Th17 cells. Together the studies of IL-17^{+} γδ T cells and Th17 cells identify an important role for endogenous SerpinB1 in restricting the number of IL-17^{+} cells and thus providing a set point that balances neutrophil number and function at sufficient levels to provide innate antimicrobial defense but prevent inflammatory disease.

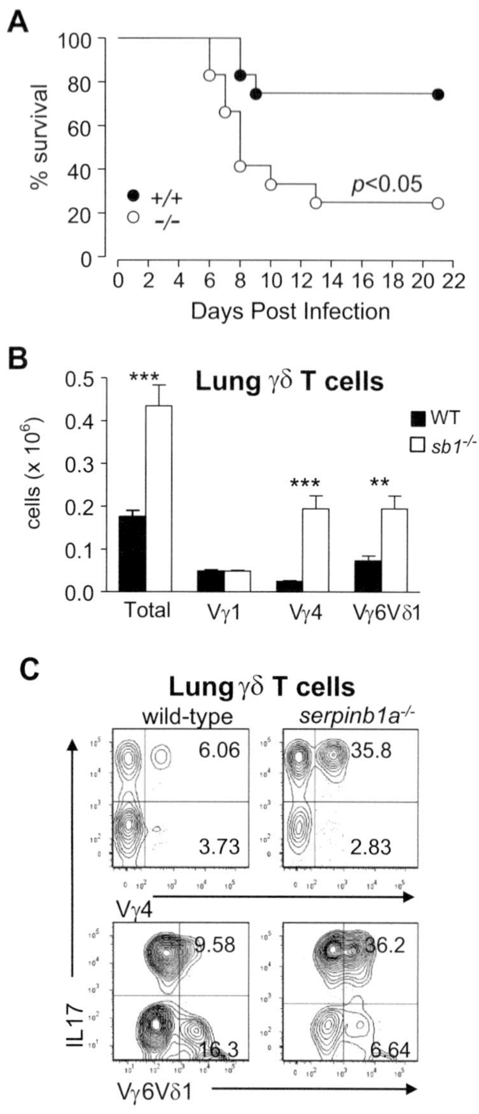

Fig. 6.3 Role of SerpinB1 in restricting IL-17 producing γδ T cells. (**a**) Survival of *serpinb1*$^{-/-}$ mice after influenza challenge. Wild-type (+/+) and *serpinb1*$^{-/-}$ mice (−/−) were inoculated with influenza virus strain. Shown are Kaplan-Meier survival curves for 12 mice per genotype. Levels of IL-17 and numbers of IL-17^{+} γδ cells were increased in lungs of *serpinb1*$^{-/-}$ mice compared with WT mice on day 2 of infection (not shown). (**b**) Selective increase of Vγ4^{+} and Vγ6/Vδ1^{+} γδ T cells in lungs of naive *serpinb1*$^{-/-}$ mice. Lung cells of naive WT and *serpinb1*$^{-/-}$ mice were stained with antibody to γδ-TCR followed by antibodies that detect Vγ1, Vγ4, and Vγ6/Vδ1. (**c**) Increased percentage of IL-17^{+} cells within the *serpinb1*$^{-/-}$ Vγ4^{+} and Vγ6/Vδ1^{+} subsets. Lung cells of naive WT and *serpinb1*$^{-/-}$ mice were cultured with PMA and ionomycin, surface stained as in (**b**) and intracellularly stained for IL-17. No IL-17^{+} cells were found within the Vγ1^{+} subset. Shown are (**b**) means ± SEM for 8 mice per group and (**c**) representative data for 4 mice per group. Reprinted with permission from (**a**) Gong et al., J Infect Dis 204: 592, 2011 and (**b**, **c**) Zhao et al., 95: 521, 2014

6.10 SerpinB1: A Global Regulator of Neutrophil Number and Function

SerpinB1, encoded by an ancient clade B serpin, inhibits serine proteases including neutrophil elastase, proteinase-3, and cathepsin G and a subset of cysteine cathepsins. We propose that the findings described above justify referring to SerpinB1 as a global regulator of neutrophil function as follows: SerpinB1 is critical for preserving the bone marrow reserve of mature neutrophils by protecting bone marrow neutrophils from cathepsin G-mediated death and protecting bone marrow and circulating neutrophils from proteinase-3-mediated caspase-3-dependent death and protecting circulating and extravasated neutrophils from death by NETosis. At sites of infection, SerpinB1 protects the host bacterial killing capacity, thus decreasing microbe-induced neutrophil death that facilitates protease-mediated injury and prolongs inflammatory cytokine production. The alternative neutrophil death mode favored by the presence of SerpinB1 is apoptosis linked to phagocytosis by macrophages, which serves to decrease inflammatory cytokines. The presence of inflammatory cytokines (IL-6, IL-1β, IL-23) supports activation of IL-17$^+$ γδ T cells and generation of Th17 cells; the extent of IL-17$^+$ T cell expansion is restricted by endogenous SerpinB1. IL-17 produced at mucosal sites by IL-17$^+$ T cells induces G-CSF production by bone marrow stromal cells and local production of chemokines and cytokines that recruit and activate neutrophils (summarized in Fig. 6.4).

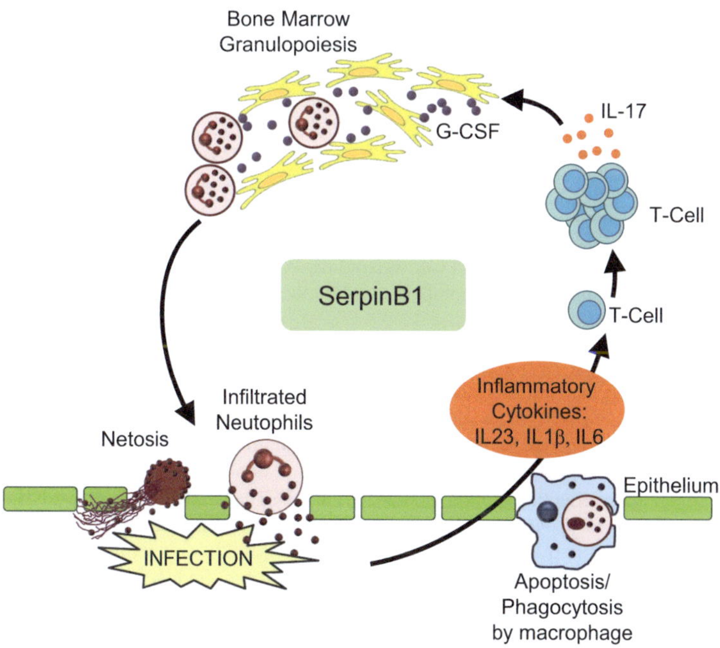

Fig. 6.4 SerpinB1: a global regulator of neutrophil homeostasis. Shown in the schematic are neutrophils generated in the bone marrow and exiting to the circulation. SerpinB1 protects bone

Acknowledgments This work was supported by the National Institutes of Health R01 HL066548, R21 AI102407, and an RRRC award from Boston Children's Hospital.

References

Akashi K, Traver D, Miyamoto T, Weissman IL (2000) A clonogenic common myeloid progenitor that gives rise to all myeloid lineages. Nature 404(6774):193–197

Baumann M, Pham CT, Benarafa C (2013) SerpinB1 is critical for neutrophil survival through cell-autonomous inhibition of cathepsin G. Blood 121(19):3900–3907, S3901–3906

Benarafa C, Remold-O'Donnell E (2005) The ovalbumin serpins revisited: perspective from the chicken genome of clade B serpin evolution in vertebrates. Proc Natl Acad Sci USA 102 (32):11367–11372

Benarafa C, Cooley J, Zeng W, Bird PI, Remold-O'Donnell E (2002) Characterization of four murine homologs of the human ov-serpin monocyte neutrophil elastase inhibitor MNEI (SERPINB1). J Biol Chem 277(44):42028–42033

Benarafa C, Priebe GP, Remold-O'Donnell E (2007) The neutrophil serine protease inhibitor serpinb1 preserves lung defense functions in Pseudomonas aeruginosa infection. J Exp Med 204(8):1901–1909

Benarafa C, LeCuyer TE, Baumann M, Stolley JM, Cremona TP, Remold-O'Donnell E (2011) SerpinB1 protects the mature neutrophil reserve in the bone marrow. J Leukoc Biol 90 (1):21–29

Bird CH, Blink EJ, Hirst CE et al (2001) Nucleocytoplasmic distribution of the ovalbumin serpin PI-9 requires a nonconventional nuclear import pathway and the export factor Crm1. Mol Cell Biol 21(16):5396–5407

Boxio R, Bossenmeyer-Pourie C, Steinckwich N, Dournon C, Nusse O (2004) Mouse bone marrow contains large numbers of functionally competent neutrophils. J Leukoc Biol 75 (4):604–611

Brinkmann V, Reichard U, Goosmann C et al (2004) Neutrophil extracellular traps kill bacteria. Science 303(5663):1532–1535

Brustle A, Brenner D, Knobbe CB et al (2012) The NF-kappaB regulator MALT1 determines the encephalitogenic potential of Th17 cells. J Clin Invest 122(12):4698–4709

Chen K, Nishi H, Travers R et al (2012) Endocytosis of soluble immune complexes leads to their clearance by FcgammaRIIIB but induces neutrophil extracellular traps via FcgammaRIIA in vivo. Blood 120(22):4421–4431

Chervenick PA, Boggs DR, Marsh JC, Cartwright GE, Wintrobe MM (1968) Quantitative studies of blood and bone marrow neutrophils in normal mice. Am J Physiol 215(2):353–360

Fig. 6.4 (continued) marrow and blood neutrophils by preventing excess death by cathepsin G- and proteinase-3-mediated pathways. SerpinB1 also protects circulating and extravasated neutrophils from excess death by NETosis, a highly inflammatory pathway. SerpinB1 protection at sites of infections includes preventing tissue injury and non-apoptotic neutrophil death, which increases injury and production of inflammatory cytokines. The alternative neutrophil death mode, apoptosis, is linked to phagocytosis by macrophages, which decreases inflammatory cytokine production. Inflammatory cytokines including IL-1β, IL-23, and IL-6 support IL-17 production by IL-17^{+} γδ T cells and Th17 cell generation and expansion. Endogenous SerpinB1 restricts expansion of γδ T cells and restricts generation of Th17 cells. IL-17 acts locally by inducing epithelial cells and fibroblasts to produce chemokines and cytokines that recruit and activate neutrophils. IL-17 also acts on bone marrow stromal cells to induce G-CSF and increase granulopoiesis

Christopher MJ, Link DC (2007) Regulation of neutrophil homeostasis. Curr Opin Hematol 14 (1):3–8

Cooley J, Takayama TK, Shapiro SD, Schechter NM, Remold-O'Donnell E (2001) The serpin MNEI inhibits elastase-like and chymotrypsin-like serine proteases through efficient reactions at two active sites. Biochemistry 40(51):15762–15770

Cooley J, Sontag MK, Accurso FJ, Remold-O'Donnell E (2011) SerpinB1 in cystic fibrosis airway fluids: quantity, molecular form and mechanism of elastase inhibition. Eur Respir J 37 (5):1083–1090

Dubin A, Travis J, Enghild JJ, Potempa J (1992) Equine leukocyte elastase inhibitor. Primary structure and identification as a thymosin-binding protein. J Biol Chem 267(10):6576–6583

Farley K, Stolley JM, Zhao P, Cooley J, Remold-O'Donnell E (2012) A serpinB1 regulatory mechanism is essential for restricting neutrophil extracellular trap generation. J Immunol 189 (9):4574–4581

Fogli LK, Sundrud MS, Goel S et al (2013) T cell-derived IL-17 mediates epithelial changes in the airway and drives pulmonary neutrophilia. J Immunol 191(6):3100–3111

Forlow SB, Schurr JR, Kolls JK, Bagby GJ, Schwarzenberger PO, Ley K (2001) Increased granulopoiesis through interleukin-17 and granulocyte colony-stimulating factor in leukocyte adhesion molecule-deficient mice. Blood 98(12):3309–3314

Fuchs TA, Abed U, Goosmann C et al (2007) Novel cell death program leads to neutrophil extracellular traps. J Cell Biol 176(2):231–241

Garcia-Romo GS, Caielli S, Vega B et al (2011) Netting neutrophils are major inducers of type I IFN production in pediatric systemic lupus erythematosus. Sci Transl Med 3(73):73ra20

Gong D, Farley K, White M, Hartshorn KL, Benarafa C, Remold-O'Donnell E (2011) Critical role of serpinB1 in regulating inflammatory responses in pulmonary influenza infection. J Infect Dis 204(4):592–600

Grigoryev SA, Bednar J, Woodcock CL (1999) MENT, a heterochromatin protein that mediates higher order chromatin folding, is a new serpin family member. J Biol Chem 274 (9):5626–5636

Hakkim A, Fuchs TA, Martinez NE et al (2011) Activation of the Raf-MEK-ERK pathway is required for neutrophil extracellular trap formation. Nat Chem Biol 7(2):75–77

Harrington LE, Hatton RD, Mangan PR et al (2005) Interleukin 17-producing CD4+ effector T cells develop via a lineage distinct from the T helper type 1 and 2 lineages. Nat Immunol 6 (11):1123–1132

Heinemann C, Heink S, Petermann F et al (2014) IL-27 and IL-12 oppose pro-inflammatory IL-23 in CD4+ T cells by inducing Blimp1. Nat Commun 5:3770

Hillier LW et al (2004) Sequence and comparative analysis of the chicken genome provide unique perspectives on vertebrate evolution. Nature 432:695–716

Hou L, Cooley J, Remold-O'Donnell E (2014) T cell intrinsic SerpinB1 and a cathepsin protease serve as yin and yang to control Th17 differentiation (LYM3P.732). J Immunol 192:64.6

Huber R, Carrell RW (1989) Implications of the three-dimensional structure of alpha 1-antitrypsin for structure and function of serpins. Biochemistry 28(23):8951–8966

Janoff A (1985) Elastase in tissue injury. Annu Rev Med 36:207–216

Jensen PH, Schuler E, Woodrow G et al (1994) A unique interhelical insertion in plasminogen activator inhibitor-2 contains three glutamines, Gln83, Gln84, Gln86, essential for transglutaminase-mediated cross-linking. J Biol Chem 269(21):15394–15398

Kaiserman D, Bird PI (2005) Analysis of vertebrate genomes suggests a new model for clade B serpin evolution. BMC Genomics 6:167

Kaiserman D, Knaggs S, Scarff KL et al (2002) Comparison of human chromosome 6p25 with mouse chromosome 13 reveals a greatly expanded ov-serpin gene repertoire in the mouse. Genomics 79(3):349–362

Keller M, Ruegg A, Werner S, Beer HD (2008) Active caspase-1 is a regulator of unconventional protein secretion. Cell 132(5):818–831

Lande R, Ganguly D, Facchinetti V et al (2011) Neutrophils activate plasmacytoid dendritic cells by releasing self-DNA-peptide complexes in systemic lupus erythematosus. Sci Transl Med 3 (73):73ra19

Li P, Li M, Lindberg MR, Kennett MJ, Xiong N, Wang Y (2010) PAD4 is essential for antibacterial innate immunity mediated by neutrophil extracellular traps. J Exp Med 207 (9):1853–1862

Loison F, Zhu H, Karatepe K et al (2014) Proteinase 3-dependent caspase-3 cleavage modulates neutrophil death and inflammation. J Clin Invest 124(10):4445–4458

Martin B, Hirota K, Cua DJ, Stockinger B, Veldhoen M (2009) Interleukin-17-producing gammadelta T cells selectively expand in response to pathogen products and environmental signals. Immunity 31(2):321–330

Massberg S, Grahl L, von Bruehl ML et al (2010) Reciprocal coupling of coagulation and innate immunity via neutrophil serine proteases. Nat Med 16(8):887–896

Miyamoto M, Prause O, Sjostrand M, Laan M, Lotvall J, Linden A (2003) Endogenous IL-17 as a mediator of neutrophil recruitment caused by endotoxin exposure in mouse airways. J Immunol 170(9):4665–4672

Narasaraju T, Yang E, Samy RP et al (2011) Excessive neutrophils and neutrophil extracellular traps contribute to acute lung injury of influenza pneumonitis. Am J Pathol 179(1):199–210

Narayan K, Sylvia KE, Malhotra N et al (2012) Intrathymic programming of effector fates in three molecularly distinct gammadelta T cell subtypes. Nat Immunol 13(5):511–518

Papayannopoulos V, Metzler KD, Hakkim A, Zychlinsky A (2010) Neutrophil elastase and myeloperoxidase regulate the formation of neutrophil extracellular traps. J Cell Biol 191 (3):677–691

Papayannopoulos V, Staab D, Zychlinsky A (2011) Neutrophil elastase enhances sputum solubilization in cystic fibrosis patients receiving DNase therapy. PLoS One 6(12), e28526

Park H, Li Z, Yang XO et al (2005) A distinct lineage of CD4 T cells regulates tissue inflammation by producing interleukin 17. Nat Immunol 6(11):1133–1141

Pham CT, Ivanovich JL, Raptis SZ, Zehnbauer B, Ley TJ (2004) Papillon-Lefevre syndrome: correlating the molecular, cellular, and clinical consequences of cathepsin C/dipeptidyl peptidase I deficiency in humans. J Immunol 173(12):7277–7281

Rees DD, Rogers RA, Cooley J, Mandle RJ, Kenney DM, Remold-O'Donnell E (1999) Recombinant human Monocyte/Neutrophil elastase inhibitor protects rat lungs against injury from cystic fibrosis airway secretions. Am J Respir Cell Mol Biol 20(1):69–78

Remold-O'Donnell E (1993) The ovalbumin family of serpin proteins. FEBS Lett 315(2):105–108

Remold-O'Donnell E, Chin J, Alberts M (1992) Sequence and molecular characterization of human monocyte/neutrophil elastase inhibitor. Proc Natl Acad Sci USA 89(12):5635–5639

Remold-O'Donnell E, Lewandrowski K (1983) Two proteinase inhibitors associated with peritoneal macrophages. J Biol Chem 258(5):3251–3257

Schwarzenberger P, Huang W, Ye P et al (2000) Requirement of endogenous stem cell factor and granulocyte-colony-stimulating factor for IL-17-mediated granulopoiesis. J Immunol 164 (9):4783–4789

Shibata K, Yamada H, Hara H, Kishihara K, Yoshikai Y (2007) Resident Vdelta1+ gammadelta T cells control early infiltration of neutrophils after Escherichia coli infection via IL-17 production. J Immunol 178(7):4466–4472

Sorensen OE, Clemmensen SN, Dahl SL et al (2014) Papillon-Lefevre syndrome patient reveals species-dependent requirements for neutrophil defenses. J Clin Invest 124(10):4539–4548

Stark MA, Huo Y, Burcin TL, Morris MA, Olson TS, Ley K (2005) Phagocytosis of apoptotic neutrophils regulates granulopoiesis via IL-23 and IL-17. Immunity 22(3):285–294

Sutton CE, Lalor SJ, Sweeney CM, Brereton CF, Lavelle EC, Mills KH (2009) Interleukin-1 and IL-23 induce innate IL-17 production from gammadelta T cells, amplifying Th17 responses and autoimmunity. Immunity 31(2):331–341

Theilgaard-Monch K, Jacobsen LC, Borup R et al (2005) The transcriptional program of terminal granulocytic differentiation. Blood 105(4):1785–1796

Wang Y, Li M, Stadler S et al (2009) Histone hypercitrullination mediates chromatin decondensation and neutrophil extracellular trap formation. J Cell Biol 184(2):205–213
Wang L, Li Q, Wu L et al (2013) Identification of SERPINB1 as a physiological inhibitor of human granzyme H. J Immunol 190(3):1319–1330
Woods DE, Cantin A, Cooley J, Kenney DM, Remold-O'Donnell E (2005) Aerosol treatment with MNEI suppresses bacterial proliferation in a model of chronic Pseudomonas aeruginosa lung infection. Pediatr Pulmonol 39(2):141–149
Yasumatsu R, Altiok O, Benarafa C et al (2006) SERPINB1 upregulation is associated with in vivo complex formation with neutrophil elastase and cathepsin G in a baboon model of bronchopulmonary dysplasia. Am J Physiol Lung Cell Mol Physiol 291(4):L619–L627
Yosef N, Shalek AK, Gaublomme JT et al (2013) Dynamic regulatory network controlling TH17 cell differentiation. Nature 496(7446):461–468
Zhao P, Hou L, Farley K, Sundrud MS, Remold-O'Donnell E (2014) SerpinB1 regulates homeostatic expansion of IL-17+ gammadelta and CD4+ Th17 cells. J Leukoc Biol 95(3):521–530

Serpins in T Lymphocyte Immunity and Blood Development

7

Philip G. Ashton-Rickardt

Contents

Abstract

T lymphocytes are required for cell-mediated immunity but also give rise to immunological diseases such as autoimmunity and transplantation rejection. Intracellular serine protease inhibitors ($\mathrm{serpins_{IC}}$) regulate the activity of key of proteases that control T lymphocyte function and development. An emerging view of $\mathrm{serpins_{IC}}$ is that they are important promoters of cellular viability through their inhibition of executioner proteases. This will be discussed in the context of the T lymphocyte survival during effector responses and the development and persistence of long-lived memory T cells. The potent anti-apoptotic properties of serpins can also work against adaptive cell immunity by protecting viruses and tumors from eradication by cytotoxic T cells (CTL). Recent insights

P.G. Ashton-Rickardt (✉)
Faculty of Medicine, Division of Immunology and Inflammation, Section of Immunobiology, Department of Medicine, Imperial College London, Exhibition Road, London SW7 2AZ, UK
e-mail: p.ashton-rickardt@imperial.ac.uk

M. Geiger et al. (eds.), *The Serpin Family*, DOI 10.1007/978-3-319-22711-5_7

from knockout mouse models demonstrate that these serpins also are required for hematological progenitor cells and so are critical for the development of lineages other than T lymphocytes. Given the emerging role of serpins in multiple aspects of lymphocyte immunity and blood development, this chapter will examine new immunotherapeutic approaches based directly on serpins or on knowledge gained from identifying their physiologically relevant protease targets.

7.1 Intracellular Serine Protease Inhibitors and Programmed Cell Death

Most serpins are secreted and constitute about 10 % of plasma proteins. In vertebrates, clade B serpins belong to a large intracellular family (Remold-O'Donnell 1993). Intracellular serpins (serpins$_{IC}$) lack cleavable N-terminal signal peptides and dwell within the nucleocytoplasmic compartment. Serpins$_{IC}$ inhibit several key proteases that trigger programmed cell death (PCD), and so they can be viewed as key regulators of cell survival. In addition to inhibiting serine proteases, some serpins$_{IC}$ are cross-class specific and inhibit cysteine proteases (Fig. 7.1). The first description of an anti-PCD function of serpins$_{IC}$ came from work on CrmA from the cowpox virus. CrmA inhibits the serine protease granzyme B (GrB), which is used by T lymphocytes to induce PCD in infected cells (Ashton-Rickardt 2010). Mammalian serpins$_{IC}$, which inhibit GrB, have also been identified in humans, proteinase inhibitor 9 (PI9, SERPINB9) (Sun et al. 1996), and mice—serine proteases inhibitor 6 (Spi6) (Sun et al. 1997). In addition to inhibiting executioner serine proteases, some cross-class-specific serpins$_{IC}$ inhibit death-inducing cysteine cathepsins. CrmA can inhibit both caspases 1 and 8 at physiological rate constants, whereas only caspase 1 is inhibited efficiently by PI9. Cross-class-specific serpins$_{IC}$ can also inhibit executioner cathepsins. Spi2A inhibits the serine protease cathepsin G and also cysteine cathepsins such as cathepsin B, V, L, K, and H after they are released from the lysosome into the cytoplasm (Liu et al. 2003). Spi2A protects from apoptotic PCD by inhibiting cathepsin B-mediated cleavage of Bid after stimulation by tumor necrosis factor receptor 1 (TNF-R1) (Liu et al. 2003) (Fig. 7.1). Spi2A can also protect from necrotic PCD induced by reactive oxygen species (ROS) or TNF-α in the absence of caspase activity, presumably because cathepsin B can induce both apoptotic and necrotic PCD (Liu et al. 2004b). Spi2A is a physiological target of NF-κB (Liu et al. 2003) and can substitute for the transcription factor in protecting cell from the lysosomal pathways of PCD (Liu et al. 2003). This cross-class specificity of Spi2A for cysteine cathepsins is also a property of another mouse serpin SQN-5 (Serpinb3a) and the human serpin SCCA1 (SERPINB3), as well as SRP-6, which protects in *C. elegans* from necrotic PCD during development (Ashton-Rickardt 2010). The mechanism by which cysteine proteases are inhibited involves the cleavage of the serpins$_{IC}$, in some cases involving as stable covalent complex and in other cases not (Ashton-Rickardt 2010).

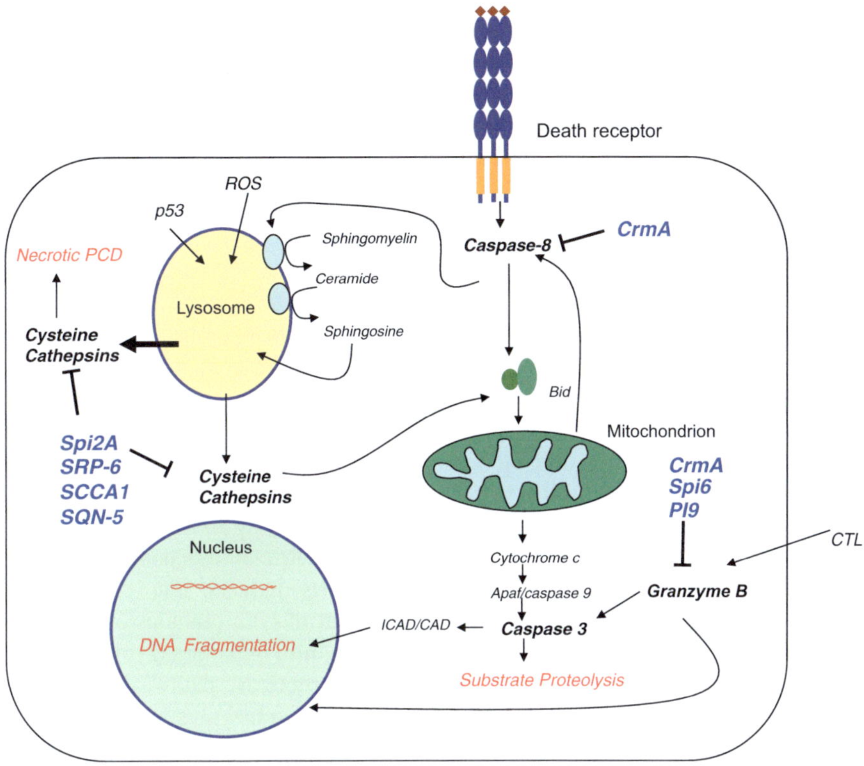

Fig. 7.1 Inhibition of death pathways by serpins. Serpins$_{IC}$ specific for executioner proteases are indicated

7.2 Self-Inflicted Damage of Cytotoxic Lymphocytes

Cytotoxic lymphocytes protect from infection by inducing the PCD of infected or transformed cells through contact-dependent, lymphocyte-mediated cytotoxicity, which can proceed through two pathways. The first involves the action of proteins present in specialized granules in cytotoxic T cells and natural killer (NK) cells (Russell and Ley 2002). In this pathway, perforin facilitates the entry of granzymes, which trigger rapid PCD in target cells. The second pathway is triggered by members of TNF-R family of which Fas is the most important (Russell and Ley 2002). Granzymes A and B are the most abundant granzymes in mice and humans and are the best characterized. When an activated CTL recognizes antigen on an infected cell, a tight junction is formed with the target cell, into which the contents of lytic granules are secreted (Russell and Ley 2002). GrB is released into the synapse between CTL and target and is endocytosed into the target cell through a mechanism that requires perforin. Once released into the cytoplasm, GrB rapidly traffics to the nucleus and initiates death by cleaving protein substrates that leads to DNA fragmentation and cell death.

Given the effectiveness of the granule exocytosis pathway of death, it is not surprising that CTL and NK cells are susceptible to self-inflicted damage. A dramatic demonstration of this is the phenotype of perforin-deficient CTL, which are defective in granule-mediated killing but also undergo less PCD in vitro and in vivo (Ashton-Rickardt 2010). The clinical relevance of perforin as a negative regulator of CTL survival comes from patients with familial hemophagocytic lymphohistiocytosis (FHL) who harbor homozygous loss-of-function defects in the perforin gene and massively increased levels of CTL (Stepp et al. 1999).

There is evidence to indicate that granzymes are turned against CTLs through both fratricide and suicide. During fratricide, CTL become the targets for killing by other CTL, which may occur if CTL express antigen on its surface (Walden and Eisen 1990). Of potentially wider relevance are observations that, in the course of delivering the lethal hit, a CTL internalizes and then reexpresses antigen, on its plasma membrane rendering it a target for fratricidal killing (Huang et al. 1999). During the killing of a target, the perforin/granzyme pathway also induces the death of the CTL itself delivering the lethal hit in a fratricide-independent manner (suicide) (Ashton-Rickardt 2010). Several reports indicate the presence of "misdirected" GrB in compartments other than that of lytic granules or the cytoplasm of a target cell. During CTL killing, up to one third of the granzymes A and B produced are secreted directly via the constitutive secretory pathway rather than from lytic granules (Lieberman 2003). This source of granzymes is thought to be responsible for bystander killing of targets, which occurs without the formation of conjugates with CTL. It has long been appreciated that granzymes can be found in the serum during CTL immune reactions, indicating a potential extracellular source of misdirected killing molecules (Lieberman 2003).

7.3 Protection of Cytotoxic Lymphocytes from Self-Inflicted Damage by Serpins

Whichever way GrB enters the cytoplasm of a CTL, what is now clear is that $\text{serpins}_{\text{IC}}$ play a key role in protecting CTL from self-inflicted damage. PI9 is an endogenous $\text{serpins}_{\text{IC}}$ specific for GrB, expressed in human CTL and NK cells (Sun et al. 1996). Expression of transgenic PI9 improved the viability of human CTL, suggesting that upregulation of PI9 may protect CTLs from self-inflicted damage (Hirst et al. 2003). Transgenic expression of the mouse homologue, Spi6, protected mouse CTL from GrB and increased the number of antigen-specific CD8 T cells after infection of mice with lymphocytic choriomeningitis virus (LCMV) (Phillips et al. 2004). However, the most convincing evidence that GrB-specific $\text{serpins}_{\text{IC}}$ protect CTLs from self-inflicted damage comes from the study of Spi6 KO mice, which revealed that Spi6 inhibits GrB in the cytoplasm after it leaks of the cytotoxic granules by acting as a classic suicide substrate (Zhang et al. 2006) (Fig. 7.2). The increase in cytoplasmic GrB activity leads to the induction of PCD and decreased

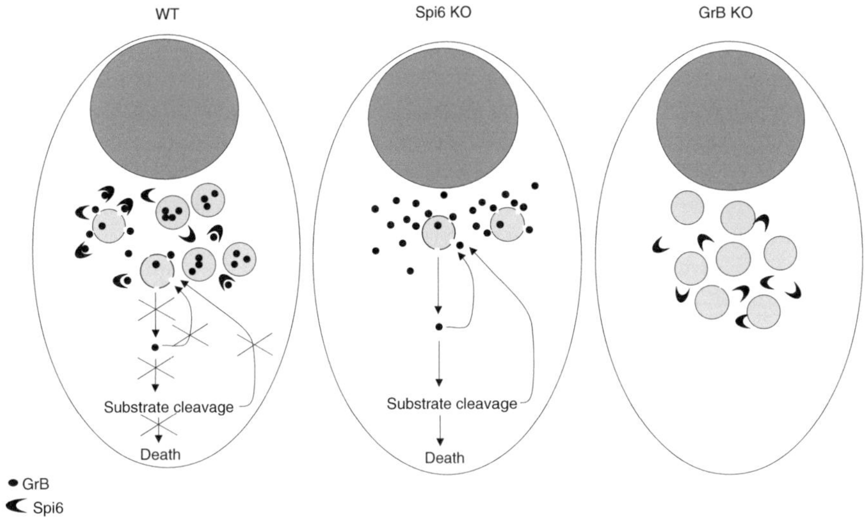

Fig. 7.2 Spi6 protects CTL from self-inflicted damage. Cartoons showing role of Spi6 in the inhibition of GrB in wild-type (WT), Spi6 KO and GrB KO CTL

clonal burst of CTL specific for LCMV and the intracellular bacterium *L. monocytogenes* resulting in impaired immunity (Zhang et al. 2006). Spi6 not only protected CTL from PCD caused by cytoplasmic GrB but also ensured the integrity of cytotoxic granules by preventing granzyme-mediated breakdown (Zhang et al. 2006) (Fig. 7.2). Spi6 KO CTL have fewer cytotoxic granules than normal, and those they do have appear to not to be arranged correctly into cytotoxic vesicles. This defect could be corrected by GrB deficiency, suggesting an active role for GrB in granule breakdown. How cytoplasmic GrB feeds back and causes the destruction of cytotoxic granules is not known, but presumably granule structural proteins could be directly attacked by GrB or by downstream products of GrB activity. Inhibition of GrB by Spi6 protects cytotoxic lymphocytes of the innate as well as the adaptive branches from self-inflicted damage. Invariant NKT (iNKT) cells are subset of cytotoxic lymphocyte, which participate in both immediate immune responses against invariant antigen (innate) as well as long-term immune responses (adaptive) (Godfrey et al. 2004). Spi6-deficient iNKT cells harbored increased levels of GrB and were susceptible to PCD after stimulation antigen (Ansari et al. 2010). As with classical CTL, activation-induced PCD resulted in impaired survival of Spi6 KO iNKT and a decreased clonal burst size after challenge in vivo.

7.4 Serpins and the Development and Persistence of Memory T Cells

Immunological memory is unique to vertebrates and is defined as the persistence of a reactive state initiated by antigen challenge (Ahmed and Gray 1996). This persistence of an immunological reactive state is dependent on memory lymphocytes, which increase the frequency of pathogen-specific cells and are functionally superior to naïve lymphocytes. After the rapid expansion in the number of antigen-specific CTL, which may be several orders of magnitude, an equally dramatic decrease in cell number occurs (Ahmed and Gray 1996). This so-called contraction phase in the immune response is driven by PCD and is an important homeostatic mechanism for regulating T cell numbers (Ahmed and Gray 1996). Those CTL that escape PCD differentiate into memory T lymphocytes (MTL), which provide long-term adaptive immunity through robust recall responses (Opferman et al. 1999). For long-term persistence, and self-renewal, MTL must also undergo a steady rate of homeostatic proliferation (HSP), which preferentially occurs in the bone marrow (Becker et al. 2005). HSP of MTL is critically dependent on survival cytokines such as interleukin-15 (IL-15), which induce the expression of anti-PCD molecules such as Bcl-2, Bcl-X_L, and Mcl-1 in MTL (Opferman et al. 2003). Studies support a linear differentiation model of MTL development, in which MTL are the progeny of CTL (Jacob and Baltimore 1999; Opferman et al. 1999). Since CTL are highly susceptible to PCD, the differentiation of MTL requires the escape from PCD, which may be facilitated by the upregulation of so-called protective genes that encode anti-PCD factors in memory cell precursors. Work from our group has shown that two different serpins$_{IC}$ expressed in CTL control the development of MTL in different ways.

7.4.1 Serine Protease Inhibitor 2A (Spi2A)

Spi2A (Seprina3g) is cross-class-specific serpins$_{IC}$ of cysteine cathepsins, which protects cells from the lysosomal pathway of PCD (Liu et al. 2003) and is upregulated in the MTL and their precursors (Liu et al. 2004a). Overexpressing *Spi2A* increased the number of MTL, and conversely, knockdown of expression by antisense message blocked the development of MTL (Liu et al. 2004a). However, the *Serpina3g* locus contains several genes, which share a very high degree of homology to Spi2A, and so it is difficult to definitely ascribe the effects on MTL development uniquely to Spi2A. The role of Spi2A in MTL self-renewal and HSP has also been studied in Spi2A KO mice. There was an age-dependent deficit and impaired self-renewal of MTL in Spi2A KO mice (Byrne et al. 2012). The defective HSP of MTL in Spi2A KO mice could be corrected by cathepsin B deficiency, demonstrating that cathepsin B was the physiological target of Spi2A in maintaining MTL. Spi2A transcription is induced by IL-15 and other survival cytokines, and so the protection from cathepsin B-mediated PCD by Spi2A is a mechanism by which the survival cytokines ensure MTL self-renewal through HSP (Fig. 7.3).

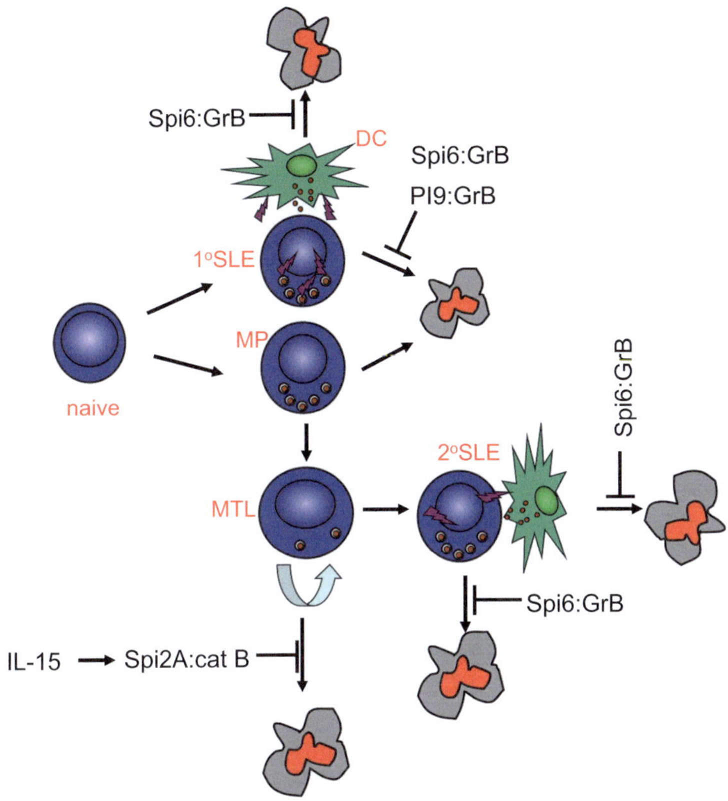

Fig. 7.3 Serpins in CTL and memory CD8 T cell development. CD8α dendritic cells (DC), primary short-lived effector (1^0SLE), secondary (2^0) SLE, memory cell precursor (MP), memory T lymphocyte (MTL)

7.4.2 Serine Protease Inhibitor 6 (Spi6)

Work with Spi6 KO mice has clearly demonstrated a role for the endogenous GrB inhibitor in protecting CTL from self-inflicted damage and controlling the clonal burst of pathogen-specific CTL (Zhang et al. 2006). However, despite an almost complete reduction in the size of the CTL pool, the size of the MTL pool was unaffected in Spi6 KO mice (Zhang et al. 2007). Therefore, the development of MTL was independent of Spi6 (Fig. 7.3). Although Spi6 was required for the survival of the majority of short-lived effector CTL, it was dispensable for the survival of the small proportion of the CTL pool that were memory cell precursors because these cells express low levels of GrB and so are not susceptible to GrB-mediated self-destruction. However, in Spi6 KO mice, the number of secondary CTL derived from MTL was severely diminished in antigen rechallenge responses (Zhang et al. 2007). Therefore, although Spi6 is not required for the development of primary MTL, it still controls memory responses by determining the size of the secondary CTL clonal burst in after reinfection. The T-bet transcription factor is required for the differentiation of CTL but not MTL (Joshi et al. 2007). GrB is a transcriptional target of T-bet, and so the dichotomy between short-lived GrB^{hi} CTL and GrB^{lo} memory cell precursors can be explained in terms of graded expression of T-bet (Joshi et al. 2007). An attractive hypothesis would be that the coordinate expression of Spi6 and GrB is part of the T-bet transcriptional program that ensures the survival of CTL. The study of Spi2A KO and Spi6 KO mice suggests that $serpins_{IC}$ have evolved separate functions in determining the size of a primary CTL (Spi6 inhibition of GrB) versus controlling long-term self-renewal (Spi2A inhibition of cathepsin B) (Fig. 7.3).

7.5 Protection of Dendritic Cells from CTL

Dendritic cells (DC) are the physiological antigen presenting cells that stimulate both naïve cells and MTL to proliferate and differentiate into CTL. In mice, the $CD11c^{++}$ $CD8\alpha^{++}$ $CD205^{+}$ DC population (CD8αDC) can best acquire a wide variety of cellular antigens (including viral proteins) from infected and dying cells and present them on self-class I major histocompatibility molecules (MHC 1) to cognate CD8 T cells in a process known as cross-presentation (Heath et al. 2004). Experiments with perforin-deficient CTL show that both primary and recall CTL eliminate cognate DC as part of a negative feedback mechanism that limits their own expansion (Ashton-Rickardt 2010). However, the fact that DC are still highly effective at priming CTL expansion implies that they have mechanisms that protect them from CTL killing. The upregulation of Spi6 in DC upon maturation or through transgene expression results in the protection of DC from granule-mediated PCD by CTL (Medema et al. 2001a). Spi6 KO mice have demonstrated that Spi6 protects CD8α DC from GrB delivered by CTL (Lovo et al. 2012) (Fig. 7.3). The impaired survival of Spi6 KO CD8α DC resulted in impaired priming and expansion of both primary and memory LCMV-specific CTL. This defect could be corrected by GrB deficiency, demonstrating that GrB was the physiological target through which Spi6 protected CD8αDC from CTL. Therefore, the negative regulation of DC priming of cellular adaptive immunity by CTL killing is mitigated by the physiological inhibition of GrB by Spi6.

7.6 Serpins and CTL-Mediated Transplantation Rejection

Although immunological memory gives long-term protective immunity to infection, it can also convert acute disease to chronic disease when the antigen is a histoincompatible transplanted tissue (allograft). The failure to extinguish or suppress T cell memory to allografts presents a major barrier to curing chronic transplant rejection. Spi6 also protects allo-specific CTL from misdirected GrB, thereby controlling the size of the CTL pool in mouse skin and cardiac allograft models (Azzi et al. 2015). As was observed for antiviral CTL, the deficit in Spi6 KO allo-specific CTL clonal burst did not translate into a deficit in the MTL pool (Zhang et al. 2007). However, it is still not known whether Spi6 controls memory responses at the level of secondary allo-specific CTL expansion as it does for antivirus memory (Zhang et al. 2007). In any case in the models tested, the deficit in the survival of Spi6 KO allo-specific CTL did not result in increased allograft survival. This may be because Spi6 is also required to protect GrB-positive T regulatory (Tregs) from death, which are important suppressors of allo-specific CTL responses (Azzi et al. 2013). Therefore, a decrease in allograft survival due to decreased levels of Spi6 KO Treg levels may counter the increase in allograft survival resulting from decreased levels of allo-specific CTL. The situation with allograft rejection illustrates the multicellular role serpin protection from GrB plays in controlling T cell immune responses.

7.7 Serpins and CTL-Driven Inflammatory Diseases

Granzymes have been implicated as tissue-damaging agents in a variety of inflammatory and autoimmune diseases, including graft versus host disease (GVHD), rheumatoid arthritis, atherosclerosis, Crohn's diseases, and systemic lupus erythematosus. Expression studies have located PI9 to nonlymphoid cells in immunoprivileged sites and at the sites of inflammatory reactions (Bladergroen et al. 2001; Hirst et al. 2003). Therefore, it has been suggested that GrB-specific $serpins_{IC}$, such as PI9 and Spi6, in addition to protecting DC, may also have a more general protective role for bystander cells in CTL-mediated inflammation. Given the wide variety of inflammatory diseases caused by CTLs, it is likely that the upregulation of PI9 or equivalent $serpins_{IC}$ in nonlymphoid tissue may have a common strategy for limiting pathological tissue damage. In addition, PI9 may also protect from inflammatory disease by inhibiting proteases other than GrB. Interleukin-1 (IL-1) beta converting enzyme (ICE; caspase 1) is required for the production of pro-inflammatory IL-1 and IL-18 cytokines. PI9 not only inhibits GrB as a classical suicide substrate, but can also inhibit caspase 1 through a similar mechanism (Annand et al. 1999). This interaction seems to be physiologically relevant in vascular smooth muscle cells where PI9 regulates the production of IL-1β and IL-18 through inhibition of caspase 1 (Young et al. 2000). A potential protective role for PI9 in inflammatory atherosclerosis is suggested by the lower expression in human atherosclerotic lesions (Young et al. 2000).

7.8 Serpins and Immunoevasion

Inhibition of executioner proteases by $serpins_{IC}$ plays a critical role in CTL-mediated immunity. However, intracellular pathogens may use serpinsIC to inhibit CTL proteases used by the immune system to kill them. For example, CrmA may allow cowpox virus in infected cells to escape both the granule-mediated and the Fas-FasL-mediated pathways of CTL killing through the inhibition of GrB and caspases. It has been suggested that inhibition of GrB by serpins$_{IC}$ may represent a novel escape mechanism by which tumors may evade CTL-mediated immuno-surveillance. This is based on the observation that overexpression of Spi6 and PI9 protects tumor cell lines from CTL-mediated lysis and that PI9 is expressed in both Hodgkin's (T and B cell) and non-Hodgkin's lymphomas and also breast, colon, and cervix carcinomas (Bladergroen et al. 2002). However, PI9 is widely expressed in normal leukocytes and normal epithelial cells (Hirst et al. 2003), and it is still yet to be determined whether the level of expression in cancer cells is any higher than the level in noncancerous control tissue (Bladergroen et al. 2002). The correlation between PI9 expression and the poor prognosis of patients with anaplastic large cell lymphoma suggests that the escape of tumors from CTL-mediated immuno-surveillance may involve protection from GrB (ten Berge et al. 2002).

7.9 Serpins in Blood Cell Development

Studies with Spi2A KO mice have revealed an unexpected role for Spi2A cyto-protection in the development of multiple lineages of blood cells. In response to anemia, erythropoietin (EPO) is produced by renal interstitial fibroblasts and then acts within bone marrow to promote erythroid progenitor expansion via its cell surface receptor (EPOR) (Wojchowski et al. 2010). ROS as generated by iron and heme accumulation in erythroblasts brings about lysosomal membrane permeability and subsequent cell death through cathepsins (Dev et al. 2013). Spi2A is required for EPO-driven hematopoiesis through protection of erythroblasts from cathepsin B-/L-mediated PCD (Dev et al. 2013) (Fig. 7.4). Pharmacological inhibition of cathepsins B and L also enhanced EPO-induced red cell formation in normal mice and in an EPO-dependent human model. Reconstitution experiments did not reveal a role for Spi2A in the self-renewal of long-term hematopoietic stem cells (LT-HSC), but rather Spi2A was required for the expansion of LT-HSC as well as

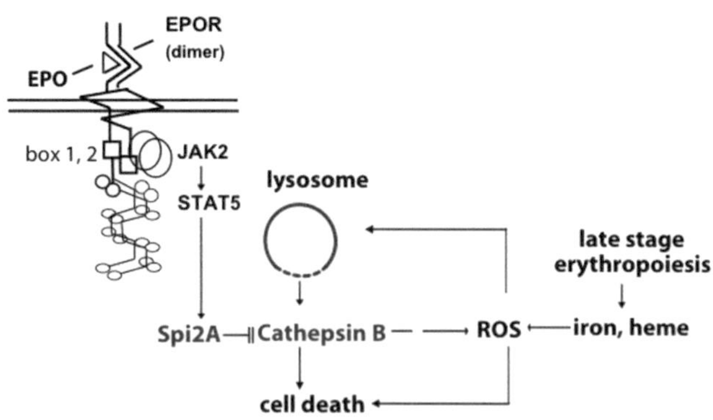

Fig. 7.4 Spi2A is an EPO-specific protective factor. Spi2A inhibits cathepsin B after it is released from lysosomes within stressed or compromised erythroblasts

granulomyelocytic and B lymphocyte progenitors after upregulation by STAT 5 transcription factor-mediated signaling through granulomyelocytic colony stimulating factor (GMCSF) or IL-7 receptors (Li et al. 2014). As with other blood cell types, the mechanism of Spi2A function is to protect from PCD causes by cysteine cathepsins released from the lysosome (Liu et al. 2003).

7.10 Therapeutic Potential of Serpins$_{IC}$

A challenge for the future is to translate our knowledge of T lymphocyte biology into new ways of harnessing cellular immunity to combat infection, cancer, and inflammatory disease. The human serpins$_{IC}$ PI9 and its mouse homologue Spi6 can influence CTL responses by both affecting the viability of DC antigen presentation cells (Medema et al. 2001b; Lovo et al. 2012) and CTL (Phillips et al. 2004; Hirst et al. 2003; Zhang et al. 2006) through the inhibition of cytoplasmic GrB. One potential adjuvant strategy is to improve the efficiency of DC priming by increasing viability. A preliminary DNA vaccination study against tumors in mice has indicated the feasibility of using *Spi6* DNA as an adjuvant in this way (Ashton-Rickardt 2010). Recent insights into how serpins$_{IC}$ control the development and function of both SLE and MTL may provide new translational opportunities. Spi6 KO mice have revealed that improving CTL viability by protecting from self-inflicted damage (Zhang et al. 2006) does not necessarily increase the size of the MTL (Zhang et al. 2007). So in order to maximize differentiation of long-term memory cells, one must suppress the death pathways specific to MTL. In this regard the suppression of cysteine cathepsins by transgenic Spi2A may be useful as an adjuvant approach. However, the lack of a structural homologue of Spi2A in humans may prevent such an adjuvant approach.

Clinical trials utilizing mesenchymal stem cells (MSC) in tissue engineering and as immunosuppressive agents have been initiated worldwide. Expression of transgenic Spi6 and PI9 protects mouse and human MSC, respectively, from killing allo-specific CTL and improved survival after transplantation (El Haddad et al. 2011a, b). Therefore, the delivery of PI9 to MSC may prove clinically useful. The impaired expansion of allo-specific Spi6 KO CTL could make Spi6 an attractive target for alleviating allograft rejection (Azzi et al. 2015). Targeted inhibition of Spi6 expression or function in allo-specific CTL would conceivably alleviate transplantation rejection. Eliminating CTL by inhibiting Spi6/PI9 may alleviate transplantation rejection and GVHD (Azzi et al. 2015), but one would need to target the approach to CTL to avoid the confounding effects of eliminating Tregs (Azzi et al. 2013). Protein therapy using T lymphocyte relevant serpins$_{IC}$ might also be useful in alleviated inflammatory disease caused by extracellular GrB.

EPO is used clinically to treat the anemia of chronic renal disease but may also worsen certain cancers and cause hypertension (Wojchowski et al. 2010). So new target downstream of EPOR may be useful to develop new drugs that will avoid EPO side effects. We have shown in mice that Spi2A facilitates the activity of EPO in anemia by protecting erythroblasts from ROS/cathepsin B-/L-driven PCD (Dev et al. 2013). However, humans lack an obvious functional homologue of Spi2A that inhibits cysteine cathepsins, and no *SERPINA* gene is upregulated by EPO in human erythroblasts.

Perhaps the most useful approach to transplanting our knowledge of serpins$_{IC}$ is to develop small molecules that are mimetics. Synthetic inhibitors of cathepsin B can boost the number of LCMV-specific MTL by inhibiting PCD (Liu et al. 2004a) and can alleviate anemia in mice by substituting for EPO (Dev et al. 2013; Wojchowski et al. 2010). Clinically useful GrB inhibitors would alleviate wide variety of inflammatory diseases (Willoughby et al. 2002). However, current cathepsin B/L and GrB inhibitors suffer from poor cell uptake and high toxicity, and so there is a great unmet need to develop new small molecules as the first step to clinically useful mimetics of serpins$_{IC}$.

Acknowledgments The author thanks his collaborators for generating the data, published and unpublished that formed the basis of this review.

References

Ahmed R, Gray D (1996) Immunological memory and protective immunity: understanding their relation. Science 272(5258):54–60

Annand RR, Dahlen JR, Sprecher CA, De Dreu P, Foster DC, Mankovich JA, Talanian RV, Kisiel W, Giegel DA (1999) Caspase-1 (interleukin-1beta-converting enzyme) is inhibited by the human serpin analogue proteinase inhibitor 9. Biochem J 342(Pt 3):655–665

Ansari AW, Temblay JN, Alyahya SH, Ashton-Rickardt PG (2010) Serine protease inhibitor 6 protects iNKT cells from self-inflicted damage. J Immunol 185(2):877–883. doi:10.4049/jimmunol.1000651, jimmunol.1000651 [pii]

Ashton-Rickardt PG (2010) Serine protease inhibitors and cytotoxic T lymphocytes. Immunol Rev 235(1):147–158. doi:10.1111/j.0105-2896.2010.00892.x

Azzi J, Ohori S, Ting C, Uehara M, Abdoli R, Smith BD, et al. (2015) Serine protease inhibitor-6 differentially affects the survival of effector and memory alloreactive CD8-T cells. American Journal of Transplantation 15:234–241. doi:10.1111/ajt.13051

Azzi J, Skartsis N, Mounayar M, Magee CN, Batal I, Ting C, Moore R, Riella LV, Ohori S, Abdoli R, Smith B, Fiorina P, Heathcote D, Bakhos T, Ashton-Rickardt PG, Abdi R (2013) Serine protease inhibitor 6 plays a critical role in protecting murine granzyme B-producing regulatory T cells. J Immunol 191(5):2319–2327. doi:10.4049/jimmunol.1300851

Becker TC, Coley SM, Wherry EJ, Ahmed R (2005) Bone marrow is a preferred site for homeostatic proliferation of memory CD8 T cells. J Immunol 174(3):1269–1273, 174/3/1269 [pii]

Bladergroen BA, Meijer CJ, ten Berge RL, Hack CE, Muris JJ, Dukers DF, Chott A, Kazama Y, Oudejans JJ, van Berkum O, Kummer JA (2002) Expression of the granzyme B inhibitor, protease inhibitor 9, by tumor cells in patients with non-Hodgkin and Hodgkin lymphoma: a novel protective mechanism for tumor cells to circumvent the immune system? Blood 99(1):232–237

Bladergroen BA, Strik MC, Bovenschen N, van Berkum O, Scheffer GL, Meijer CJ, Hack CE, Kummer JA (2001) The granzyme B inhibitor, protease inhibitor 9, is mainly expressed by dendritic cells and at immune-privileged sites. J Immunol 166:3218–3225

Byrne SM, Aucher A, Alyahya S, Elder M, Olson ST, Davis DM, Ashton-Rickardt PG (2012) Cathepsin B controls the persistence of memory CD8+ T lymphocytes. J Immunol 189(3):1133–1143. doi:10.4049/jimmunol.1003406, jimmunol.1003406 [pii]

Dev A, Byrne SM, Verma R, Ashton-Rickardt PG, Wojchowski DM (2013) Erythropoietin-directed erythropoiesis depends on serpin inhibition of erythroblast lysosomal cathepsins. J Exp Med. doi:10.1084/jem.20121762, jem.20121762 [pii]

El Haddad N, Heathcote D, Moore R, Yang S, Azzi J, Mfarrej B, Atkinson M, Sayegh MH, Lee JS, Ashton-Rickardt PG, Abdi R (2011a) Mesenchymal stem cells express serine protease inhibitor to evade the host immune response. Blood 117(4):1176–1183. doi:10.1182/blood-2010-06-287979, blood-2010-06-287979 [pii]

El Haddad N, Moore R, Heathcote D, Mounayar M, Azzi J, Mfarrej B, Batal I, Ting C, Atkinson M, Sayegh MH, Ashton-Rickardt PG, Abdi R (2011b) The novel role of SERPINB9 in cytotoxic protection of human mesenchymal stem cells. J Immunol 187(5):2252–2260. doi:10.4049/jimmunol.1003981, jimmunol.1003981 [pii]

Forsyth S, Horvath A, Coughlin P (2003) A review and comparison of the murine alpha1-antitrypsin and alpha1-antichymotrypsin multigene clusters with the human clade A serpins. Genomics 81(3):336–345, S0888754302000411 [pii]

Godfrey DI, MacDonald HR, Kronenberg M, Smyth MJ, Van Kaer L (2004) NKT cells: what's in a name? Nat Rev Immunol 4(3):231–237. doi:10.1038/nri1309

Heath WR, Belz GT, Behrens GM, Smith CM, Forehan SP, Parish IA, Davey GM, Wilson NS, Carbone FR, Villadangos JA (2004) Cross-presentation, dendritic cell subsets, and the generation of immunity to cellular antigens. Immunol Rev 199:9–26. doi:10.1111/j.0105-2896.2004.00142.xIMR142

Hirst CE, Buzza MS, Bird CH, Warren HS, Cameron PU, Zhang M, Ashton-Rickardt PG, Bird PI (2003) The intracellular granzyme b inhibitor, proteinase inhibitor 9, is up-regulated during accessory cell maturation and effector cell degranulation, and its overexpression enhances CTL potency. J Immunol 170:805–815

Huang JF, Yang Y, Sepulveda H, Shi W, Hwang I, Peterson PA, Jackson MR, Sprent J, Cai Z (1999) TCR-Mediated internalization of peptide-MHC complexes acquired by T cells. Science 286(5441):952–954

Jacob J, Baltimore D (1999) Modelling T-cell memory by genetic marking of memory T cells in vivo. Nature 399(6736):593–597

Joshi NS, Cui W, Chandele A, Lee HK, Urso DR, Hagman J, Gapin L, Kaech SM (2007) Inflammation directs memory precursor and short-lived effector CD8(+) T cell fates via the graded expression of T-bet transcription factor. Immunity 27(2):281–295

Li L, Byrne SM, Rainville N, Su S, Jachimowicz E, Aucher A, Davis DM, Ashton-Rickardt PG, Wojchowski DM (2014) Brief report: serpin Spi2A as a novel modulator of hematopoietic progenitor cell formation. Stem Cells 32(9):2550–2556. doi:10.1002/stem.1778

Lieberman J (2003) The ABCs of granule-mediated cytotoxicity: new weapons in the arsenal. Nat Rev Immunol 3:361–370

Liu N, Phillips T, Zhang M, Wang Y, Opferman JT, Shah R, Ashton-Rickardt PG (2004a) Serine protease inhibitor 2A is a protective factor for memory T cell development. Nat Immunol 5(9): 919–926

Liu N, Raja SM, Zazzeroni F, Metkar SS, Shah R, Zhang M, Wang Y, Bromme D, Russin WA, Lee JC, Peter ME, Froelich CJ, Franzoso G, Ashton-Rickardt PG (2003) NF-kappaB protects from the lysosomal pathway of cell death. EMBO J 22(19):5313–5322

Liu N, Wang Y, Ashton-Rickardt PG (2004b) Serine protease inhibitor 2A inhibits caspase-independent cell death. FEBS Lett 569(1-3):49–53

Lovo E, Zhang M, Wang L, Ashton-Rickardt PG (2012) Serine protease inhibitor 6 is required to protect dendritic cells from the kiss of death. J Immunol 188(3):1057–1063. doi:10.4049/jimmunol.1102667, jimmunol.1102667 [pii]

Medema JP, de Jong J, Peltenburg LT, Verdegaal EM, Gorter A, Bres SA, Franken KL, Hahne M, Albar JP, Melief CJ, Offringa R (2001a) Blockade of the granzyme B/perforin pathway through overexpression of the serine protease inhibitor PI-9/SPI-6 constitutes a mechanism for immune escape by tumors. Proc Natl Acad Sci USA 98(20):11515–11520

Medema JP, Schuurhuis DH, Rea D, van Tongeren J, de Jong J, Bres SA, Laban S, Toes RE, Toebes M, Schumacher TN, Bladergroen BA, Ossendorp F, Kummer JA, Melief CJ, Offringa R (2001b) Expression of the serpin serine protease inhibitor 6 protects dendritic cells from cytotoxic T lymphocyte-induced apoptosis: differential modulation by T helper type 1 and type 2 cells. J Exp Med 194(5):657–667
Opferman JT, Letai A, Beard C, Sorcinelli MD, Ong CC, Korsmeyer SJ (2003) Development and maintenance of B and T lymphocytes requires antiapoptotic MCL-1. Nature 426(6967): 671–676
Opferman JT, Ober BT, Ashton-Rickardt PG (1999) Linear differentiation of cytotoxic effectors into memory T lymphocytes. Science 283(5408):1745–1748
Phillips T, Opferman JT, Shah R, Liu N, Froelich CJ, Ashton-Rickardt PG (2004) A role for the granzyme B inhibitor serine protease inhibitor 6 in CD8+ memory cell homeostasis. J Immunol 173(6):3801–3809
Remold-O'Donnell E (1993) The ovalbumin family of serpin proteins. FEBS Lett 315(2):105–108
Russell JH, Ley TJ (2002) Lymphocyte-mediated cytotoxicity. Annu Rev Immunol 20:323–370
Stepp SE, Dufourcq-Lagelouse R, Le Deist F, Bhawan S, Certain S, Mathew PA, Henter J-I, Bennett M, Fischer A, de Saint Basile G, Kumar V (1999) Perforin gene defects in familial hemophagocytic lymphohistiocytosis. Science 286:1957–1959
Sun J, Bird CH, Sutton V, McDonald L, Coughlin PB, De Jong TA, Trapani JA, Bird PI (1996) A cytosolic granzyme B inhibitor related to the viral apoptotic regulator cytokine response modifier a is present in cytotoxic lymphocytes. J Biol Chem 271(44):27802–27809
Sun J, Ooms L, Bird CH, Sutton VR, Trapani JA, Bird PI (1997) A new family of 10 murine ovalbumin serpins includes two homologs of proteinase inhibitor 8 and two homologs of the granzyme B inhibitor (proteinase inhibitor 9). J Biol Chem 272(24):15434–15441
ten Berge RL, Meijer CJ, Dukers DF, Kummer JA, Bladergroen BA, Vos W, Hack CE, Ossenkoppele GJ, Oudejans JJ (2002) Expression levels of apoptosis-related proteins predict clinical outcome in anaplastic large cell lymphoma. Blood 99(12):4540–4546
Walden PR, Eisen HN (1990) Cognate peptides induce self-destruction of $CD8^+$ cytolytic T lymphocytes. Proc Natl Acad Sci USA 87:9015–9019
Willoughby CA, Bull HG, Garcia-Calvo M, Jiang J, Chapman KT, Thornberry NA (2002) Discovery of potent, selective human granzyme B inhibitors that inhibit CTL mediated apoptosis. Bioorg Med Chem Lett 12(16):2197–2200
Wojchowski DM, Sathyanarayana P, Dev A (2010) Erythropoietin receptor response circuits. Curr Opin Hematol 17(3):169–176. doi:10.1097/MOH.0b013e328338008b
Young JL, Sukhova GK, Foster D, Kisiel W, Libby P, Schonbeck U (2000) The serpin proteinase inhibitor 9 is an endogenous inhibitor of interleukin 1beta-converting enzyme (caspase-1) activity in human vascular smooth muscle cells. J Exp Med 191(9):1535–1544
Zhang M, Byrne S, Liu N, Wang Y, Oxenius A, Ashton-Rickardt PG (2007) Differential survival of cytotoxic T cells and memory cell precursors. J Immunol 178(6):3483–3491
Zhang M, Park SM, Wang Y, Shah R, Liu N, Murmann AE, Wang CR, Peter ME, Ashton-Rickardt PG (2006) Serine protease inhibitor 6 protects cytotoxic T cells from self-inflicted injury by ensuring the integrity of cytotoxic granules. Immunity 24(4):451–461

Plasminogen Activator Inhibitor Type-2 (PAI-2)/SerpinB2: A Unique Multifunctional Serpin

8

Tierra A. Johnson, Marguerite S. Buzza, Ekemini A.U. Riley, and Toni M. Antalis

Contents

T.A. Johnson • M.S. Buzza, Ph.D. • E.A.U. Riley, Ph.D.
Department of Physiology, Center for Vascular and Inflammatory Diseases, University of Maryland School of Medicine, Biopark Building One, Room 235, 800 West Baltimore Street, Baltimore, MD 21201, USA
e-mail: tierra.johnson@umaryland.edu; mbuzza@som.umaryland.edu; eudof001@umaryland.edu

T.M. Antalis, Ph.D. (✉)
Department of Physiology, Center for Vascular and Inflammatory Diseases, University of Maryland School of Medicine, Biopark Building One, Room 220, 800 West Baltimore Street, Baltimore, MD 21201, USA
e-mail: tantalis@som.umaryland.edu

M. Geiger et al. (eds.), *The Serpin Family*, DOI 10.1007/978-3-319-22711-5_8

Abstract

Plasminogen activator inhibitor type-2 (PAI-2) or SerpinB2 is a clade B serine protease inhibitor (serpin) with unique properties that was originally identified as an inhibitor of the secreted serine protease, urokinase-plasminogen activator (uPA). The plasminogen activating cascade is well recognized to play an important fibrinolytic role in maintaining vascular patency and for remodeling of the extracellular matrix. On the other hand, PAI-2 is present predominantly as a nucleocytoplasmic protein, and it has been shown to have a number of pleiotropic activities associated with its intracellular localization. Upregulation of PAI-2 is a major stress response in multiple cell types and recent data have started to illuminate PAI-2 functions in the cell, having immunomodulatory, cytoprotective, and anti-inflammatory activities.

8.1 Introduction

Plasminogen activator inhibitor type 2 (PAI-2), also known as SerpinB2, was one of the first discovered clade B serpins. PAI-2 was first identified in human placental trophoblasts (Kawano et al. 1970; Wun and Reich 1987) and was later isolated and cloned independently by different groups (Ye et al. 1987; Webb et al. 1987; Schleuning et al. 1987; Antalis et al. 1988). As a member of the clade B serpins, PAI-2 is structurally similar to ovalbumin (Remold-O'Donnell 1993), having a common genomic structure and a lack of a conventional secretory signal (Jensen 1987). SerpinB2 was originally identified as an inhibitor of urokinase-plasminogen activator (uPA), although a number of uPA-independent, intracellular activities associated with cell survival, innate and adaptive immunity, cell differentiation, and autophagy have also been recognized (Fig. 8.1). A number of insights into the activities and functions of PAI-2 have come from recent biochemical and cell biological analyses, animal models, and human patient studies. However, our current understanding is far from complete. This chapter will attempt to shed light on this interesting serpin and discuss recent findings that implicate a role for PAI-2 as a unique moderator of inflammation and its resolution.

8.2 PAI-2 Structure and Cellular Localization

The official gene symbol for PAI-2 is *SERPINB2* (human) or *SerpinB2* (other species), as commissioned by the HUGO Gene Nomenclature Committee (Silverman et al. 2001). All human clade B serpins are found clustered in two chromosomal loci, 6p25 ($n = 3$) or 18q21 ($n = 10$), suggesting functional similarities and evolution from a common ancestor (Kaiserman and Bird 2005). *SERPINB2* is located in the serpin cluster on chromosome 18, mapping specifically to 18q21 (Ye et al. 1989). Similarly, mouse clade B serpins also cluster at specific chromosomal loci and the mouse clade B cluster at 1D was reported to have nearly

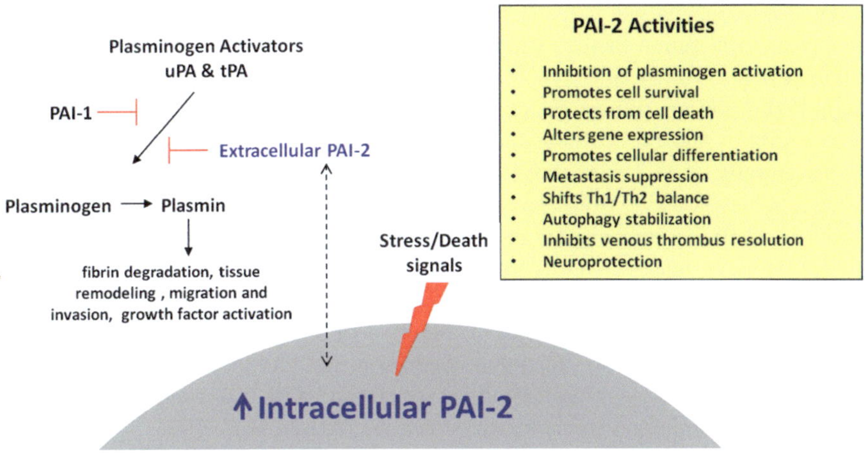

Fig. 8.1 PAI-2 activities. Intracellular PAI-2 is induced by cell stress pathways and its expression is associated with a number of pleiotropic activities associated with cell survival, immunity, and cell differentiation. Extracellular PAI-2 is an inhibitor of uPA, a protease which plays an important fibrinolytic role in vascular patency, cell migration, and extracellular matrix remodeling

complete conservation of gene number, order, and orientation relative to those of the human 18q21 locus (Askew et al. 2004).

SERPINB2 is 16.9 kilobases (kb) in length and has 8 exons (Ye et al. 1987). There are two common PAI-2 alleles that differ by six nucleotides in exons 1, 4, and 8. The human PAI-2 protein is synthesized as a single chain, 415 amino acid protein (Law et al. 2006). Mouse PAI-2 cDNA encodes for a protein that is 80 % homologous to human PAI-2 (Kruithof et al. 1995). As such, studies on mouse PAI-2 function and *SerpinB2* regulation have provided insight into the function of human PAI-2 and the regulation of *SERPINB2*. Recently, the 5′ UTR of both mouse *SerpinB2* and human *SERPINB2* were aligned and compared (Udofa et al. 2013). The presence of several repetitive sequence elements delineated five broadly homologous regions between the human and murine promoters and several *cis*-acting regulatory elements defined in the human *SERPINB2* proximal promoter that are conserved in the murine *SerpinB2* promoter.

The crystal structure of human PAI-2 protein revealed it is comprised of nine α-helices, three β-sheets, and a reactive center loop (RCL) (Harrop et al. 1999), typical of the serpins (Law et al. 2006; Gettins 2002). The RCL is cleaved upon interaction with uPA or tPA serine proteases at PAI-2 P_1-RCL(Arg^{380}) and becomes inserted into beta sheet A of the serpin. PAI-2 is synthesized without a typical N-terminal signal sequence and is found predominantly located in the intracellular space. PAI-2 has a nucleocytoplasmic distribution within the cell and is found in two forms: a nonglycosylated intracellular 47 kDa protein and a glycosylated extracellular 60 kDa protein. A small percentage of the PAI-2 protein is able to enter the secretory pathway, become glycosylated, and then secreted due to the presence of an inefficient internal hydrophobic region that can facilitate spontaneous facultative translocation through the Golgi complex—resulting in bi-topical localization of the serpin within both the extracellular and intracellular compartments (von Heijne et al. 1991; Medcalf and Stasinopoulos 2005). Interestingly PAI-2 and maspin are the only two clade B serpins that are found located in both the intra- and extracellular spaces.

Structurally, PAI-2 has another distinctive feature: an interhelical loop that bridges helices C and D of the protein (CD loop). The PAI-2 CD loop is the longest of all the clade B serpins and is believed to serve as a protein binding domain that allows PAI-2 to interact with a host of other proteins (Jensen et al. 1996; Darnell et al. 2003; Tonnetti et al. 2008; Fan et al. 2004; Boncela et al. 2011; Gan et al. 2008; Kasyapa et al. 2006; Katic et al. 2014), thereby extending its functional significance in the cell. There are also glutamine residues in the CD loop that are subject to crosslinking by tissue transglutaminase or factor XIII, allowing PAI-2 to be crosslinked to fibrin and trophoblastic structures (Jensen et al. 1993, 1994). PAI-2 can also undergo spontaneous polymerization (Wilczynska et al. 2003). In fact, it is the only serpin to polymerize under physiological conditions, largely dependent on the redox status of the cell and the CD loop (Wilczynska et al. 2003).

8.3 PAI-2 Expression and Induction

SERPINB2 is very tightly regulated and its expression is restricted to very few cell and tissue types, particularly macrophages, keratinocytes, and fibroblasts. Early studies also showed constitutive PAI-2 expression by trophoblasts during pregnancy and its presence in the serum of pregnant women (Astedt et al. 1998; Coolman et al. 2012; Hui et al. 2012). Notably, *SERPINB2* expression is induced by a number of pro-inflammatory agents, namely, viral (Antalis et al. 1998; Krishnamurti et al. 1989; Tyner et al. 2005), bacterial (Gan et al. 2008; Montemurro et al. 2001; Mirlashari et al. 2001; Xiao et al. 2001; Losick and Isberg 2006), and parasitic pathogens (Bystrom et al. 2004; Swartz et al. 2004), growth factors, and cytokines (Kruithof et al. 1995; Dickinson et al. 1995; Shafren et al. 1999), thus implicating it in a wide variety of physiological and pathophysiological processes. PAI-2 expression is upregulated in a range of cells including cancer cells, monocytes, macrophages, fibroblasts, endothelial cells, and dendritic cells, and it is often induced during inflammatory conditions, cell differentiation, and wound healing (Kruithof et al. 1995; Medcalf and Stasinopoulos 2005; Schroder et al. 2011). Recently PAI-2 was also shown to be induced in human vascular smooth muscle cells after stimulation of nuclear calcium increase with leukotriene C(4) (Eaton et al. 2012).

PAI-2 is one of the most abundantly induced proteins in macrophages in response to bacterial lipopolysaccharide (LPS), with induction reported over 105-fold (Medcalf 2011). LPS signals through the TLR4 receptor resulting in the concomitant activation of several MAP kinase (MAPK) pathways as well as NF-κB activation (Akira and Takeda 2004; Guha and Mackman 2001), leading to the upregulation of target genes, including PAI-2 (Wygrecka et al. 2004; Park et al. 2005). PAI-2 expression is also upregulated when hepatic stellate cells (HSC) in the liver are exposed to hypoxia (Copple et al. 2011). Recent studies link PAI-2 to inflammatory diseases such as asthma (Woodruff et al. 2007), lupus (Palafox-Sanchez et al. 2009), antiphospholipid syndrome (Vazquez-Del Mercado et al. 2007), periodontal disease (Toyman et al. 2015), the regulation of adaptive immunity after schistosome infection (Schroder et al. 2010a), and recurrent coronary event risk in patients with high HDL and C-reactive protein levels (Corsetti et al. 2013).

8.4 Transcriptional Regulation

Both transcriptional and posttranscriptional mechanisms regulate *SERPINB2* expression (Medcalf 2007, 2011). *SERPINB2* was found to be transcriptionally regulated by an inducible proximal promoter, an upstream silencer (PAUSE-1), and a distal transactivator region (Stringer et al. 2012). Two AP-1-like sites in the proximal *SERPINB2* promoter contribute to both constitutive and PMA-stimulated transcription of PAI-2 (Cousin et al. 1991). Further, the PAI-2 proximal promoter is critically dependent on a CCAAT enhancer binding (C/EBP) element which binds the transcription factor C/EBP-β for constitutive and LPS-inducible transcription of *SerpinB2* (Udofa et al. 2013). PAI-2 is also reported to be an Ets-1-regulated gene in fibroblasts stimulated with bFGF (Hahne et al. 2011), and its transcription in response to gastrin stimulation is dependent on proteasome B1 which is recruited to the PAI-2 promotor (O'Hara et al. 2013).

8.5 Biological and Physiological Functions of PAI-2

Many serpins have evolved inhibitory and non-inhibitory functions, and PAI-2 appears to be no exception. Extracellular PAI-2 was originally identified as an inhibitor of uPA. With respect to intracellular PAI-2, there are a myriad of biological functions that have been associated with its expression which are independent of uPA inhibition, such as cytoprotection, innate immunity, differentiation, modulation of cell stress, and others [reviewed in Kruithof et al. (1995), Medcalf and Stasinopoulos (2005), Medcalf (2011)]. While these functions are uPA independent, this does not rule out the possibility that PAI-2 could be inhibiting another protease target to mediate the aforementioned processes. Below we briefly describe the various biological functions associated with extracellular and intracellular PAI-2 expression.

8.5.1 Insights from Mouse Models of PAI-2 Deficiency

PAI-2-deficient mice develop normally (Dougherty et al. 1999). When challenged, PAI-2 deficiency in mice is associated increased HIV-1 gene expression in activated PAI-2$^{-/-}$ macrophages (Darnell et al. 2006), susceptibility to cell death in response to *Legionella pneumophila* infection (Losick and Isberg 2006), impairment of nutritionally induced adipose tissue development (Lijnen et al. 2007), and lower tumor burden during skin carcinogenesis (Tonnetti et al. 2008). PAI-2-deficient mice exhibit impaired responses to infections (Schroder et al. 2010a, b; Lin and Wang 1992), and recently PAI-2 deficiency has been associated with enhanced venous thrombus resolution (Siefert et al. 2014).

8.5.2 Inhibition of Plasminogen Activation

Generation of active plasmin, which is important for many necessary biological activities including the degradation of fibrin (fibrinolysis) during blood clot dissolution, the activation of collagenase, and the cleavage of thrombospondin, laminin, and von Willebrand factor during cell migration, wound healing, and extracellular matrix turnover, occurs through the proteolytic cleavage of its zymogen, plasminogen, by tissue-type and urokinase-type plasminogen activators (tPA, and uPA) (Vassalli et al. 1991; Plow et al. 1995). These plasminogen activators may in turn be inhibited by plasminogen activator inhibitor type 1 (PAI-1/SerpinE1) and PAI-2. Despite the fact that PAI-1 and PAI-2 share similar protease targets, they are phylogenetically dissimilar [reviewed in Kruithof et al. (1995)]. PAI-2 is a less effective inhibitor of uPA than is PAI-1 (Law et al. 2006), and PAI-$2^{-/-}$ mice fail to demonstrate the fibrinolytic defects seen in PAI-$1^{-/-}$ mice (Dougherty et al. 1999).

uPA is an extracellular protein, whereas the majority of PAI-2 is found intracellularly. This difference in localization has led to uncertainty as to the *physiological* relevance of PAI-2-mediated inhibition of uPA. Direct inhibition of uPA by PAI-2 has only been shown using purified recombinant proteins or addition of recombinant PAI-2 to cell culture systems in vitro (Baker et al. 1990). It has been speculated that under conditions of inflammation, nonglycosylated PAI-2 could be secreted from dying cells as an immediate source of uPA inhibition (Medcalf et al. 1988). Recently, PAI-2 expressed by B16 melanoma cells was reported to effectively reach the extracellular milieu on the surface of microparticles, in a manner similar to that reported for transglutaminase-2 (van den Akker et al. 2012), where it was able to inhibit uPA (Schroder et al. 2014). In addition, it was recently found that nonglycosylated PAI-2 can be secreted from endothelial cells activated with LPS (Boncela et al. 2013). The mechanism involves brefeldin A-dependent delivery of PAI-2 to the plasma membrane in trans-Golgi-derived vesicles that fuse with the plasma membrane to release contents. Together, these mechanisms may have implications for the release of high local concentrations of SerpinB2 at inflammation sites to effect uPA inhibition.

8.5.3 Venous Thrombus Resolution

Recent studies show that PAI-2 functions as a modulator of venous thrombus resolution in vivo (Siefert et al. 2014). PAI-$2^{-/-}$ mice displayed enhanced venous thrombus resolution as compared to wild-type mice independent of initial thrombus formation. This was accompanied by a 12-fold elevation in active uPA levels in PAI-2$^{-/-}$ thrombi with no significant effect on MMP-2 and MMP-9 activities. Since uPA plays a critical role in endogenous thrombus formation via uPA-mediated lysis (Singh et al. 2003; Gossage et al. 2006), the results may suggest that the increased uPA activity in the absence of PAI-2 enhanced thrombus resolution via a uPA-mediated mechanism. However, the mechanism appears more complex since thrombi formed in PAI-2$^{-/-}$ mice also displayed a simultaneous reduction in PAI-1, which could also be responsible for the elevated uPA levels found in the thrombus. Murine PAI-1 deficiency resulted in a decrease in thrombus resolution similar to PAI-2 deficiency but has an additional negative effect on thrombus formation and also altered MMP activity (Siefert et al. 2014). PAI-2-deficient thrombi were also found to have increased levels of the neutrophil chemoattractant, CXCL2, which was associated with early enhanced neutrophil recruitment. These data identify PAI-2 as a novel regulator of venous thrombus resolution, where its expression inhibits resolution and decreases active uPA in the thrombus.

8.5.4 Gene Expression and Cell Signaling

PAI-2 has been reported to alter gene expression in several cell types. Ectopic PAI-2 expression transcriptionally downregulated the expression of intercellular adhesion molecule 1 (ICAM-1) in HeLa cells, thus protecting these cells from lytic infection (Shafren et al. 1999). ICAM-1(or CD54) is a transmembrane glycoprotein expressed constitutively on many cell types and it has been shown that ICAM-1 may support the adhesion of cancer cells to mesothelium (Ksiazek et al. 2010). This could also explain the loss of cell-cell adhesion observed in PAI-2-expressing THP-1 cells (Yu et al. 2002). PAI-2 expression mediates antiviral responses by the upregulation of antiviral genes such as IFN-stimulated gene factor 3 (ISGF3) and oligo(A) synthetase (OAS) (Antalis et al. 1998) and the downregulation of picornavirus receptors decay-accelerating factor (DAF) and coxsackie-adenovirus receptor (CAR) (Shafren et al. 1999). PAI-2 expression is also implicated in cytokine gene expression as its expression has been reported to alter the secretory profile of cytokines in macrophages (Greten et al. 2007; Hsu et al. 2008; Mosser and Edwards 2008; Zhao et al. 2013).

8.5.5 Cytoprotection

This is the most well-known function associated with intracellular PAI-2 expression. The prosurvival and cytoprotective function of PAI-2 has been studied in a range of cell types (Gan et al. 1995, 2008; Antalis et al. 1998; Dickinson et al. 1995, 1998; Park et al. 2005; Greten et al. 2007; Kumar and Baglioni 1991; Varro et al. 2004; Delhase et al. 2012) with most studies being performed in macrophages. PAI-2 is strongly induced in macrophages in response to multiple "danger" signals and demonstrates cytoprotective properties in various experimental contexts. The cytoprotective activity of PAI-2 appears to be at the level of the control of life or death cell fate decisions (Tonnetti et al. 2008) and is selective in that PAI-2 expression shows protection from only some cytotoxic agents, such as LPS, TNFα, and certain viruses, and is dependent on high expression of PAI-2. Cytoprotection also appears to be cell type and cell context dependent—the complement of activated cell survival and/or death pathways is relevant. A report using lentiviral delivery of PAI-2 did not find that PAI-2 expression protected against TNFα-induced apoptosis (Fish and Kruithof 2006). The reason for this is not known but may be related to the possible influence of lentiviral transduction on cell activation and signaling pathways (Tan et al. 2006; Chen et al. 2004). Of note, PAI-2 cytoprotective activity appears independent of uPA, but dependent on the PAI-2 CD loop (Dickinson et al. 1998) and the PAI-2 P_1-RCL (Dickinson et al. 1995), indicating that PAI-2 requires both its protein interaction capabilities and serpin inhibitory activity to confer cytoprotection. A protease target for PAI-2 in this context remains unidentified, although there is some evidence that PAI-2 can modulate calpain-mediated cleavage of the retinoblastoma protein (Rb) (Tonnetti et al. 2008). In contrast, in endothelial cells, PAI-2 has been implicated in inhibition of proteasome activity, including degradation of the proteasome substrate p53 (Boncela et al. 2011).

Recent studies establish PAI-2 and transglutaminase 2 as downstream mediators in the antiapoptotic response triggered upon activation of the IκB kinase-related kinase TANK-binding kinase 1 (TBK1) (Delhase et al. 2012). Genetic studies in mice identified TBK1 as a regulatory molecule that promotes survival downstream of TNFα. It was reported that TBK1 triggers an antiapoptotic response via inducible expression of PAI-2 that limits caspase-3 activation through stabilization of transglutaminase-2, which cross-links and inactivates procaspase-3. This study may provide insight into the mechanism by which PAI-2 mediates its cytoprotective activities.

8.5.6 Immunomodulation of Innate and Adaptive Immunity

Macrophages are key producers of inflammatory mediators, cytokines and chemokines, that are the main cellular effectors of the innate and adaptive immune responses. Macrophages become activated or polarized depending on the cytokine microenvironment. Typically, macrophages undergo classical (M1) activation when stimulated by LPS or other Th1 cytokines such as IFNγ. Alternative (M2) activation occurs in response to Th2 cytokines such as IL-4 and IL-13 (Zhao et al. 2013). PAI-2 is one of the most highly induced genes in M1-activated macrophages (Wygrecka et al. 2004). PAI-2 expression is important for macrophage survival in response to LPS (Park et al. 2005) and also antagonizes the secretion of the pro-inflammatory cytokine IL-1β (Greten et al. 2007; Hsu et al. 2008). PAI-$2^{-/-}$ mice show enhanced Th1 responses following immunization with ovalbumin and complete Freund's adjuvant (OVA/CFA), a Th1 immunogen (Schroder et al. 2010a).

More recently, PAI-2 has been further implicated in the modulation of adaptive immunity, specifically Th1/Th2 immune responses. For example, PAI-2 deficiency was found to promote some Th1 responses and decrease some Th2 responses following *Schistosoma* infection (Schroder et al. 2010b). Following nematode infection, PAI-2$^{-/-}$ mice exhibit a lack of Th2 responses as evidenced by decreased intestinal IL-13 production following nematode infection (Zhao et al. 2013). In addition, PAI-2$^{-/-}$ mice exhibit decreased macrophage recruitment to the intestine following nematode infection, possibly due to a decreased intestinal CCL2 expression in PAI-2$^{-/-}$ mice (Zhao et al. 2013). In pigtail macaques infected with the macrophage-tropic lentivirus SIV_{mac251}, PAI-2 mRNA expression was found to be induced in peripheral blood mononuclear cells. PAI-2$^{-/-}$ mice infected with a chimeric HIV lentivirus (EcoHIV) demonstrated a lower IgG1 response, induction of IL-4 and decrease clearance of the infection, suggesting PAI-2 expression promotes a Th2 response (Major et al. 2013). Together, these results indicate that PAI-2 contributes an immunomodulatory role to host immunity and resolution of infectious pathogens.

8.5.7 Differentiation

PAI-2 expression is highly induced in monocytes in response to PMA (Genton et al. 1987), a classic macrophage differentiation agent, and increasing levels of PAI-2 are found in the differentiating layers of the epidermis (Lavker et al. 1998). At present it is unclear whether PAI-2 is solely a property of more differentiated cells or whether it plays a functional role in cellular differentiation. However, in vitro studies using cultured keratinocytes show PAI-2 expression retards cell proliferation and induces the expression of other keratinocyte differentiation markers (Jang et al. 2010). PAI-2 also has a retinoic acid (RA)-like response element in its promoter (Schuster et al. 1994). Retinoids such as *all-trans*-retinoic acid (ATRA) and *N*-(4-hydroxyphenyl) retinamide (4-HPR, fenretinide) are used clinically to induce terminal cellular differentiation (Leszczyniecka et al. 2001). RA stimulates PAI-2 mRNA expression in human blood mononuclear cells (Montemurro et al. 1999), promyelocytic cells (Schuster et al. 1993; Tapiovaara et al. 1994), and epidermal keratinocytes (Braungart et al. 2001), indicating that PAI-2 has the properties of a differentiation antigen.

8.5.8 Inflammation and Autophagy Stabilization

Autophagy (or macroautophagy) is an evolutionarily conserved, adaptive cellular process that is frequently activated in response to inflammatory and ER-stress-inducing stimuli and is involved in a range of processes including cell survival and cell death (Platini et al. 2010), differentiation (Zhang et al. 2012), immunity and inflammation (Levine et al. 2011), and cancer development (White 2012). The autophagic process involves the formation of cytosolic, double membrane vesicles (autophagosomes) which engulf old or damaged organelles and fuse with lysosomes (autophagolysosomes) to degrade their luminal contents. The process ensures organelle turnover and adaptation to increased energy demands of the cell. In a recent study, shRNA-mediated-PAI-2 depletion in mouse bone marrow-derived macrophages was reported to enhance the activity of the TLR-activated NOD-like receptor pyrin domain-containing 3 (NLRP3) inflammasome complex, which results in increased caspase-1 activation and IL-1β production (Chuang et al. 2013). This study determined that upon TLR engagement in macrophages, PAI-2 expression is induced, where it complexes with the autophagic protein Beclin-1 and HSP90, to protect Beclin-1 from ubiquitin-proteasome-dependent degradation. In PAI-2 expressing macrophages, the elevated levels of Beclin-1 protein promote autophagy which leads to increased NLRP3 degradation and reduced inflammasome activity. Interestingly, in the same study it was shown that expression of wild-type PAI-2 in PAI-2-deficient human monocytic THP-1 cells reduced inflammasome activation, whereas expression of the PAI-2 P_1-R380A mutant did not stabilize Beclin-1 or result in reduced inflammasome activity, suggesting this function may be dependent upon protease inhibitory activity (Chuang et al. 2013). Collectively, these results reveal a novel function of PAI-2 as a regulator of autophagy in macrophages that suppresses inflammasome function.

8.5.9 Neuroprotection

Recent studies implicate PAI-2 activities in neuroprotection. PAI-2 has long been known to be expressed in neurons and is potently and rapidly increased following kainite (a glutamate analogue) injection in vivo (Sharon et al. 2002). A microarray screen for calcium-dependent genes required for neuroprotection recently identified PAI-2 as one of nine "activity-regulated inhibitor of death (AID)" genes that was highly induced in hippocampal neurons. PAI-2 demonstrated strong survival promoting activity both in cell culture and in an animal model of neurodegeneration (Zhang et al. 2009). PAI-2 also has been reported to participate in neuronal cell migration (Katic et al. 2014). PAI-2 was identified as one of the binding partners for cell adhesion molecule close homologue of L1 (CHL1). In humans, CHL1 is linked to mental retardation, schizophrenia, epilepsy, and autism spectrum disorders and functions in neural cell proliferation, migration, differentiation, and survival. Differential molecular interactions with CHL1 underlie the diverse functions of this protein (Katic et al. 2014). During postnatal cerebellar development, the interaction of vitronectin with the extracellular domain of CHL1 induces integrin-mediated neuritogenesis, whereas interactions of CHL1 with plasminogen activator inhibitor-2 (PAI-2), plasminogen activators, and integrins promote cerebellar granule cell migration.

8.5.10 Tumor Biology

Given the diverse activities attributed to PAI-2, it is not surprising that the role of PAI-2 in tumor biology is complex. PAI-2 expression has been associated with both tumor suppression and tumor development, depending on the tumor type, tissue context, and its localization within cells or in the extracellular environment.

Tumor Suppression PAI-2 expression has long been associated with metastasis suppression; however, there is currently no consensus on mechanisms by which this activity is mediated [reviewed in Kruithof et al. (1995), Croucher et al. (2008)]. A metastasis suppressor is defined as a molecule whose expression results in the inhibition of a cancer cell's ability to metastasize while having little to no effect on primary tumor growth (Bodenstine and Welch 2008; Rinker-Schaeffer et al. 2006). Clinically, PAI-2 expression in breast, lung, and pancreatic cancers is associated with positive prognostic indicators such as an increase in relapse-free survival, disease-free survival, and/or mean overall survival (Croucher et al. 2008). In human ovarian cancer, increased expression of intracellular PAI-2 in metastatic tumor cells is correlated with an increased overall survival, whereas increased soluble PAI-2 found in ascites is associated with poor prognosis (Chambers et al. 1995, 1997).

There are also several in vitro and in vivo experimental cancer models which indicate that PAI-2 expression mediates both anti-invasive and antimetastatic outcomes (Schroder et al. 2014; Varro et al. 2004; Laug et al. 1993; Mueller et al. 1995; Shimizu et al. 2003; Praus et al. 1999). While the mechanisms of PAI-2-mediated metastasis suppression are not clearly defined, PAI-2 has been implicated in the suppression of cellular growth pathways and inhibition of the activity of plasminogen activators (Darnell et al. 2003; Tonnetti et al. 2008; Schroder et al. 2011; Yu et al. 2002; Croucher et al. 2008). Recently, it was also reported that lower PAI-2 mRNA and protein levels correlated with chemoresistance in head and neck squamous cell carcinoma (HNSCC) cell lines (Huang et al. 2014). Importantly, PAI-2 downregulation was also found to be a strong predictor of reduced overall survival in patients with HNSCC who received cisplatin-based chemotherapy. This is the first report of a role for PAI-2 in preventing acquired resistance to cisplatin in patients with HNSCC. Further, using an in vivo model of human prostate cancer, it was found that G-protein-coupled receptor kinase 3 (GRK3) stimulated angiogenesis is at least in part mediated through downregulation of PAI-2 (Li et al. 2014), suggesting PAI-2 may normally contribute to the suppression of tumor angiogenesis in this model. In another recent study PAI-2 expression was found to be suppressed by the tumor-promoting molecules PKD2 and PKD3, suggesting PAI-2 loss may contribute to prostate tumor progression in this system (Zou et al. 2012).

Tumor Promotion In contrast, in other cancers such as endometrial, bladder, and colorectal cancers, PAI-2 expression is associated with a poor prognosis (Chambers et al. 1995; Nordengren et al. 2002; Casslen et al. 1994; Ganesh et al. 1994; Champelovier et al. 2002). The mechanisms by which PAI-2 may promote tumor progression are yet to be defined and may involve multiple functional activities of PAI-2. The transcription of PAI-2 was recently shown to be essential for EGF-induced mammary cell migration (Tarcic et al. 2012). PAI-2 expression has also been implicated in the development of skin carcinogenesis induced by DMBA/TPA in murine models (Tonnetti et al. 2008; Zhou et al. 2001). Recently studies show that loss of dual-specificity phosphatase 5 (DUSP5) results in an increase in activated HRas/ERK-mediated PAI-2 gene transcription, which led to increased skin cancer susceptibility (Rushworth et al. 2014), suggesting that DUSP5 suppresses carcinogenesis through suppression of PAI-2 expression. Recent gene expression data implicates PAI-2 expression as an important regulator of leukemia stem cell (LSC) survival and growth in vivo (Ashton et al. 2012). PAI-2 was identified as one of a group of genes synergistically dysregulated by cooperating oncogenes that regulate LSC growth and survival. Furthermore, suppression of PAI-2 expression using a lentiviral RNAi-based approach in isolated hematopoietic stem cells showed reduced bone marrow engraftment and frequency of LSC development when injected into recipient mice.

In the case of the brain, tumor PAI-2 has been shown to function as a prometastatic factor. In a recent study, uPA and tPA produced by astrocytes in the reactive brain stroma were identified to function as a host defense mechanism against metastatic outgrowth and invasion through the activation of plasminogen to plasmin. Plasmin in turn inhibits invading cancer cells via activation of soluble FasL to induce cancer cell apoptosis and inactivates L1CAM which is required for tumor cell spreading along capillaries (Valiente et al. 2014; Erler 2014). To counter the inhibitory activity of plasmin, they found that PAI-2 and neuroserpin are produced by the cancer cells to shield the cells against the action of sFasL and allow them to proliferate in a perivascular niche (Valiente et al. 2014; Annand et al. 1999). These data highlight the complex interplay between invading tumor cells and the host tissue.

Tumor Therapy The uPA binding properties of PAI-2 have been exploited for targeted therapy of tumor cells. uPA is well recognized to play a key role in tumor invasion and metastasis by facilitating the ability of both tumor and tumor-associated cells to generate plasmin for matrix remodeling and the breakdown of tissue barriers (Andreasen et al. 1997). Binding of cell surface, receptor-bound uPA by PAI-2 results in receptor-mediated endocytosis of the uPAR/uPA:PAI-2 complex (Al Ejeh et al. 2004; Croucher et al. 2006). PAI-2-cytotoxin conjugates are being developed (Vine et al. 2012; Ranson et al. 2012) with enhanced pharmacokinetic properties (Vine et al. 2015) to specifically target uPA-positive cancers with additional prospective utility as a noninvasive diagnostic imaging agent.

8.5.11 Summary

It is apparent that PAI-2 executes a myriad of activities, for which the molecular mechanisms are recently beginning to be elucidated. PAI-2 interacts with a variety of intracellular and extracellular protein partners, modulating their functions, which may explain in part the variety of pleiotropic activities attributed to PAI-2. Whether many of these PAI-2 activities are dependent on the inhibitory serpin mechanism and protease targets remains to be determined. Studies examining the functional activity of PAI-2 mutants could provide some further insight. Cells constantly respond to stressors by activating a number of adaptive stress response pathways in an attempt to resolve cellular damage and establish homeostasis/cell survival. PAI-2 appears to be part of a cytoprotective pathway activated in response to cell stress. PAI-2 modulates cellular gene expression and protein activities, directly affecting immune cell polarization and cell adhesion and migration, which has implications for innate immune responses, inflammation, and cancer. Certainly PAI-2 constitutes a druggable target and would be amenable to therapeutic intervention for a variety of human diseases.

References

Akira S, Takeda K (2004) Toll-like receptor signalling. Nat Rev Immunol 4:499–511

Al Ejeh F, Croucher D, Ranson M (2004) Kinetic analysis of plasminogen activator inhibitor type-2: urokinase complex formation and subsequent internalisation by carcinoma cell lines. Exp Cell Res 297:259–271

Andreasen PA, Kjoller L, Christensen L, Duffy MJ (1997) The urokinase-type plasminogen activator system in cancer metastasis: a review. Int J Cancer 72:1–22

Annand RR, Dahlen JR, Sprecher CA, De Dreu P, Foster DC, Mankovich JA, Talanian RV, Kisiel W, Giegel DA (1999) Caspase-1 (interleukin-1beta-converting enzyme) is inhibited by the human serpin analogue proteinase inhibitor 9. Biochem J 342(Pt 3):655–665

Antalis TM, Clark MA, Barnes T, Lehrbach PR, Devine PL, Schevzov G, Goss NH, Stephens RW, Tolstoshev P (1988) Cloning and expression of a cDNA coding for a human monocyte-derived plasminogen activator inhibitor. Proc Natl Acad Sci USA 85:985–989

Antalis TM, La Linn M, Donnan K, Mateo L, Gardner J, Dickinson JL, Buttigieg K, Suhrbier A (1998) The serine proteinase inhibitor (serpin) plasminogen activation inhibitor type 2 protects against viral cytopathic effects by constitutive interferon alpha/beta priming. J Exp Med 187:1799–1811

Ashton JM, Balys M, Neering SJ, Hassane DC, Cowley G, Root DE, Miller PG, Ebert BL, McMurray HR, Land H, Jordan CT (2012) Gene sets identified with oncogene cooperativity analysis regulate in vivo growth and survival of leukemia stem cells. Cell Stem Cell 11:359–372

Askew DJ, Askew YS, Kato Y, Turner RF, Dewar K, Lehoczky J, Silverman GA (2004) Comparative genomic analysis of the clade B serpin cluster at human chromosome 18q21: amplification within the mouse squamous cell carcinoma antigen gene locus. Genomics 84:176–184

Astedt B, Lindoff C, Lecander I (1998) Significance of the plasminogen activator inhibitor of placental type (PAI-2) in pregnancy. Semin Thromb Hemost 24:431–435

Baker MS, Bleakley P, Woodrow GC, Doe WF (1990) Inhibition of cancer cell urokinase plasminogen activator by its specific inhibitor PAI-2 and subsequent effects on extracellular matrix degradation. Cancer Res 50:4676–4684
Bodenstine TM, Welch DR (2008) Metastasis suppressors and the tumor microenvironment. Cancer Microenviron 1:1–11
Boncela J, Przygodzka P, Papiewska-Pajak I, Wyroba E, Cierniewski CS (2011) Association of plasminogen activator inhibitor type 2 (PAI-2) with proteasome within endothelial cells activated with inflammatory stimuli. J Biol Chem 286:43164–43171
Boncela J, Przygodzka P, Wyroba E, Papiewska-Pajak I, Cierniewski CS (2013) Secretion of SerpinB2 from endothelial cells activated with inflammatory stimuli. Exp Cell Res 319:1213–1219
Braungart E, Magdolen V, Degitz K (2001) Retinoic acid upregulates the plasminogen activator system in human epidermal keratinocytes. J Invest Dermatol 116:778–784
Bystrom J, Wynn TA, Domachowske JB, Rosenberg HF (2004) Gene microarray analysis reveals interleukin-5-dependent transcriptional targets in mouse bone marrow. Blood 103:868–877
Casslen B, Bossmar T, Lecander I, Astedt B (1994) Plasminogen activators and plasminogen activator inhibitors in blood and tumour fluids of patients with ovarian cancer. Eur J Cancer 30A:1302–1309
Chambers SK, Gertz RE Jr, Ivins CM, Kacinski BM (1995) The significance of urokinase- type plasminogen activator, its inhibitors, and its receptor in ascites of patients with epithelial ovarian cancer. Cancer 75:1627–1633
Chambers SK, Ivins CM, Carcangiu ML (1997) Expression of plasminogen activator inhibitor-2 in epithelial ovarian cancer: a favorable prognostic factor related to the actions of CSF-1. Int J Cancer 74:571–575
Champelovier P, Boucard N, Levacher G, Simon A, Seigneurin D, Praloran V (2002) Plasminogen- and colony-stimulating factor-1-associated markers in bladder carcinoma: diagnostic value of urokinase plasminogen activator receptor and plasminogen activator inhibitor type-2 using immunocytochemical analysis. Urol Res 30:301–309
Chen X, He J, Chang LJ (2004) Alteration of T cell immunity by lentiviral transduction of human monocyte-derived dendritic cells. Retrovirology 1:37
Chuang SY, Yang CH, Chou CC, Chiang YP, Chuang TH, Hsu LC (2013) TLR-induced PAI-2 expression suppresses IL-1beta processing via increasing autophagy and NLRP3 degradation. Proc Natl Acad Sci USA 110:16079–16084
Coolman M, Timmermans S, de Groot CJ, Russcher H, Lindemans J, Hofman A, Geurts-Moespot AJ, Sweep FC, Jaddoe VV, Steegers EA (2012) Angiogenic and fibrinolytic factors in blood during the first half of pregnancy and adverse pregnancy outcomes. Obstet Gynecol 119:1190–1200
Copple BL, Bai S, Burgoon LD, Moon JO (2011) Hypoxia-inducible factor-1alpha regulates the expression of genes in hypoxic hepatic stellate cells important for collagen deposition and angiogenesis. Liver Int 31:230–244
Corsetti JP, Salzman P, Ryan D, Moss AJ, Zareba W, Sparks CE (2013) Plasminogen activator inhibitor-2 polymorphism associates with recurrent coronary event risk in patients with high HDL and C-reactive protein levels. PLoS One 8, e68920
Cousin E, Medcalf RL, Bergonzelli GE, Kruithof EK (1991) Regulatory elements involved in constitutive and phorbol ester-inducible expression of the plasminogen activator inhibitor type 2 gene promoter. Nucleic Acids Res 19:3881–3886
Croucher D, Saunders DN, Ranson M (2006) The urokinase/PAI-2 complex: a new high affinity ligand for the endocytosis receptor low density lipoprotein receptor-related protein. J Biol Chem 281:10206–10213
Croucher DR, Saunders DN, Lobov S, Ranson M (2008) Revisiting the biological roles of PAI2 (SERPINB2) in cancer. Nat Rev Cancer 8:535–545
Darnell GA, Antalis TM, Johnstone RW, Stringer BW, Ogbourne SM, Harrich D, Suhrbier A (2003) Inhibition of retinoblastoma protein degradation by interaction with the serpin plasminogen activator inhibitor 2 via a novel consensus motif. Mol Cell Biol 23:6520–6532

Darnell GA, Schroder WA, Gardner J, Harrich D, Yu H, Medcalf RL, Warrilow D, Antalis TM, Sonza S, Suhrbier A (2006) SerpinB2 is an inducible host factor involved in enhancing HIV-1 transcription and replication. J Biol Chem 281:31348–31358

Delhase M, Kim SY, Lee H, Naiki-Ito A, Chen Y, Ahn ER, Murata K, Kim SJ, Lautsch N, Kobayashi KS, Shirai T, Karin M, Nakanishi M (2012) TANK-binding kinase 1 (TBK1) controls cell survival through PAI-2/serpinB2 and transglutaminase 2. Proc Natl Acad Sci USA 109:E177–E186

Dickinson JL, Bates EJ, Ferrante A, Antalis TM (1995) Plasminogen activator inhibitor type 2 inhibits tumor necrosis factor alpha-induced apoptosis. Evidence for an alternate biological function. J Biol Chem 270:27894–27904

Dickinson JL, Norris BJ, Jensen PH, Antalis TM (1998) The C-D interhelical domain of the serpin plasminogen activator inhibitor-type 2 is required for protection from TNF-alpha induced apoptosis. Cell Death Differ 5:163–171

Dougherty KM, Pearson JM, Yang AY, Westrick RJ, Baker MS, Ginsburg D (1999) The plasminogen activator inhibitor-2 gene is not required for normal murine development or survival. Proc Natl Acad Sci USA 96:686–691

Eaton A, Nagy E, Pacault M, Fauconnier J, Back M (2012) Cysteinyl leukotriene signaling through perinuclear CysLT(1) receptors on vascular smooth muscle cells transduces nuclear calcium signaling and alterations of gene expression. J Mol Med (Berl) 90:1223–1231

Erler JT (2014) Cancer: disabling defences in the brain. Nature 508:46–47

Fan J, Zhang YQ, Li P, Hou M, Tan L, Wang X, Zhu YS (2004) Interaction of plasminogen activator inhibitor-2 and proteasome subunit, beta type 1. Acta Biochim Biophys Sin (Shanghai) 36:42–46

Fish RJ, Kruithof EK (2006) Evidence for serpinB2-independent protection from TNF-alpha-induced apoptosis. Exp Cell Res 312:350–361

Gan H, Newman GW, Remold HG (1995) Plasminogen activator inhibitor type 2 prevents programmed cell death of human macrophages infected with Mycobacterium avium, serovar 4. J Immunol 155:1304–1315

Gan H, Lee J, Ren F, Chen M, Kornfeld H, Remold HG (2008) Mycobacterium tuberculosis blocks crosslinking of annexin-1 and apoptotic envelope formation on infected macrophages to maintain virulence. Nat Immunol 9:1189–1197

Ganesh S, Sier CF, Griffioen G, Vloedgraven HJ, de Boer A, Welvaart K, van de Velde CJ, van Krieken JH, Verheijen JH, Lamers CB (1994) Prognostic relevance of plasminogen activators and their inhibitors in colorectal cancer. Cancer Res 54:4065–4071

Genton C, Kruithof EK, Schleuning WD (1987) Phorbol ester induces the biosynthesis of glycosylated and nonglycosylated plasminogen activator inhibitor 2 in high excess over urokinase-type plasminogen activator in human U-937 lymphoma cells. J Cell Biol 104:705–712

Gettins PG (2002) Serpin structure, mechanism, and function. Chem Rev 102:4751–4804

Gossage JA, Humphries J, Modarai B, Burnand KG, Smith A (2006) Adenoviral urokinase-type plasminogen activator (uPA) gene transfer enhances venous thrombus resolution. J Vasc Surg 44:1085–1090

Greten FR, Arkan MC, Bollrath J, Hsu LC, Goode J, Miething C, Goktuna SI, Neuenhahn M, Fierer J, Paxian S, van Rooijen N, Xu Y, O'Cain T, Jaffee BB, Busch DH, Duyster J, Schmid RM, Eckmann L, Karin M (2007) NF-kappaB is a negative regulator of IL-1beta secretion as revealed by genetic and pharmacological inhibition of IKKbeta. Cell 130:918–931

Guha M, Mackman N (2001) LPS induction of gene expression in human monocytes. Cell Signal 13:85–94

Hahne JC, Okuducu AF, Fuchs T, Florin A, Wernert N (2011) Identification of ETS-1 target genes in human fibroblasts. Int J Oncol 38:1645–1652

Harrop SJ, Jankova L, Coles M, Jardine D, Whittaker JS, Gould AR, Meister A, King GC, Mabbutt BC, Curmi PM (1999) The crystal structure of plasminogen activator inhibitor 2 at 2.0 A resolution: implications for serpin function. Struct Fold Des 7:43–54

Hsu LC, Ali SR, McGillivray S, Tseng PH, Mariathasan S, Humke EW, Eckmann L, Powell JJ, Nizet V, Dixit VM, Karin M (2008) A NOD2-NALP1 complex mediates caspase-1-dependent IL-1beta secretion in response to Bacillus anthracis infection and muramyl dipeptide. Proc Natl Acad Sci USA 105:7803–7808

Huang Z, Li H, Huang Q, Chen D, Han J, Wang L, Pan C, Chen W, House MG, Nephew KP, Guo Z (2014) SERPINB2 down-regulation contributes to chemoresistance in head and neck cancer. Mol Carcinog 53:777–786

Hui C, Lili M, Libin C, Rui Z, Fang G, Ling G, Jianping Z (2012) Changes in coagulation and hemodynamics during pregnancy: a prospective longitudinal study of 58 cases. Arch Gynecol Obstet 285:1231–1236

Jang S, Yang TH, An EJ, Yoon HK, Sohn KC, Cho AY, Ryu EK, Park YS, Yoon TY, Lee JH, Kim CD (2010) Role of plasminogen activator inhibitor-2 (PAI-2) in keratinocyte differentiation. J Dermatol Sci 59:25–30

Jensen PH (1987) Structure and function of plasminogen activator inhibitor-2: an intracellular serine proteinase inhibitor modulating apoptosis. Int J Oncol 11:557–570

Jensen PH, Lorand L, Ebbesen P, Gliemann J (1993) Type-2 plasminogen-activator inhibitor is a substrate for trophoblast transglutaminase and factor XIIIa. Transglutaminase-catalyzed cross-linking to cellular and extracellular structures. Eur J Biochem 214:141–146

Jensen PH, Schuler E, Woodrow G, Richardson M, Goss N, Hojrup P, Petersen TE, Rasmussen LK (1994) A unique interhelical insertion in plasminogen activator inhibitor-2 contains three glutamines, Gln83, Gln84, Gln86, essential for transglutaminase-mediated cross-linking. J Biol Chem 269:15394–15398

Jensen PH, Jensen TG, Laug WE, Hager H, Gliemann J, Pepinsky B (1996) The exon 3 encoded sequence of the intracellular serine proteinase inhibitor plasminogen activator inhibitor 2 is a protein binding domain. J Biol Chem 271:26892–26899

Kaiserman D, Bird PI (2005) Analysis of vertebrate genomes suggests a new model for clade B serpin evolution. BMC Genomics 6:167

Kasyapa CS, Kunapuli P, Hawthorn L, Cowell JK (2006) Induction of the plasminogen activator inhibitor-2 in cells expressing the ZNF198/FGFR1 fusion kinase that is involved in atypical myeloproliferative disease. Blood 107:3693–3699

Katic J, Loers G, Kleene R, Karl N, Schmidt C, Buck F, Zmijewski JW, Jakovcevski I, Preissner KT, Schachner M (2014) Interaction of the cell adhesion molecule CHL1 with vitronectin, integrins, and the plasminogen activator inhibitor-2 promotes CHL1-induced neurite outgrowth and neuronal migration. J Neurosci 34:14606–14623

Kawano T, Morimoto K, Uemura Y (1970) Partial purification and properties of urokinase inhibitor from human placenta. J Biochem 67:333–342

Krishnamurti C, Wahl LM, Alving BM (1989) Stimulation of plasminogen activator inhibitor activity in human monocytes infected with dengue virus. Am J Trop Med Hyg 40:102–107

Kruithof EK, Baker MS, Bunn CL (1995) Biological and clinical aspects of plasminogen activator inhibitor type 2. Blood 86:4007–4024

Ksiazek K, Mikula-Pietrasik J, Catar R, Dworacki G, Winckiewicz M, Frydrychowicz M, Dragun D, Staniszewski R, Jorres A, Witowski J (2010) Oxidative stress-dependent increase in ICAM-1 expression promotes adhesion of colorectal and pancreatic cancers to the senescent peritoneal mesothelium. Int J Cancer 127:293–303

Kumar S, Baglioni C (1991) Protection from tumor necrosis factor-mediated cytolysis by overexpression of plasminogen activator inhibitor type-2. J Biol Chem 266:20960–20964

Laug WE, Cao XR, Yu YB, Shimada H, Kruithof EK (1993) Inhibition of invasion of HT1080 sarcoma cells expressing recombinant plasminogen activator inhibitor 2. Cancer Res 53:6051–6057

Lavker RM, Risse B, Brown H, Ginsburg D, Pearson J, Baker MS, Jensen PJ (1998) Localization of plasminogen activator inhibitor type 2 (PAI-2) in hair and nail: implications for terminal differentiation. J Invest Dermatol 110:917–922

Law RH, Zhang Q, McGowan S, Buckle AM, Silverman GA, Wong W, Rosado CJ, Langendorf CG, Pike RN, Bird PI, Whisstock JC (2006) An overview of the serpin superfamily. Genome Biol 7:216

Leszczyniecka M, Roberts T, Dent P, Grant S, Fisher PB (2001) Differentiation therapy of human cancer: basic science and clinical applications. Pharmacol Ther 90:105–156

Levine B, Mizushima N, Virgin HW (2011) Autophagy in immunity and inflammation. Nature 469:323–335

Li W, Ai N, Wang S, Bhattacharya N, Vrbanac V, Collins M, Signoretti S, Hu Y, Boyce FM, Gravdal K, Halvorsen OJ, Nalwoga H, Akslen LA, Harlow E, Watnick RS (2014) GRK3 is essential for metastatic cells and promotes prostate tumor progression. Proc Natl Acad Sci USA 111:1521–1526

Lijnen HR, Frederix L, Scroyen I (2007) Deficiency of plasminogen activator inhibitor-2 impairs nutritionally induced murine adipose tissue development. J Thromb Haemost 5:2259–2265

Lin BT, Wang JY (1992) Cell cycle regulation of retinoblastoma protein phosphorylation. Ciba Found Symp 170:227–241

Losick VP, Isberg RR (2006) NF-kappaB translocation prevents host cell death after low-dose challenge by Legionella pneumophila. J Exp Med 203:2177–2189

Major LD, Partridge TS, Gardner J, Kent SJ, de Rose R, Suhrbier A, Schroder WA (2013) Induction of SerpinB2 and Th1/Th2 modulation by SerpinB2 during lentiviral infections in vivo. PLoS One 8, e57343

Medcalf RL (2007) Fibrinolysis, inflammation, and regulation of the plasminogen activating system. J Thromb Haemost 5(Suppl 1):132–142

Medcalf RL (2011) Plasminogen activator inhibitor type 2: still an enigmatic serpin but a model for gene regulation. Methods Enzymol 499:105–134

Medcalf RL, Stasinopoulos SJ (2005) The undecided serpin. The ins and outs of plasminogen activator inhibitor type 2. FEBS J 272:4858–4867

Medcalf RL, Kruithof EK, Schleuning WD (1988) Plasminogen activator inhibitor 1 and 2 are tumor necrosis factor/cachectin-responsive genes. J Exp Med 168:751–759

Mirlashari MR, Hoiby EA, Holst J, Lyberg T (2001) Outer membrane vesicles from Neisseria meningitidis: effects on tissue factor and plasminogen activator inhibitor-2 production in human monocytes. Thromb Res 102:375–380

Montemurro P, Barbuti G, Conese M, Gabriele S, Petio M, Colucci M, Semeraro N (1999) Retinoic acid stimulates plasminogen activator inhibitor 2 production by blood mononuclear cells and inhibits urokinase-induced extracellular proteolysis. Br J Haematol 107:294–299

Montemurro P, Barbuti G, Dundon WG, Del Giudice G, Rappuoli R, Colucci M, De Rinaldis P, Montecucco C, Semeraro N, Papini E (2001) Helicobacter pylori neutrophil-activating protein stimulates tissue factor and plasminogen activator inhibitor-2 production by human blood mononuclear cells. J Infect Dis 183:1055–1062

Mosser DM, Edwards JP (2008) Exploring the full spectrum of macrophage activation. Nat Rev Immunol 8:958–969

Mueller BM, Yu YB, Laug WE (1995) Overexpression of plasminogen activator inhibitor 2 in human melanoma cells inhibits spontaneous metastasis in scid/scid mice. Proc Natl Acad Sci USA 92:205–209

Nordengren J, Fredstorp LM, Bendahl PO, Brunner N, Ferno M, Hogberg T, Stephens RW, Willen R, Casslen B (2002) High tumor tissue concentration of plasminogen activator inhibitor 2 (PAI-2) is an independent marker for shorter progression-free survival in patients with early stage endometrial cancer. Int J Cancer 97:379–385

O'Hara A, Howarth A, Varro A, Dimaline R (2013) The role of proteasome beta subunits in gastrin-mediated transcription of plasminogen activator inhibitor-2 and regenerating protein1. PLoS One 8, e59913

Palafox-Sanchez CA, Vazquez-Del Mercado M, Orozco-Barocio G, Garcia-De la Torre I, Torres-Carrillo N, Torres-Carrillo NM, Illades-Aguiar B, Munoz-Valle JF (2009) A functional Ser

(413)/Ser(413) PAI-2 polymorphism is associated with susceptibility and damage index score in systemic lupus erythematosus. Clin Appl Thromb Hemost 15:233–238
Park JM, Greten FR, Wong A, Westrick RJ, Arthur JS, Otsu K, Hoffmann A, Montminy M, Karin M (2005) Signaling pathways and genes that inhibit pathogen-induced macrophage apoptosis--CREB and NF-kappaB as key regulators. Immunity 23:319–329
Platini F, Perez-Tomas R, Ambrosio S, Tessitore L (2010) Understanding autophagy in cell death control. Curr Pharm Des 16:101–113
Plow EF, Herren T, Redlitz A, Miles LA, Hoover-Plow JL (1995) The cell biology of the plasminogen system. FASEB J 9:939–945
Praus M, Wauterickx K, Collen D, Gerard RD (1999) Reduction of tumor cell migration and metastasis by adenoviral gene transfer of plasminogen activator inhibitors. Gene Ther 6:227–236
Ranson M, Berghofer P, Vine KL, Greguric I, Shepherd R, Katsifis A (2012) Different radiolabelling methods alter the pharmacokinetic and biodistribution properties of plasminogen activator inhibitor type 2 (PAI-2) forms. Nucl Med Biol 39:833–839
Remold-O'Donnell E (1993) The ovalbumin family of serpin proteins. FEBS Lett 315:105–108
Rinker-Schaeffer CW, O'Keefe JP, Welch DR, Theodorescu D (2006) Metastasis suppressor proteins: discovery, molecular mechanisms, and clinical application. Clin Cancer Res 12:3882–3889
Rushworth LK, Kidger AM, Delavaine L, Stewart G, van Schelven S, Davidson J, Bryant CJ, Caddye E, East P, Caunt CJ, Keyse SM (2014) Dual-specificity phosphatase 5 regulates nuclear ERK activity and suppresses skin cancer by inhibiting mutant Harvey-Ras (HRasQ61L)-driven SerpinB2 expression. Proc Natl Acad Sci USA 111:18267–18272
Schleuning WD, Medcalf RL, Hession C, Rothenbuhler R, Shaw A, Kruithof EK (1987) Plasminogen activator inhibitor 2: regulation of gene transcription during phorbol ester-mediated differentiation of U-937 human histiocytic lymphoma cells. Mol Cell Biol 7:4564–4567
Schroder WA, Le TT, Major L, Street S, Gardner J, Lambley E, Markey K, MacDonald KP, Fish RJ, Thomas R, Suhrbier A (2010a) A physiological function of inflammation-associated SerpinB2 is regulation of adaptive immunity. J Immunol 184:2663–2670
Schroder WA, Gardner J, Le TT, Duke M, Burke ML, Jones MK, McManus DP, Suhrbier A (2010b) SerpinB2 deficiency modulates Th1Th2 responses after schistosome infection. Parasite Immunol 32:764–768
Schroder WA, Major L, Suhrbier A (2011) The role of SerpinB2 in immunity. Crit Rev Immunol 31:15–30
Schroder WA, Major LD, Le TT, Gardner J, Sweet MJ, Janciauskiene S, Suhrbier A (2014) Tumor cell-expressed SerpinB2 is present on microparticles and inhibits metastasis. Cancer Med 3:500–513
Schuster WA, Medcalf RL, Kruithof EK (1993) Retinoic acid potentiates phorbol ester-mediated induction of urokinase and plasminogen activator inhibitor type 2 in human myeloid leukemic cell lines. Endocrinology 133:1724–1730
Schuster WA, Medcalf RL, Kruithof EKO (1994) Localization and characterization of a retinoic acid response-like element in the plasminogen activator inhibitor-2 gene promoter. Fibrinolysis 8:113–9
Shafren DR, Gardner J, Mann VH, Antalis TM, Suhrbier A (1999) Picornavirus receptor downregulation by plasminogen activator inhibitor type 2. J Virol 73:7193–7198
Sharon R, Abramovitz R, Miskin R (2002) Plasminogen mRNA induction in the mouse brain after kainate excitation: codistribution with plasminogen activator inhibitor-2 (PAI-2) mRNA. Brain Res Mol Brain Res 104:170–175
Shimizu T, Sato K, Suzuki T, Tachibana K, Takeda K (2003) Induction of plasminogen activator inhibitor-2 is associated with suppression of invasive activity in TPA-mediated differentiation of human prostate cancer cells. Biochem Biophys Res Commun 309:267–271
Siefert SA, Chabasse C, Mukhopadhyay S, Hoofnagle MH, Strickland DK, Sarkar R, Antalis TM (2014) Enhanced venous thrombus resolution in plasminogen activator inhibitor type-2-deficient mice. J Thromb Haemost 12:1706–1716

Silverman GA, Bird PI, Carrell RW, Church FC, Coughlin PB, Gettins PG, Irving JA, Lomas DA, Luke CJ, Moyer RW, Pemberton PA, Remold-O'Donnell E, Salvesen GS, Travis J, Whisstock JC (2001) The serpins are an expanding superfamily of structurally similar but functionally diverse proteins. Evolution, mechanism of inhibition, novel functions, and a revised nomenclature. J Biol Chem 276:33293–33296

Singh I, Burnand KG, Collins M, Luttun A, Collen D, Boelhouwer B, Smith A (2003) Failure of thrombus to resolve in urokinase-type plasminogen activator gene-knockout mice: rescue by normal bone marrow-derived cells. Circulation 107:869–875

Stringer B, Udofa EA, Antalis TM (2012) Regulation of the human plasminogen activator inhibitor type 2 gene: cooperation of an upstream silencer and transactivator. J Biol Chem 287:10579–10589

Swartz JM, Bystrom J, Dyer KD, Nitto T, Wynn TA, Rosenberg HF (2004) Plasminogen activator inhibitor-2 (PAI-2) in eosinophilic leukocytes. J Leukoc Biol 76:812–819

Tan PH, Xue SA, Manunta M, Beutelspacher SC, Fazekasova H, Alam AK, McClure MO, George AJ (2006) Effect of vectors on human endothelial cell signal transduction: implications for cardiovascular gene therapy. Arterioscler Thromb Vasc Biol 26:462–467

Tapiovaara H, Matikainen S, Hurme M, Vaheri A (1994) Induction of differentiation of promyelocytic NB4 cells by retinoic acid is associated with rapid increase in urokinase activity subsequently downregulated by production of inhibitors. Blood 83:1883–1891

Tarcic G, Avraham R, Pines G, Amit I, Shay T, Lu Y, Zwang Y, Katz M, Ben-Chetrit N, Jacob-Hirsch J, Virgilio L, Rechavi G, Mavrothalassitis G, Mills GB, Domany E, Yarden Y (2012) EGR1 and the ERK-ERF axis drive mammary cell migration in response to EGF. FASEB J 26:1582–1592

Tonnetti L, Netzel-Arnett S, Darnell GA, Hayes T, Buzza MS, Anglin IE, Suhrbier A, Antalis TM (2008) SerpinB2 protection of retinoblastoma protein from calpain enhances tumor cell survival. Cancer Res 68:5648–5657

Toyman U, Tuter G, Kurtis B, Kivrak E, Bozkurt S, Yucel AA, Serdar M (2015) Evaluation of gingival crevicular fluid levels of tissue plasminogen activator, plasminogen activator inhibitor 2, matrix metalloproteinase-3 and interleukin 1-beta in patients with different periodontal diseases. J Periodontal Res 50:44–51

Tyner JW, Uchida O, Kajiwara N, Kim EY, Patel AC, O'Sullivan MP, Walter MJ, Schwendener RA, Cook DN, Danoff TM, Holtzman MJ (2005) CCL5-CCR5 interaction provides antiapoptotic signals for macrophage survival during viral infection. Nat Med 11:1180–1187

Udofa EA, Stringer BW, Gade P, Mahony D, Buzza MS, Kalvakolanu DV, Antalis TM (2013) The transcription factor C/EBP-beta mediates constitutive and LPS-inducible transcription of murine SerpinB2. PLoS One 8, e57855

Valiente M, Obenauf AC, Jin X, Chen Q, Zhang XH, Lee DJ, Chaft JE, Kris MG, Huse JT, Brogi E, Massague J (2014) Serpins promote cancer cell survival and vascular co-option in brain metastasis. Cell 156:1002–1016

van den Akker J, van Weert A, Afink G, Bakker EN, van der Pol E, Boing AN, Nieuwland R, Van Bavel E (2012) Transglutaminase 2 is secreted from smooth muscle cells by transamidation-dependent microparticle formation. Amino Acids 42:961–973

Varro A, Noble PJ, Pritchard DM, Kennedy S, Hart CA, Dimaline R, Dockray GJ (2004) Helicobacter pylori induces plasminogen activator inhibitor 2 in gastric epithelial cells through nuclear factor-kappaB and RhoA: implications for invasion and apoptosis. Cancer Res 64:1695–1702

Vassalli JD, Sappino AP, Belin D (1991) The plasminogen activator/plasmin system. J Clin Invest 88:1067–1072

Vazquez-Del Mercado M, Garcia-Cobian TA, Munoz Valle JF, Torres-Carrillo N, Martin-Marquez BT, Arana-Argaez VE, Best-Aguilera CR, Martinez-Garcia EA, Petri MH, Nunez-Atahualpa L, Delgado-Rizo V (2007) Genotype Ser413/Ser of PAI-2 polymorphism Ser413/Cys is associated with anti-phospholipid syndrome and systemic lupus erythematosus in a familial case: comparison with healthy controls. Scand J Rheumatol 36:206–210

Vine KL, Indira CV, Locke JM, Matesic L, Lee J, Skropeta D, Bremner JB, Ranson M (2012) Targeting urokinase and the transferrin receptor with novel, anti-mitotic N-alkylisatin cytotoxin conjugates causes selective cancer cell death and reduces tumor growth. Curr Cancer Drug Targets 12:64–73

Vine KL, Lobov S, Chandran VI, Harris NL, Ranson M (2015) Improved pharmacokinetic and biodistribution properties of the selective urokinase inhibitor PAI-2 (SerpinB2) by site-specific PEGylation: implications for drug delivery. Pharm Res 32(3):1045–54

von Heijne G, Liljestrom P, Mikus P, Andersson H, Ny T (1991) The efficiency of the uncleaved secretion signal in the plasminogen activator inhibitor type 2 protein can be enhanced by point mutations that increase its hydrophobicity. J Biol Chem 266:15240–15243

Webb AC, Collins KL, Snyder SE, Alexander SJ, Rosenwasser LJ, Eddy RL, Shows TB, Auron PE (1987) Human monocyte Arg-Serpin cDNA. Sequence, chromosomal assignment, and homology to plasminogen activator-inhibitor. J Exp Med 166:77–94

White E (2012) Deconvoluting the context-dependent role for autophagy in cancer. Nat Rev Cancer 12:401–410

Wilczynska M, Lobov S, Ohlsson PI, Ny T (2003) A redox-sensitive loop regulates plasminogen activator inhibitor type 2 (PAI-2) polymerization. EMBO J 22:1753–1761

Woodruff PG, Boushey HA, Dolganov GM, Barker CS, Yang YH, Donnelly S, Ellwanger A, Sidhu SS, Dao-Pick TP, Pantoja C, Erle DJ, Yamamoto KR, Fahy JV (2007) Genome-wide profiling identifies epithelial cell genes associated with asthma and with treatment response to corticosteroids. Proc Natl Acad Sci USA 104:15858–15863

Wun TC, Reich E (1987) An inhibitor of plasminogen activation from human placenta. Purification and characterization. J Biol Chem 262:3646–3653

Wygrecka M, Markart P, Ruppert C, Kuchenbuch T, Fink L, Bohle RM, Grimminger F, Seeger W, Gunther A (2004) Compartment- and cell-specific expression of coagulation and fibrinolysis factors in the murine lung undergoing inhalational versus intravenous endotoxin application. Thromb Haemost 92:529–540

Xiao Y, Li H, Bunn C, Bartold PM (2001) The expression of plasminogen activator system in a rat model of periodontal wound healing. J Periodontol 72:849–857

Ye RD, Wun TC, Sadler JE (1987) cDNA cloning and expression in Escherichia coli of a plasminogen activator inhibitor from human placenta. J Biol Chem 262:3718–3725

Ye RD, Ahern SM, Le Beau MM, Lebo RV, Sadler JE (1989) Structure of the gene for human plasminogen activator inhibitor-2. The nearest mammalian homologue of chicken ovalbumin. J Biol Chem 264:5495–5502

Yu H, Maurer F, Medcalf RL (2002) Plasminogen activator inhibitor type 2: a regulator of monocyte proliferation and differentiation. Blood 99:2810–2818

Zhang SJ, Zou M, Lu L, Lau D, Ditzel DA, Delucinge-Vivier C, Aso Y, Descombes P, Bading H (2009) Nuclear calcium signaling controls expression of a large gene pool: identification of a gene program for acquired neuroprotection induced by synaptic activity. PLoS Genet 5, e1000604

Zhang Y, Morgan MJ, Chen K, Choksi S, Liu ZG (2012) Induction of autophagy is essential for monocyte-macrophage differentiation. Blood 119:2895–2905

Zhao A, Yang Z, Sun R, Grinchuk V, Netzel-Arnett S, Anglin IE, Driesbaugh KH, Notari L, Bohl JA, Madden KB, Urban JF Jr, Antalis TM, Shea-Donohue T (2013) SerpinB2 is critical to Th2 immunity against enteric nematode infection. J Immunol 190:5779–5787

Zhou HM, Bolon I, Nichols A, Wohlwend A, Vassalli JD (2001) Overexpression of plasminogen activator inhibitor type 2 in basal keratinocytes enhances papilloma formation in transgenic mice. Cancer Res 61:970–976

Zou Z, Zeng F, Xu W, Wang C, Ke Z, Wang QJ, Deng F (2012) PKD2 and PKD3 promote prostate cancer cell invasion by modulating NF-kappaB- and HDAC1-mediated expression and activation of uPA. J Cell Sci 125:4800–4811

Maspin Is a Multifunctional Tumor-Suppressing Serpin

9

Heidi Y. Shi, Jeremy S. Zhang, and Ming Zhang

Contents

Abstract

Maspin (**MA**mmary **S**er**PIN**) is a member of the family of **SER**ine **P**rotease **IN**hibitors or serpins. It is non-inhibitory and is grouped in the ovalbumin serpin (ov-serpin) subfamily because it exhibits significant sequence similarity to chicken ovalbumin. Initially identified from human normal mammary epithelial cells, maspin was named after its source of identification (mammary epithelia).

H.Y. Shi • J.S. Zhang • M. Zhang, Ph.D (✉)
Department of Biochemistry and Molecular Genetics, Robert H. Lurie Comprehensive Cancer Center, Feinberg School of Medicine, Northwestern University, 303 E. Superior Street, Chicago, IL 60611, USA
e-mail: m-zhang@northwestern.edu

M. Geiger et al. (eds.), *The Serpin Family*, DOI 10.1007/978-3-319-22711-5_9

Subsequent studies by our laboratory and many colleagues concluded that maspin is not only expressed in mammary epithelial cells but also in many other tissues and cell types. Functionally, maspin was first reported to be a breast tumor-suppressing serpin. Maspin has been shown to be a potent metastasis inhibitor in breast and prostate cancers by numerous groups in the last 20 years. In addition, maspin possesses multiple functions in other cancers, vascular diseases, and physiological processes. Numerous mouse models have been developed to determine its role in normal biology and disease development. In this review, we intend to summarize some of the key findings of maspin research in the last two decades. These studies offer great potential for developing maspin-based therapeutics against cancer and other diseases. We focus discussion on maspin's biological functions, gene regulation, and animal models of maspin research.

9.1 Maspin Expression

In contrast to its indicated name, maspin is not a mammary-specific gene, but is expressed in a variety of epithelial cells, including those of the prostate, colon, epidermis, ovary, pancreas, and lung. The pattern and level of maspin expression varies in different epithelial cells and tissues. In the last 20 years, over 300 research articles have been published regarding the differential expression of maspin in various epithelial tissues and cancers. A majority of these reports show that maspin is downregulated in various cancers, including those from breast, prostate, stomach, colon, and lung cancers. In breast cancer, maspin is found to be expressed in normal mammary epithelial cells, higher in myoepithelial cells than luminal cells (Zou et al. 1994). Its expression is reduced in breast cancer patients along with the progression from ductal carcinoma in situ (DCIS) to more aggressive carcinoma and metastasis (Maass et al. 2000, 2001a). For example, in one of the early studies, Maass et al. examined a large panel of breast tissues for maspin expression using a maspin monoclonal antibody. They found that myoepithelia invariably showed strong maspin expression. A significant stepwise decrease in maspin expression occurred in the sequence from DCIS to invasive cancer and lymph node metastasis (Maass et al. 2001a). In prostate cancer, maspin level was found to be associated with high recurrence-free survival, lower tumor grade, and metastasis (Machtens et al. 2001). Others found that maspin downregulation is specifically correlated with the progression of various cancer subtypes, including ovarian epithelial carcinoma (Abd El-Wahed 2005), breast cancer with low erbB-2 expression (Sopel et al. 2005), thyroid cancer (Ito et al. 2004), and acinar cell pancreatic cancer (Oh et al. 2002). High maspin expression was also found to be correlated with better prognosis and better survival for those with bladder cancer (Acikalin et al. 2012), gastric cancer (He et al. 2007), colon cancer (Boltze 2005), thyroid cancer (Boltze et al. 2003), oral squamous cell carcinoma (Yoshizawa et al. 2009; Xia et al. 2000), gallbladder carcinoma (Baghel et al. 2014), and glioma (Wang et al. 2007a).

There are also conflicting reports regarding the effect of maspin expression on clinical outcomes and prognostic implications (Maass et al. 2001b; Sood et al. 2002;

Nakashima et al. 2006; Tsuji et al. 2007; Umekita et al. 2011). For example, maspin overexpression was reported to be associated with poor prognosis in ovarian and pancreas cancers (Maass et al. 2001b; Sood et al. 2002). Since the evaluation of maspin expression in human samples is based on immunohistochemistry, it is not surprising that the immunostaining results are dependent on the antibodies and scoring systems used in the assays. Indeed, many reports used different antibodies and immunostaining protocols. Moreover, it becomes clear that at least some of the discrepancies are due to subcellular localization of maspin (Solomon et al. 2006; Mohsin et al. 2003; Lonardo et al. 2006). For example, Solomon et al. showed that cytoplasmic maspin is associated with poor prognosis, while nuclear maspin is associated with favorable prognosis (Solomon et al. 2006). It is important to point out that maspin is present in different compartments of the cell, which can affect its function. Thus, it is possible that in ovarian cancer, nuclear maspin and not cytoplasmic maspin may behave as a tumor suppressor. There are still many unanswered questions. Is cytoplasmic maspin cleaved by certain protease(s)? Is cytoplasmic maspin bound by other protein partners and is it not able to function as a tumor suppressor in ovarian cancer cells? More research is needed to understand why maspin behaves differently in various subcellular compartments. Finally, there is also a possibility that in certain cancers, the maspin gene may display polymorphisms or mutations which render the protein nonfunctional (e.g., not possessing tumor-suppressing activity). Jang et al. reported that in 89 % of gastric cancer samples, there is a maspin polymorphism at amino acid 176 from proline to serine (Jang et al. 2008). They showed that this change abolishes maspin's tumor-suppressing activity in their mouse model of cancer. Interestingly, the Hendrix group also showed that the P176S mutant maspin also abolishes maspin's ability to inhibit tumor cell invasion in breast cancer cells (Bodenstine et al. 2012). Whether this and other polymorphisms or mutations in maspin are present in various cancers remains to be discovered in the future. For the reasons described above, one needs to take caution in analyzing maspin's gene expression and its functions.

In addition to epithelial cells, maspin is also expressed in stromal cells of the cornea (Bailey et al. 2006), endoderm of embryos (Gao et al. 2004), and endothelial cells of the vessel (Qin and Zhang 2010). Their expressions clearly play diverse roles in normal biology and development. This is still an understudied area, but it will become a new frontier of maspin research with the development of novel tissue-specific maspin gene knockouts and maspin overexpressing transgenic mice.

9.2 Regulation of Maspin Gene Expression

Maspin was initially identified to be highly expressed in human normal mammary epithelial cells but not in aggressive breast cancer cells. What drives maspin expression in normal mammary epithelial cells, and why is the maspin gene downregulated in breast cancer cells? These questions prompted us to study maspin gene's transcriptional regulation. We started this work by identifying maspin gene promoter from the human genomic DNA library. This turned out to be a daunting task since maspin's first exon is separated 7.6 kB away from the second exon (ATG containing), of which the original cDNA sequence matched with the part of the genomic DNA sequence. However, at that time the gene structure of maspin was totally unknown, and the human genome project had not yet been completed until many years later. After over a year's effort, we isolated the genomic DNA fragment containing the maspin promoter from a human DNA phage library and mapped the maspin transcription start site and identified the key elements within the promoter that transcription factors bind in order to activate maspin gene transcription in human normal mammary epithelial cells (Zhang et al. 1997a). Specifically, these observations showed that the Ets and Ap1 sites in the maspin promoter are active in regulating its expression in normal mammary epithelial cells, but are inactive in tumor cells (Zhang et al. 1997a). Consequently, the enhancing function by Ets and Ap1 is decreased in tumor cells and abolished in invasive tumor cells. This study demonstrated that the loss of maspin expression during tumor progression results at least in part from the absence of transcriptional activation through Ets and Ap1 sites. It also laid the foundation for identifying transcription activators, repressors, and other factors that are responsible for regulating maspin gene expression.

9.2.1 Positive Upregulation

After we identified the Ets and Ap1 elements as the binding sites for transcription activators in maspin promoter, Yamada et al. showed that one of the members of the Ets family, PDEF or PSE (prostate-derived Ets factor), is able to bind to the Ets site of maspin promoter and activate maspin gene expression in prostate epithelial cells (Yamada et al. 2000). We tested this PDEF for maspin transactivation in our breast cancer cells by using a maspin-luciferase reporter system. Indeed, PDEF is able to activate maspin gene expression in both MCF7 and MCF231 breast tumor cells. Feldman et al. also showed that PDEF activated maspin gene expression in breast cancer cells (Feldman et al. 2003). We further demonstrated that PDEF also activates a cell cycle inhibitor p21 gene expression through an Ets binding site in its promoter (Schaefer et al. 2010). Interestingly, both maspin and p21 are themselves activated by p53. Zou et al. identified a consensus-binding site for p53 in the published maspin promoter and showed that maspin is a p53 inducible gene (Zou et al. 2000).

In addition to PDEF, AP-1, and p53, several studies reported that maspin can be transactivated by TGF-β (Wang et al. 2007b), PTEN (Eitel et al. 2009; Zhang 2009; Maekawa et al. 2008), ATF-2 (Maekawa et al. 2008), and p63 (Kim et al. 2004). In the case of TGF-β, transactivation of maspin requires both TGF-β and wild-type p53 and is dependent on the binding of Smad2/Smad3 and p53 to the maspin promoter (Wang et al. 2007b). Similarly, PTEN transactivation of maspin expression also requires the cooperation of PTEN with p53 (Eitel et al. 2009). Maekawa et al. identified a CRE-like element in the maspin promoter through which ATF-2 transcription factor binds (Maekawa et al. 2008). Finally, p63 isoform TA63-γ, a member of the p53 family proteins, was shown to induce maspin expression in lung cancer cells (Kim et al. 2004).

9.2.2 Negative Downregulation

At the same time we identified Ets and Ap1 as positive regulatory elements of maspin expression, we also identified a negative regulatory site named HRE (**H**ormone **R**esponse **E**lement) located at −297 bp upstream of the maspin transcription start site. Upon binding by androgen receptor, the HRE site is responsible for the downregulation of maspin expression in prostate epithelial cells (Zhang et al. 1997b). In support of our initial findings, Zou et al. showed that androgen depletion through castration increased maspin expression in prostate cancer (Zou et al. 2002).

A seminal study by the Karin laboratory demonstrated that maspin is transcriptionally repressed by NFκB signal, linking inflammation and cancer metastasis through maspin. Specifically, Luo et al. showed that IκB kinase α (IKKα) functions to repress maspin, thus promoting prostate cancer metastasis (Luo et al. 2007). They used a mouse model harboring a mutation that prevented IKKα activation. In this *Ikkα*$^{AA/AA}$ mouse line, the serine residues whose phosphorylation is required for IKKα activation have been replaced with alanines. To investigate the role of IKKα in cancer progression, these mice were crossed with prostate oncogenic mice called TRAMP (**TR**ansgenic **A**denocarcinoma **M**ouse **P**rostate), which developed metastatic prostate cancer with a high frequency. Luo and colleagues demonstrated that while single mutant (WT/TRAMP) mice develop prostate cancer fairly early and begin dying at approximately 22 weeks of age, bitransgenic mice (*Ikkα*$^{AA/AA}$ and *TRAMP*) prolong tumor onset and also delay mortality. Interestingly, while the IKKα AA mutation also affected the growth rate of primary cancer, its most profound effect was on metastasis inhibition. After examining over 30 known tumor suppressor genes using qRT-PCR to evaluate differences in their gene expression levels between these genotypes, they found that the only gene exhibiting striking and consistent differences was maspin. Having established that IKKα exerts its pro-metastatic effects via transcriptional repression of the maspin gene, they sought to elucidate the mechanism behind this intriguing process. Their further investigations concluded that phosphorylation of IKKα, triggered by RANKL binding to RANK, induces the nuclear localization of IKKα, where it is then able to interact with the maspin promoter and cause transient repression of maspin expression. Subsequently, the maspin gene is permanently silenced by methylation, committing the prostate cancer cells to their metastatic fate (Luo et al. 2007; Affara and Coussens 2007). This pivotal study links inflammation and cancer development with maspin gene expression at the molecular level.

Another important negative regulation of maspin expression is through epigenetic modification by promoter methylation. Futscher group first demonstrated that the maspin gene is methylated at the promoter, which represses its gene expression (Futscher et al. 2002). They showed that in normal cells expressing maspin, the promoter is unmethylated and the promoter region has acetylated histones and an accessible chromatin structure. Conversely, cells that do not express maspin possess a completely methylated promoter that is occupied by hypoacetylated histones, resulting in an inaccessible chromatin structure and repression of transcription. Notably, this repression was found to be relieved by inhibiting DNA methylation. Their study suggests that methylation is a primary impediment to maspin expression and consequently determines its cell-type specificity (Futscher et al. 2002;

Costello and Vertino 2002). Unsurprisingly, DNA methylation also plays a critical role in controlling maspin expression during cancer progression. This is highlighted by the fact that aberrant methylation of the maspin promoter is associated with the silencing of maspin gene expression in breast cancer cells (Domann et al. 2000). Maspin expression could be recovered in these cancer cell lines when they were treated with 5-aza-2′-deoxycytidine, a C5-DNA methyltransferase inhibitor. Moreover, Maass et al. described the reactivation of maspin expression in a series of maspin-negative breast cancer cell lines (Maass et al. 2002). All of the seven maspin-negative breast cancer cell lines (MCF7, ZR-75-1, SK-BR-3, T-47D, MDA-MB231, MDA-MB468, and BT-20) showed induction of maspin promoter activity in a promoter reporter assay. In addition, the treatment of 5-aza-2-′-deoxycytidine, trichostatin A, or a combination of both led to the re-expression of maspin in the maspin-negative breast cancer cell lines. In addition, the restoration of p53 levels in breast cancer cell lines can partially compensate for the aberrant cytosine methylation (Oshiro et al. 2003). Treatment with 5-aza-2-′-deoxycytidine along with the p53 infection further restored maspin expression to near wild-type levels (Oshiro et al. 2003). This reactivation is a result of wt p53 binding to its consensus DNA-binding sites within the maspin and DSC3 promoters, stimulating histone acetylation and enhancing chromatin accessibility for their promoters. These findings indicate that DNA methylation and/or histone deacetylation are partially responsible for the silencing of maspin gene expression in breast cancer cells. The re-expression of maspin by pharmacological intervention potentially offers a promising new target as a therapeutic option in breast cancer. A combination of transcription factors, transcription factor binding elements, and DNA methylation coalesce to form a very complicated and intricate control system that contributes to ensure the proper expression of the maspin gene. Alteration of any of these can have a dramatic impact on the cells; an increase in methylation or a decrease in transactivation due to mutations of critical enzymes and transcription factors can disrupt maspin expression and enable cells to acquire the necessary properties to become angiogenic, invasive, and metastatic.

9.2.3 Other Regulatory Factors

There are other factors that regulate maspin's gene expression. These factors act together with transcription factors or co-activators to regulate maspin gene transcription. Jiang et al. showed that γ-linolenic acid induces maspin mRNA expression, while linoleic acid reduces maspin expression (Jiang et al. 1997). Li et al. showed that MnSOD is able to induce maspin expression in MCF7 breast tumor cells, which suppresses tumor invasion and migration (Li et al. 1998; Duan et al. 2003). Maspin was also shown to be induced by nitric oxide in breast tumor cells, and introduction of eNOS in MCF7 breast tumor cells directly induced maspin expression (Khalkhali-Ellis and Hendrix 2003). Finally, we and other groups have shown that antiestrogen inhibitor tamoxifen is able to induce maspin expression through upregulation of maspin promoter activity (Shao et al. 2000; Khalkhali-Ellis et al. 2004; Liu et al. 2004).

Clearly, maspin expression is tightly regulated at numerous levels. While there is a great deal of crosstalk between several of these regulatory mechanisms, each pathway is also distinct in its own right. Thus, each merits careful investigation and represents a potential avenue to be explored as a therapeutic target for cancer treatment and prevention. New reagents and drugs that can upregulate maspin gene expression and suppress tumor progression may be discovered in the near future.

9.3 Biological Functions

Maspin is a multifaceted protein that plays key roles in a variety of biological processes. It was initially characterized as a tumor-suppressing serpin but later found to have other properties, including migration inhibition, apoptosis induction, and angiogenesis inhibition. While maspin resides predominantly in the cytoplasm of the cell, it is also localized to other cellular compartments and is secreted (Khalkhali-Ellis and Hendrix 2007). Due to its broad localization pattern, maspin has been implicated in numerous pathways and processes involved in both tumor progression and normal development.

9.3.1 Extracellular Maspin

9.3.1.1 Inhibition of Tumor Cell Migration

One of the early discoveries was that exogenous human maspin could act on tumor cell surfaces and inhibit their migration and invasion (Zou et al. 1994; Sheng et al. 1994, 1996; Sager et al. 1994; Seftor et al. 1998). Mouse maspin could also inhibit mouse mammary tumor cell migration and invasion with similar potency (Zhang et al. 1997c). These observations were also confirmed in other kinds of cancer cells (Jiang et al. 1997; Sheng et al. 1998; Ngamkitidechakul et al. 2001) and noncancerous cells (e.g., stromal, corneal, endothelial) (Ngamkitidechakul et al. 2001). Using a time-lapse microscopy, it was shown that recombinant maspin added to the culture medium of breast tumor cells could inhibit tumor cell migration with a half-life time of 12 h (Sheng et al. 1996; Sager et al. 1996). Also, anti-maspin antibody was able to block maspin's inhibitory effect, allowing tumor cells to regain their ability to migrate (Sheng et al. 1994; Seftor et al. 1998). One of the mechanisms of maspin-mediated inhibition of migration involves inhibition of small GTPase activity. Our laboratory demonstrated that recombinant maspin could decrease tumor cell Rac1 and CDC42 but not Rho1 activities (Shi et al. 2007).

9.3.1.2 Increased Cell Adhesion

Maspin is well known to increase cell adhesion, and this property may partially explain why tumor cells are less mobile after maspin treatment. The Hendrix group first demonstrated that the addition of recombinant maspin to breast cancer cells increases their expression of $\alpha5\beta1$ and $\alpha3\beta1$ integrins (Seftor et al. 1998), which corresponds to the increased cell adhesion to fibronectin. Our laboratory, on the other hand, showed that maspin from normal immortalized human mammary epithelial cells (MCF10A) increased cell adhesion to self-deposited laminin matrix (α3β1) and that silencing maspin gene expression in MCF10A drastically reduced cell adhesion to laminin5 matrix (Cella et al. 2006). We showed that maspin interacts with integrin β1 in a complex. Furthermore, we identified one region in maspin protein (139–225 aa) that is partially responsible for maspin-mediated cell adhesion.

In early reports, the Sheng group showed that maspin binds to tPA and uPA/uPAR (Sheng et al. 1998; Yin et al. 2006). Binding of maspin to uPA/uPAR induced the internalization of uPA/uPAR complex on the cell surface (Yin et al. 2006), which may explain why maspin partially inhibits uPA or tPA activity, even though maspin is a non-inhibitory serpin. Since uPAR is also known to interact with integrin β1 (Wei et al. 1996, 2005), my laboratory performed experiments to test whether maspin could act as a molecular bridge bringing integrin β1 and uPA/uPAR together (Endsley et al. 2011). Indeed, we found that maspin is in a complex with both integrin β1 and uPA/uPAR. Maspin binds to the wild-type MEF cell surface but not to $uPAR^{-/-}$ MEF cells (Endsley et al. 2011). Anti-uPAR antibody blocking uPA binding to uPAR could in fact inhibit maspin interaction with uPA/uPAR complex. Our study further showed that three amino acids structurally close in two regions, 190–202 and 260–275 aa, are able to form a functional domain to increase cell adhesion (Endsley et al. 2011). Interesting, Gettins's laboratory previously showed that maspin binds to uPA through a site close to the maspin RSL region (Al-Ayyoubi et al. 2007). Since the 260–275 region is structurally close to the maspin RSL, the three amino acid functional domain that we identified may be the site where maspin binds uPA/uPAR. Others showed that maspin increases cell adhesion through integrin β1 in vascular smooth muscle cells, and a G-helix of maspin acts as a potential site for integrin β1 binding (Bass et al. 2009; Ravenhill et al. 2010). Therefore, it is likely that maspin may interact with integrin β1 and uPA/uPAR through a G-helix and the functional domain of three amino acids that we identified.

9.3.1.3 Angiogenesis Inhibition

Angiogenesis, the formation of new blood vessels, is vital to many physiological and pathological processes. A few non-inhibitory serpin members have been shown to regulate angiogenesis (Kisker et al. 2001; Dawson et al. 1999; Zheng et al. 2013). For example, pigment epithelium-derived factor or PEDF is a non-inhibitory serpin with known functions in pigment cell differentiation and is also a very potent anti-angiogenic factor (Dawson et al. 1999). A cleaved product of antithrombin, an inhibitory serpin, is shown to directly inhibit angiogenesis (Kisker et al. 2001). The question of whether maspin can inhibit angiogenesis was not known until my laboratory performed experiments to address this in early 2000. We were the first to demonstrate that maspin is a potent angiogenesis inhibitor (Zhang et al. 2000a). We showed that maspin acts directly on cultured endothelial cells to impede their migration toward basic fibroblastic growth factor (bFGF) and vascular endothelial growth factor (VEGF), which serve as critical chemoattractants during angiogenesis (Zhang et al. 2000a). Specifically, maspin blocked HUVEC endothelial cell migration induced by VEGF and bFGF in a dose-dependent manner with an ED_{50} of 0.2–0.3 μM. Additionally, maspin was shown to prevent endothelial cells from forming tubes in a Matrigel assay. We also demonstrated that maspin effectively inhibited neovascularization in rat cornea in vivo and reduces the density of tumor-associated microvessels in athymic mice transplanted with human prostate tumors (Zhang et al. 2000a). Following our initial report, Cher et al. provided further support in regard to maspin's effect on angiogenesis in cancer progression (Cher et al. 2003). They showed that maspin-expressing transfectant cells derived from prostate cancer cell line DU145 were inhibited in in vitro extracellular matrix and collagen degradation assays. They injected the maspin-transfected DU145 cells into human fetal bone fragments, which were previously implanted in immunodeficient mice. Overall, their studies showed that maspin expression decreases tumor growth and angiogenesis (Cher et al. 2003).

Regarding the mechanism of maspin action, we showed that maspin treatment of HUVEC endothelial cells increases their adhesion to various matrices (Qin and Zhang 2010). Maspin-mediated enhancement of cell adhesion was dependent on the activation of integrin β_1, which subsequently leads to the pattern change of vinculin and phalloidin, indicating that maspin affects cell adhesion and cytoskeleton rearrangement through the integrin signal transduction pathway. We also showed that maspin increases ILK activity and phosphorylated FAK level in HUVECs. In addition to increased cell adhesion, we showed that maspin also decreases HUVEC focal adhesion disassembly, which leads to the retardation in endothelial cell migration (Qin and Zhang 2010).

9.3.1.4 Maspin Secretion

One interesting question arises regarding maspin's action on the cell membrane: how is maspin secreted from intracellular locations to the extracellular space? Some tissues and cells can secrete a large amount of maspin. For example, Twinning group showed that maspin is secreted to the tears of human and mouse eye in large quantities (Ngamkitidechakul et al. 2001, 2003). Their study showed that both corneal stromal and epithelial cells express and secrete maspin efficiently (Ngamkitidechakul et al. 2001). We concluded that maspin is secreted from MCF10A mammary epithelial cells and HUVECs because we were able to detect maspin from conditioned medium of these cells (Qin and Zhang 2010). The secreting ability of maspin varies depending on the passages of cell lines. Teoh et al. found that maspin was not secreted in their MCF10A cells (Teoh et al. 2010). However, we showed that MCF10A cells from old passage displayed diminished levels of maspin secretion than those from early passage, suggesting that maspin secretion is dependent on cell differentiation and growing conditions (Endsley et al. 2011; Mikus et al. 1993). Similar findings are observed in corneal stromal cells. Since maspin protein does not contain the signal peptide, the question of how maspin is secreted remains unsolved. Interestingly, one serpin with high homology to maspin, PAI-2, was shown to use a facultative secretion signal encoded by the first 50 amino acids (Mikus et al. 1993). It is possible that maspin may also use the hydrophobic peptide at its N-terminal region, which is homologous to PAI-2's N-terminal region, for its secretion. Another possibility is that maspin may be secreted through the nontraditional pathway. Dean et al. reported in a poster abstract in a 2013 AACR meeting that human prostate epithelial cells secrete maspin as a soluble and exosome-associated protein. The Twining group showed that corneal epithelium-secreted maspin has eight serine and threonine sites of phosphorylation detected by mass spectrometry (Narayan et al. 2011). Whether these phosphorylation events are related to the exosome-mediated secretion is not known though. Finally, one previous study clearly showed that maspin is a major component of the exosome-secreted products following p53-induced stress responses (Yu et al. 2006). Yu et al. showed that activation of p53 by DNA damage in the cells induced maspin and TSAP6 expression and their secretion in the exosome (Yu et al. 2006). They then hypothesized that maspin and TSAP6 may play a role in the exosome-mediated secretion possibly involved in cell-cell communication. This is consistent with the role of extracellular maspin in cell adhesion, which controls a variety of cellular processes including cell differentiation, proliferation, and migration.

9.3.1.5 Other Extracellular Partners of Maspin Interaction

Maspin is also known to bind other extracellular partners. The Hendrix laboratory identified cathepsin D as an extracellular partner of maspin (Khalkhali-Ellis and Hendrix 2007). Cathepsin D is a lysosomal aspartyl protease that degrades the extracellular matrices. Binding of maspin to cathepsin D was shown to prevent matrix incorporation of cathepsin D, thus reducing matrix degradation (Khalkhali-Ellis and Hendrix 2007). Blacque et al. showed that maspin binds to collagens I and III, but not any other collagen subtypes (Blacque and Worrall 2002).

9.3.2 Cytoplasmic Maspin

9.3.2.1 Enhanced Apoptosis

The majority of maspin is located in the cytoplasm (Pemberton et al. 1997). The role of maspin in the cytoplasm has been extensively studied for a long time. My laboratory was the first to demonstrate that intracellular maspin functions to induce mammary epithelial cell apoptosis (Zhang et al. 1999). We showed that maspin overexpression in mammary epithelial cells under a whey acidic protein (WAP) promoter in a WAP-maspin transgenic mouse induces mammary cell apoptosis, which reduces mammary alveolar structures (Zhang et al. 1999). When WAP-maspin transgenic mice were mated with WAP-SV40 T-antigen tumor mice, the bitransgenic mice showed maspin overexpression and reduced tumor growth (Zhang et al. 2000b). This was accompanied by an increased rate of apoptosis of precancerous and carcinomatous mammary epithelial cells (Zhang et al. 2000b). Subsequently, the Sheng laboratory showed that overexpression of maspin in MDA-231 human breast tumor cells sensitized them to apoptosis induced by staurosporine treatment (Jiang et al. 2002). We further studied the mechanisms involved in maspin-mediated apoptosis under stress (Jiang et al. 2002; Latha et al. 2005). We showed that this effect is due to maspin's activation of caspase pathways, as maspin-expressing tumor cells exhibited a reduced level of anti-apoptotic Bcl-2 protein and an increased level of pro-apoptotic Bax protein (Zhang et al. 2005). Specifically, maspin alters the delicate balance of these proteins via selective control of Bcl-2 and Bax stability. It is important to note that intracellular maspin, but not secreted maspin, is necessary to sensitize breast cancer cells to apoptosis (Bailey et al. 2006; Jiang et al. 2002; Latha et al. 2005), again highlighting the fact that maspin's tumor-suppressive properties are intimately linked to its cytoplasmic location. Further studies showed that maspin is able to migrate into the inner membrane of the mitochondria, which is accompanied by the release of cytochrome c from the mitochondria to the cytoplasm (Latha et al. 2005). The question of what controls maspin migration to mitochondria is still under investigation by my laboratory.

Maspin is also capable of inducing apoptosis in other cell types. For example, when maspin is delivered and overexpressed in the endothelial cells in vivo, it actively induces endothelial cell apoptosis (Li et al. 2005). In that particular study, the intravascular administration of adenovirus-maspin to mice bearing mammary tumors disrupted tumor-induced angiogenesis by promoting endothelial cell apoptosis. These findings present an elegant example of the ways in which maspin's apoptosis-inducing property can be used to inhibit tumor-induced angiogenesis and cancer progression.

9.3.2.2 ROS Scavenging

The Sheng group first showed that maspin interacts with GST and glutathione peroxidase, which is accompanied by increased GST activity and decreased levels of reactive oxygen species (ROS) (Yin et al. 2005). Interestingly, increased cellular ROS was shown to induce the interaction of maspin with GST, suggesting that maspin and GST cooperate to control oxidative stress in the cells (Yin et al. 2005). However, the mechanism by which maspin controls cellular ROS is not fully understood from this early study. Further investigation by my laboratory demonstrated that intracellular maspin protein undergoes posttranslational modification following oxidative stress (H_2O_2) (Nawata et al. 2011). Specifically, maspin adopted an oxidized, disulfide-bonded structure when maspin overexpressing MCF10A cells were exposed to high levels of H_2O_2, suggesting that maspin may regulate cellular ROS through its cysteine residues (Nawata et al. 2011). Under the oxidized state, maspin is no longer bound to the GST (Nawata et al. 2011). Recently, my laboratory has discovered that maspin regulates ROS through its protein surface cysteines (Mahajan et al. 2013). Protein cysteine residues are highly susceptible to various types of oxidation. Sulfenic acid [–SOH] is the initial product of cysteine oxidation and is a key intermediate in the functional modulation of enzymes and proteins. Under variant biochemical circumstances, it may serve to mediate redox signaling (Poole et al. 2004; Poole and Nelson 2008). Although sulfenic acid is likely to represent a dynamic and transient oxidation product, its unique chemistry allows it to be captured by dimedone-based labeling reagents before progressing to a potentially more complex arrays of disulfide-bonded or oxidized products (e.g., sulfinic and sulfonic acid) (Nelson et al. 2010; Klomsiri et al. 2010). Maspin has eight cysteine residues; three of the cysteines are located at positions C183, C205, and C323 and are fully structurally exposed on maspin's protein surface. We showed that these exposed cysteine residues in maspin act as potent scavengers/quenchers of ROS (Mahajan et al. 2013). These cysteines formed sulfenic acids upon oxidative stress, which can be detected by the formation of stable maspin-dimedone complex (ref). Ablation of these cysteine residues in maspin results in a significant increase in total ROS production in TM40D mouse mammary cells. Also, cells containing the triple cysteine mutant of maspin show elevated ERK1/2 activity (a downstream target of ROS) and enhanced proliferation and colony formation (Mahajan et al. 2013). These findings establish a novel mechanism by which maspin utilizes its cysteine thiols to inhibit oxidative stress and cell growth.

9.3.2.3 Other Cytoplasmic Binding Proteins

Cytoplasmic maspin was also shown to bind other proteins in the cytoplasm. For example, the Hendrix group showed that maspin binds cytoplasmic IRF6 and sequesters it from nuclear translocation and/or proteosomal degradation (Bailey et al. 2005a, b, 2008). Since IRF6 is a transcriptional factor, the binding of maspin with IRF6 in the cytoplasm may serve to downregulate IRF6-mediated transcription in the nucleus (Bailey et al. 2008). Maspin was also found to bind to Hsp90 (Yin et al. 2005). Hsp90 is a cellular chaperone involved in assisting protein folding. Many oncoproteins and pro-survival factors interact with Hsp90 through ATP hydrolysis. Blocking ATP hydrolysis by a reagent such as 17-AAG (17-(allylamino)-17-demethoxygeldanamycin) sensitizes tumor cells to apoptosis. Lockett et al. hypothesized that maspin may regulate tumor cell apoptosis through its binding with Hsp90 (Lockett et al. 2006).

9.3.3 Nuclear Maspin

9.3.3.1 HDAC1 Inhibition

The Sheng group first demonstrated that maspin interacted with HDAC1, one of the most abundant histone deacetylases, in a yeast two-hybrid assay. Li et al. confirmed this interaction and provided the first evidence that maspin exerts an inhibitory effect on HDAC1 (Li et al. 2006) through its effect on HDAC1-regulated genes such as Bax, p21, and CK18 (Lockett et al. 2006; Li et al. 2006). HDAC1 plays a critical role in transcriptional repression by facilitating the histone deacetylation that leads to chromatin condensation, thus preventing the access of transcription factors to DNA. Notably, HDAC1-mediated epigenetics play an integral role in both development and tumor progression. The fact that maspin interacts with HDAC1 suggests that maspin may control gene expression as a master regulator by influencing the acetylation state of many transcription factors. Indeed, subsequent studies by Sheng's group showed that maspin leads to changes in the expression level of a large number of proteins and that these changes are often microenvironment specific (Kaplun et al. 2012). They performed three different assays to compare gene expression levels between maspin overexpressed cells and control cells: identification of statistically enriched gene ontology groups, detection of overrepresented transcription factors binding sites in promoters of differentially expressed genes, and searching of key nodes of regulatory networks controlling these transcription factors. Their data confirmed that maspin is an endogenous inhibitor of HDAC1 and that the effect of maspin is primarily mediated by TGF-β, β-catenin/E-cadherin pathways, and network key nodes such as Abl kinase, p62, IL1, and caspase 6 and 8 (Kaplun et al. 2012).

9.3.3.2 Chromatin Binding

Goulet et al. demonstrated that nuclear localization of maspin is required for its tumor-suppressing activity and that maspin binds to the chromatin (Goulet et al. 2011). Using two invasive cancer cell lines, they showed that reintroduction of maspin in these tumor cells prevented tumor metastasis and that maspin was found to be in the nucleus in cultured cells and clinical samples. However, when maspin was excluded from the nucleus by adding a nuclear export signal, the metastasis-suppressive effect was abolished (Goulet et al. 2011, 2012). They further confirmed that nuclear maspin is associated with chromatin at the unique promoter regions of CSF1 and ERRA, which are two genes critical in breast cancer progression (Goulet et al. 2012). However, it is not clear from their study why maspin binds to chromatin with specificity for certain gene promoters.

9.3.3.3 Other Nuclear Partners

Maspin was also reported to interact with other nuclear proteins. For example, maspin was shown to interact with GC-binding factor 2 (GCF2), a transcriptional repressor whose expression is induced following tissue injury (Bailey et al. 2006). In addition to transcriptional regulators involved in the stress response, the broad-range transcription factor TFIID was also identified as a putative maspin-interacting protein (Yin et al. 2005). Maspin's interaction with these binding partners suggests that it has a wide-ranging role in transcriptional regulation.

9.4 Animal Models of Maspin Study

9.4.1 Maspin Transgenic Mice

To characterize maspin's in vivo activity during mammary development and tumor progression, we generated maspin transgenic mice that overexpress maspin gene in mammary epithelial cells. This was accomplished by placing the maspin gene under the control of a mammary-specific whey acidic protein (WAP) promoter (Zhang et al. 1999). Subsequently, these WAP-maspin transgenic mice were used to demonstrate that the overexpression of maspin in mammary epithelial cells during pregnancy inhibits mammary gland development and induces apoptosis. These mice exhibited impaired differentiation as well as a reduced number of lobular alveolar structures during late pregnancy. In addition, at midpregnancy, alveolar cells from the transgenic mammary glands displayed an increased rate of apoptosis without any changes in the rate of proliferation. These experiments reveal that maspin has a profound biological influence on the growth and function of mammary epithelial cells in vivo (Zhang et al. 1999).

The majority of cancer deaths result not from the primary tumor but from metastasis. Because metastasis is an extremely complicated, multistep process, it is usually best studied in an animal model. In order to facilitate these studies, it is critical to establish an appropriate animal model in which the tumor cells are invasive and possess the ability to metastasize to other organs (Shi et al. 2003). WAP-simian virus (SV) 40 T-antigen (WAP-TAg) transgenic mice develop mammary tumors with 100 % frequency with a high possibility of developing tumor metastasis (Li et al. 1996; Tzeng et al. 1993). The SV40 TAg initiates tumorigenesis through the inactivation of both p53- and the pRb-related family of proteins (Li et al. 2000). Thus, the WAP-TAg transgenic mice possess specific features of some human breast cancers. Since the primary goal of generating WAP-maspin transgenic mice was to test the role of maspin overexpression against mammary tumor growth and metastasis, we crossed maspin transgenic mice with the oncogenic WAP-TAg mice (Zhang et al. 2000b). The results from this study show that overexpression of maspin through a transgenic approach can significantly block tumor progression. For example, bitransgenic maspin/TAg mammary tumors grow considerably slower than their TAg controls. Growth inhibition is accompanied by an increase in apoptosis as well as a reduction in microvessel density. Significantly, maspin overexpression increases apoptosis of both normal mammary epithelial cells during pregnancy and malignant breast cancer cells. This study also revealed that the overexpression of maspin dramatically reduced tumor metastasis to the lung. Taken together, these data demonstrate that the targeted overexpression of maspin can inhibit tumor progression in vivo, likely through a combination of increased apoptosis, decreased angiogenesis, and inhibition of tumor cell migration (Zhang et al. 2000b). These studies have provided a wealth of information regarding the key roles that maspin plays in the prevention of tumor growth, invasion, and metastasis. They offer promising insight into therapeutic treatment and prevention of cancer and demonstrate the benefits of utilizing mouse models to study human diseases.

9.4.2 Maspin Knockout Mice

We also took a different approach for studying the loss of maspin in normal tissue development and cancer progression by using maspin gene knockout mice (Gao et al. 2004). We cloned mouse maspin genomic DNA and designed a vector to selectively knock out the maspin gene in the mice. To do this, we first generated maspin heterozygous knockout mouse. The heterozygous $Mp^{+/-}$ mice progeny appeared to be morphologically normal after birth. However, when they were crossed, no homozygous maspin deletion progeny were obtained at birth. Among the 571 live-born offspring, 343 were heterozygotes and 228 were wild-type mice. This indicates that the homozygous maspin null mice are lethal during embryonic development. We then proceeded to determine when the homozygous lethality exactly happens during embryonic development. We first examined the preimplantation stage of embryonic development. $Mp^{+/-}$ mice were intercrossed; embryos at the 2-cell, 4-cell, and blastocyst stages were compared. Both $Mp^{-/-}$ and $Mp^{+/+}$ embryos were observed by the late blastocyst stage, which indicates that the $Mp^{-/-}$ embryos do not die before implantation. To further determine the time of embryonic death after implantation, in utero embryos ranging from E4.5 to late gestation stages were microdissected and genotyped by either PCR or immunostaining. Our data indicated that $Mp^{-/-}$ embryos had died by E5.5. Maspin protein appeared in the visceral endoderm (VE) at E5.5. This expression was restricted to both the embryonic and the extraembryonic visceral endoderm at E6.5 and E7.5. No maspin expression was observed in the parietal endoderm, the ectoplacental cone, the chorion, and the amnion. In vitro analysis using maspin-deleted ES cells in an embryoid body formation assay supports this idea of maspin having an essential role in early embryonic development. As an early event of embryonic development, the generation of the visceral endoderm provides the embryo with nutritional and hematopoietic functions. Since maspin is specifically expressed in visceral endoderm cells after implantation, its deletion seems to be destructive for the formation of the endodermal cell layer and for the morphogenesis of the epiblast, thus preventing further embryo development.

Since the homozygous maspin knockout mice were embryonic lethal at peri-implantation stage (Gao et al. 2004), which precluded the analysis of mammary phenotype at a later stage, we examined the effect of maspin on mammary development in maspin heterozygous mice (Shi et al. 2004). We showed that maspin heterozygous mice have a defect in their ductal development at pubertal stages. However, this defect arises from a reduced level of systemic progesterone. We showed that the level of progesterone in maspin heterozygous mice was consistently reduced during puberty. When both maspin heterozygous and wild-type mice were ovariectomized, treatment with progesterone by itself stimulated ductal development. Treatment of wild-type mice with progesterone inhibitor RU486 resulted in decreased ductal growth, as was observed in maspin heterozygous mice, and the inhibitory effect was dose dependent. Furthermore, we demonstrated that progesterone, like estrogen, also plays an important role in early mammary ductal development (Shi et al. 2004). Finally, the maspin heterozygous epithelial cells displayed increased levels of ROS compared to the wild-type epithelial cells (Shi et al. 2004), and they appear to have a higher proliferation rate than the wild-type controls (unpublished data).

We also utilized maspin heterozygous mice to study the effect of maspin haploinsufficiency on prostate development and cancer progression. We showed that the loss of one copy of maspin in $Mp^{+/-}$ heterozygous mice leads to the development of prostate hyperplastic lesions and that this effect is mediated through decreased levels of CDK inhibitors p21 and p27 (Shao et al. 2008). A very interesting phenomenon we found is that the partial loss of maspin in epithelial cells in $Mp^{+/-}$ prostate not only causes hyperproliferation of epithelial cells but also induces adjacent stromal cell hyperplasia. This happens in both adult and aged prostates as well as in neonatal prostates during early ductal morphogenesis. Clearly, maspin secreted from epithelial cells can act on the adjacent stromal cells to inhibit cell proliferation in a paracrine manner. However, since smooth muscle cells also express maspin, maspin could also act on smooth muscle cells in an autocrine manner too. Thus, the observed stromal cell hyperplasia in $Mp^{+/-}$ mice may arise from reduced maspin expression in both epithelial and smooth muscle cells in $Mp^{+/-}$ mice (Sung and Chung 2002). The findings that prostate smooth muscle cells express maspin and that the loss of one copy of maspin causes stromal hyperplasia are highly significant since it highlights the importance of epithelial-stromal interaction during prostate tumor progression. It is very possible that such feedback interaction between epithelial and smooth muscle cells may in fact promote prostate cancer progression. Finally, the results from our maspin knockout mouse study also provided direct evidence that maspin acts as a prostate tumor suppressor gene in vivo in a mouse model. This novel property that maspin inhibits cell proliferation may be exploited for therapeutic treatment of prostate cancer in the future.

9.4.3 Maspin Conditional Knockout Mice

Teoh et al. recently reported the generation of a maspin conditional knockout mouse line (Teoh et al. 2014). The deletion construct was built by recombination-mediated genetic engineering using a chromosome BAC clone as the source for maspin genomic DNA. Two LoxP flox sites are located in intron 3 and intron 4. The construct was electroporated to C57BL/6-derived ES cells, and selected ES cells were injected into BALB/c blastocysts to generate chimeric embryos. They used CMV-Cre to perform the LoxP-mediated deletion of maspin exon 4 and generate the $maspin^{-/-}$ knockout mice. Their data showed that Cre-mediated deletion generated an mRNA carrying the predicted deletion and also the introduction of a stop codon 11 residues downstream. This flox-deleted maspin knockout mouse is viable and has no overt anatomical and physiological abnormalities. This result is in contrast to our previous report that maspin conventional knockout mouse is embryonic lethal and that maspin heterozygous knockout mice display developmental and physiological abnormalities (Teoh et al. 2014; Gao et al. 2004).

Why is there such a discrepancy? First of all, the targeting constructs in these two reports are different. While our group's previous construct targeted the deletion of maspin exon 7, the domain responsible for many of the biological functions of maspin as described above, the newly reported construct targeted exon 4 to generate truncated maspin by inserting a stop codon after LoxP site with the hope that this may result in a maspin "null" mouse. One concern with this model is that this new targeting method may yield an alternatively spliced maspin gene product missing exon 4 and the inserted stop codon and LoxP site, but the encoded maspin protein may still possess some or all of the maspin functions. In addition, a 133 residue truncated maspin is generated that may retain some activities. These possibilities can be tested using the right tools. Unfortunately, the two antibodies used by Teoh et al. are not very specific. For example, antibody 13G11 showed the absence of "maspin" from knockout mammary tissue but the presence of the "maspin band" in the knockout prostate and thymus (Teoh et al. 2014). Antibody H130 on the other hand showed the absence of maspin from the knockout prostate, but not in the knockout thymus. Teoh et al. explained that this might be due to antibody cross-reactivity. However, our view is that this raises doubts of whether maspin is fully deleted, since the nonspecific immunodetection signals cannot be used for the confirmation of their maspin knockout mouse. If alternative splicing does occur in these conditional knockout mice, the animals generated will have a truncated maspin missing exon 4, which may explain why their conditional knockout mice show no developmental and physiological abnormalities.

Teoh et al. also suggested that the lethality defect of our maspin conventional knockout mice might arise from an off-target effect of the neo cassette close to PHLPP1, which is nearly 100 kb away from maspin gene. This speculation is inconsistent with the evidence that PHLPP1 knockout mice are viable, while the previously characterized maspin knockout mice are embryonic lethal (Gao et al. 2004; Masubuchi et al. 2010).

My laboratory also generated a new maspin conditional knockout mouse strain, targeting the complete deletion of maspin product (unpublished data). The designed recombination-mediated deletion will not give rise to any maspin product since the start codon containing exon 2 and exon 3 are deleted. In our preliminary study, we used Tie2-Cre to delete endothelial maspin during early embryonic development. Our data showed that embryonic deletion of maspin in endothelial cells by Tie 2-Cre causes embryonic lethality of mouse embryos (unpublished data). Analysis of some embryos at E9.5 revealed the defective vasculature. Our data indicates that both conventional and tissue-specific deletion of maspin can cause mouse embryonic lethality. More research with this new conditional maspin knockout mouse strain will be performed in the future.

9.5 Developing Maspin for Therapeutic Intervention

There is mounting evidence confirming that maspin acts as a tumor-suppressing serpin, inhibiting tumor progression, metastasis, and angiogenesis. Maspin's tumor-suppressing properties prompted us to develop maspin for therapeutic application. These preclinical studies described below are mostly carried out in mouse cancer models using the maspin gene, protein, and/or peptides.

9.5.1 Maspin-Mediated Gene Therapy Against Breast Cancer

Cancer gene therapy requires both a good animal model and an effective delivery system (McCormick 2001). Genetically engineered mouse (GEM) mammary tumor models are widely used for cancer gene therapy. However, there are limitations with using GEM for cancer therapy, particularly variations in the number of tumors and the time of tumor development in different animals. These variations affect the evaluation of treatment efficacy since all reagents are supposed to be delivered to the groups of animals at the same time. To circumvent these problems, we established a syngeneic PyV-mT tumor transplantation model for maspin gene delivery study by transplanting the mammary tumor cells (PyV-mT cells) to syngeneic FVB female mammary fat pads bilaterally. Tumors developed 100 % in transplanted sites and they showed the characteristic of solid adenocarcinoma. The unique feature of this transplantation model is that implanted tumors grow out uniformly and are metastatic to the lung. We then delivered the maspin gene to tumor-bearing mice using DNA/liposome complexes for increased delivery and gene expression (Shi et al. 2002). To determine whether the system we chose could effectively deliver gene to the breast in mice, we injected a reporter DNA construct (CAT) with liposome to a group of wild-type mice through their tail veins. We found that the mammary gland is one of the tissue organs with the highest delivery efficiency (Shi et al. 2002). To test the therapeutic value of maspin against breast tumor progression, two groups of mice were implanted with PyV-mT tumor cells. After the tumors were developed in the mammary gland orthotopically, mice were treated with maspin/liposome or with control plasmid DNA (placebo) systemically for over 1 month. Primary mammary tumor growth was monitored during the treatment and at the end point. Our data showed that the overall tumor growth rate was significantly decreased for maspin-treated tumors compared to the controls. To analyze maspin's effect on tumor metastasis, lung tissues from the two groups were harvested at the end point. Tissues were serially sectioned for histology. Histological analysis of these sections revealed a significant decrease of lung tumor metastasis in the maspin-treated group (44.5 %) compared to the control (100 %). The total numbers of tumor foci were also significantly reduced (185 %) in maspin-treated samples. TUNEL assays revealed that this significant inhibition of both primary tumor growth and metastasis is mediated by increased apoptosis in the maspin-treated tumors. Furthermore, we demonstrated that maspin/liposome treatment does not result in any toxic side effect after a long period of treatment in wild-type mice (Shi et al. 2002). Female wild-type mice receiving long-term maspin gene therapy maintained normal reproductive function and did not exhibit any gross abnormalities in tissue organs upon biopsy. Thus, maspin/liposome treatment offers an effective therapy with low toxicity for breast cancer. Based on this study, we anticipate that new maspin gene-mediated therapies can be developed for cancer treatment with high efficacy delivery methods in the near future.

9.5.2 Using Maspin Protein and Peptides for Anti-angiogenesis Therapy

Tumor growth and metastasis requires neovascular formation, a process also termed angiogenesis (Folkman 1986, 1989). Angiogenesis supplies tumors with nutrients and aid through the removal of metabolic wastes (Folkman 1995a; Fidler and Ellis 1994). Most solid tumors cannot grow beyond a few millimeters without angiogenesis (Folkman 1995b; Boudreau and Myers 2003). Targeting neovessels in cancer with anti-angiogenesis inhibitors has been considered a smart approach to eradicate cancers. To determine if maspin protein can be used for anti-angiogenesis therapy against cancer, we delivered maspin to mice bearing human prostate cancer cells in an athymic mouse xenograft model (Zhang et al. 2000a). LNCaP prostate tumor cells were implanted subcutaneously on the bidorsal back of nude mice. Tumor growth and neovascularization were monitored following systemic treatment with recombinant maspin. Through CD31 immunostaining, we found that maspin-treated tumors contained significantly fewer vessels than mock-treated controls. To determine whether maspin's effects on the tumor-induced vasculature were maintained during a more prolonged treatment, the above experiment was replicated with tumors harvested after 7–8 weeks. Thirty-two tumor sites were treated with maspin and 37 with GST. When examined at week 8, the growth of 53 % of the maspin-treated tumors had been completely inhibited. The remaining 15 maspin-treated tumors were reduced in size by on average 3.43-fold when compared to mock control treated tumors. To examine if the reduced size of maspin-treated tumors coincided with reduced neovascularization, 20 representative tumors from either maspin-treated (10 sites) or GST-treated tumors (10 sites) were used to quantify the density of microvessels after immunostaining with CD31 antibody. The density of vessels in maspin-treated tumors was reduced by 2.6-fold on average compared to those in control tumors, and this difference was highly significant. We also compared the treated and control tumors of similar sizes. A reduction of vessel density was likewise observed in the maspin-treated samples (Zhang et al. 2000a). These data confirm that maspin is an effective inhibitor of tumor angiogenesis and could therefore be developed into a potent anti-angiogenesis/anticancer therapy.

Protein therapeutics like using maspin protein still faces challenges relating to high cost and short retention time in tissues. Small molecular peptides can be clinical alternatives, provided they offer comparable potency. Our collaborators and my laboratory have recently developed new maspin peptide nanostructures comprised of maspin peptide epitope bound to a nanoparticle molecular domain that promotes self-assembly through hydrogen bonding and hydrophobic interactions. This molecule, known as peptide amphiphile (PA), consists of a hydrophobic alkyl tail, a peptide sequence capable of β-sheet formation, a flexible peptide sequence as a linker, and the maspin peptide epitope at one terminus of the molecule. The maspin-nanoparticle PAs can self-assemble in aqueous solution to form high aspect-ratio nanofibers comprised of a hydrophobic core and stabilized by β-sheet formation down their long axis (Guler and Stupp 2007; Velichko et al. 2008). These nanofibers can therefore display on their surfaces a high density of biological signals (Storrie et al. 2007), stabilize epitope secondary structure (Solomon et al. 2006), and improve therapeutic retention in tissue as a result of longer half-lives associated with their supramolecular structure (Ghanaati et al. 2009; Geng et al. 2007). In a recent report, we demonstrated that these maspin-mimetic PA nanostructures possess anti-angiogenic activity at concentrations that are significantly lower than those necessary for the maspin G-helix peptide (Zha et al. 2015). Furthermore, in vivo assays in the chick chorio-allantoic membrane have demonstrated that maspin-mimetic nanostructures are highly effective in inhibiting angiogenesis (Zha et al. 2015). Therefore, we anticipate that in the future, maspin peptide-based nanostructures may be developed for the treatment of cancer or other diseases involving abnormal blood vessel formation.

9.6 Future Direction

Since the discovery of maspin in 1994 as a breast tumor suppressor, more than 500 articles have been published regarding maspin expression and its biological functions in cancers and other diseases. While the majority of these publications support a role of maspin as a bona fide tumor suppressor in the early stages of tumor progression and in tumor metastasis, some reports raise concerns pointing to the opposite conclusions. This suggests that we are still far from understanding the biological functions of maspin. We are encouraged by these debates and are even more determined to continue our research of maspin because of this. With this said, it has become clear that maspin gene expression and subcellular locations of maspin protein are context dependent. Maspin's biological functions may also vary in different kinds of tissues. Thus, we need to investigate maspin's function in the context of maspin expression, mutation, and subcellular location. Also, there is currently an urgent need to develop new in vivo animal models to study the physiological functions of maspin. These in vivo studies will likely shed more lights regarding roles of maspin in normal tissue development and cancer. In addition, studies are required to further elucidate the mechanisms by which maspin translocates to different subcellular compartments and functions in various biological processes. Finally, more research will be done focusing on the ways in which this crucial tumor suppressor can be utilized in cancer prevention, diagnosis, and treatment.

Acknowledgments Many of the studies discussed in this book chapter are endless efforts from people (postdoctoral fellows, students, and visiting scholars) who worked in the Zhang laboratory and from colleagues who collaborated with Dr. Zhang on maspin project in the last 20 years. I wish to take this opportunity to thank these people for their dedication, contribution, and support. I also wish to thank my colleagues, who worked with me at Harvard Medical School, Baylor College of Medicine, and Northwestern University for their friendship and encouragement.

The maspin studies in the Zhang laboratory have been supported generously by research funds from the National Cancer Institute-USA and the United States Department of Defense.

References

Abd El-Wahed MM (2005) Expression and subcellular localization of maspin in human ovarian epithelial neoplasms: correlation with clinicopathologic features. J Egypt Natl Canc Inst 17:173–183

Acikalin D, Oner U, Can C, Acikalin MF, Colak E (2012) Predictive value of maspin and Ki-67 expression in transurethral resection specimens in patients with T1 bladder cancer. Tumori 98:344–350

Affara NI, Coussens LM (2007) IKKalpha at the crossroads of inflammation and metastasis. Cell 129:25–26

Al-Ayyoubi M, Schwartz BS, Gettins PG (2007) Maspin binds to urokinase-type and tissue-type plasminogen activator through exosite-exosite interactions. J Biol Chem 282:19502–19509

Baghel K, Kazmi HR, Raj S, Chandra A, Srivastava RN (2014) Elevated expression of maspin mRNA as a predictor of survival in stage II and III gallbladder cancer cases. Asian Pac J Cancer Prev 15:343–347

Bailey CM et al (2005a) Mammary serine protease inhibitor (Maspin) binds directly to interferon regulatory factor 6: identification of a novel serpin partnership. J Biol Chem 280:34210–34217

Bailey CM et al (2005b) Maspin binds directly to interferon regulatory factor 6: identification of a novel serpin partnership. J Biol Chem 240(40):34210–34217

Bailey CM, Khalkhali-Ellis Z, Seftor EA, Hendrix MJ (2006) Biological functions of maspin. J Cell Physiol 209:617–624

Bailey CM, Abbott DE, Margaryan NV, Khalkhali-Ellis Z, Hendrix MJ (2008) Interferon regulatory factor 6 promotes cell cycle arrest and is regulated by the proteasome in a cell cycle-dependent manner. Mol Cell Biol 28:2235–2243

Bass R, Wagstaff L, Ravenhill L, Ellis V (2009) Binding of extracellular maspin to beta1 integrins inhibits vascular smooth muscle cell migration. J Biol Chem 284:27712–27720

Blacque OE, Worrall DM (2002) Evidence for a direct interaction between the tumour suppressor serpin maspin, and types I and III collagen. J Biol Chem 277(13):10783–10788

Bodenstine TM et al (2012) Maspin: molecular mechanisms and therapeutic implications. Cancer Metastasis Rev 31:529–551

Boltze C (2005) Loss of maspin is a helpful prognosticator in colorectal cancer: a tissue microarray analysis. Pathol Res Pract 200:783–790

Boltze C et al (2003) Maspin in thyroid cancer: its relationship with p53 and clinical outcome. Oncol Rep 10:1783–1787

Boudreau N, Myers C (2003) Breast cancer-induced angiogenesis: multiple mechanisms and the role of the microenvironment. Breast Cancer Res 5:140–146

Cella N, Contreras A, Latha K, Rosen JM, Zhang M (2006) Maspin is physically associated with [beta]1 integrin regulating cell adhesion in mammary epithelial cells. FASEB J 20:1510–1512

Cher ML et al (2003) Maspin expression inhibits osteolysis, tumor growth, and angiogenesis in a model of prostate cancer bone metastasis. Proc Natl Acad Sci USA 100:7847–7852

Costello JF, Vertino PM (2002) Methylation matters: a new spin on maspin. Nat Genet 31:123–124

Dawson DW et al (1999) Pigment epithelium-derived factor: a potent inhibitor of angiogenesis. Science 285:245–248

Domann FE, Rice JC, Hendrix MJ, Futscher BW (2000) Epigenetic silencing of maspin gene expression in human breast cancers. Int J Cancer 85:805–810

Duan H et al (2003) MnSOD up-regulates maspin tumor suppressor gene expression in human breast and prostate cancer cells. Antioxid Redox Signal 5:677–688

Eitel JA et al (2009) PTEN and p53 are required for hypoxia induced expression of maspin in glioblastoma cells. Cell Cycle 8:896–901

Endsley MP et al (2011) Maspin, the molecular bridge between the plasminogen activator system and beta1 integrin that facilitates cell adhesion. J Biol Chem 286:24599–24607

Feldman RJ, Sementchenko VI, Gayed M, Fraig MM, Watson DK (2003) Pdef expression in human breast cancer is correlated with invasive potential and altered gene expression. Cancer Res 63:4626–4631
Fidler IJ, Ellis LM (1994) The implications of angiogenesis for the biology and therapy of cancer metastasis [comment]. Cell 79:185–188
Folkman J (1986) How is blood vessel growth regulated in normal and neoplastic tissue? G.H.A. Clowes memorial Award lecture. Cancer Res 46:467–473
Folkman J (1989) Successful treatment of an angiogenic disease [editorial]. N Engl J Med 320:1211–1212
Folkman J (1995a) Tumor angiogenesis. In: Mendelsohn J, Israel MA, Liotta LA (eds) The molecular basis of cancer. WB Sauenders Co., Philadelphia, PA, pp 206–232
Folkman J (1995b) Angiogenesis in cancer, vascular, rheumatoid and other disease. Nat Med 1:27–31
Futscher BW et al (2002) Role for DNA methylation in the control of cell type specific maspin expression. Nat Genet 31:175–179
Gao F, Shi HY, Daughty C, Cella N, Zhang M (2004) Maspin plays an essential role in early embryonic development. Development 131:1479–1489
Geng Y et al (2007) Shape effects of filaments versus spherical particles in flow and drug delivery. Nat Nanotechnol 2:249–255
Ghanaati S et al (2009) Dynamic in vivo biocompatibility of angiogenic peptide amphiphile nanofibers. Biomaterials 30:6202–6212
Goulet B et al (2011) Nuclear localization of maspin is essential for its inhibition of tumor growth and metastasis. Lab Invest 91:1181–1187
Goulet B, Chan G, Chambers AF, Lewis JD (2012) An emerging role for the nuclear localization of maspin in the suppression of tumor progression and metastasis. Biochem Cell Biol 90:22–38
Guler MO, Stupp SI (2007) A self-assembled nanofiber catalyst for ester hydrolysis. J Am Chem Soc 129:12082–12083
He Y et al (2007) Expression of maspin protein and correlation between maspin and P53 proteins in gastric carcinoma. Zhonghua Yi Xue Za Zhi 87:909–912
Ito Y et al (2004) Maspin expression is directly associated with biological aggressiveness of thyroid carcinoma. Thyroid 14:13–18
Jang HL et al (2008) Maspin polymorphism associated with apoptosis susceptibility and in vivo tumorigenesis. Int J Mol Med 22:333–338
Jiang WG, Hiscox S, Horrobin DF, Bryce RP, Mansel RE (1997) Gamma linolenic acid regulates expression of maspin and the motility of cancer cells. Biochem Biophys Res Commun 237:639–644
Jiang N, Meng Y, Zhang S, Mensah-Osman E, Sheng S (2002) Maspin sensitizes breast carcinoma cells to induced apoptosis. Oncogene 21:4089–4098
Kaplun A, Dzinic S, Bernardo M, Sheng S (2012) Tumor suppressor maspin as a rheostat in HDAC regulation to achieve the fine-tuning of epithelial homeostasis. Crit Rev Eukaryot Gene Expr 22:249–258
Khalkhali-Ellis Z, Hendrix MJ (2003) Nitric oxide regulation of maspin expression in normal mammary epithelial and breast cancer cells. Am J Pathol 162:1411–1417
Khalkhali-Ellis Z, Hendrix MJ (2007) Elucidating the function of secreted maspin: inhibiting cathepsin D-mediated matrix degradation. Cancer Res 67:3535–3539
Khalkhali-Ellis Z et al (2004) Regulating the tumor suppressor gene maspin in breast cancer cells: a potential mechanism for the anticancer properties of tamoxifen. Clin Cancer Res 10:449–454
Kim S, Han J, Kim J, Park C (2004) Maspin expression is transactivated by p63 and is critical for the modulation of lung cancer progression. Cancer Res 64:6900–6905
Kisker O et al (2001) Generation of multiple angiogenesis inhibitors by human pancreatic cancer. Cancer Res 61:7298–7304

Klomsiri C et al (2010) Use of dimedone-based chemical probes for sulfenic acid detection evaluation of conditions affecting probe incorporation into redox-sensitive proteins. Methods Enzymol 473:77–94
Latha K, Zhang W, Cella N, Shi HY, Zhang M (2005) Maspin mediates increased tumor cell apoptosis upon induction of the mitochondrial permeability transition. Mol Cell Biol 25:1737–1748
Li M, Hu J, Heermeier K, Hennighausen L, Furth PA (1996) Expression of a viral oncoprotein during mammary gland development alters cell fate and function: induction of p53-independent apoptosis is followed by impaired milk protein production in surviving cells. Cell Growth Differ 7:3–11
Li JJ, Colburn NH, Oberley LW (1998) Maspin gene expression in tumor suppression induced by overexpressing manganese-containing superoxide dismutase cDNA in human breast cancer cells. Carcinogenesis 19:833–839
Li M, Lewis B, Capuco AV, Laucirica R, Furth PA (2000) WAP-TAg transgenic mice and the study of dysregulated cell survival, proliferation, and mutation during breast carcinogenesis. Oncogene 19:1010–1019
Li Z, Shi HY, Zhang M (2005) Targeted expression of maspin in tumor vasculatures induces endothelial cell apoptosis. Oncogene 24(12):2008–2019
Li X, Yin S, Meng Y, Sakr W, Sheng S (2006) Endogenous inhibition of histone deacetylase 1 by tumor-suppressive maspin. Cancer Res 66:9323–9329
Liu Z, Shi HY, Nawaz Z, Zhang M (2004) Tamoxifen induces the expression of maspin through estrogen receptor-alpha. Cancer Lett 209:55–65
Lockett J, Yin S, Li X, Meng Y, Sheng S (2006) Tumor suppressive maspin and epithelial homeostasis. J Cell Biochem 97:651–660
Lonardo F et al (2006) Maspin nuclear localization is linked to favorable morphological features in pulmonary adenocarcinoma. Lung Cancer 51:31–39
Luo JL et al (2007) Nuclear cytokine-activated IKKalpha controls prostate cancer metastasis by repressing Maspin. Nature 446:690–694
Maass N et al (2000) Maspin – a novel protease inhibitor with tumor-suppressing activity in breast cancer. Acta Oncol 39:931–934
Maass N et al (2001a) Down regulation of the tumor suppressor gene maspin in breast carcinoma is associated with a higher risk of distant metastasis. Clin Biochem 34:303–307
Maass N et al (2001b) Expression of the tumor suppressor gene Maspin in human pancreatic cancers. Clin Cancer Res 7:812–817
Maass N et al (2002) Hypermethylation and histone deacetylation lead to silencing of the maspin gene in human breast cancer. Biochem Biophys Res Commun 297:125–128
Machtens S et al (2001) Expression of the p53 and Maspin protein in primary prostate cancer: correlation with clinical features. Int J Cancer 95:337–342
Maekawa T et al (2008) ATF-2 controls transcription of Maspin and GADD45 alpha genes independently from p53 to suppress mammary tumors. Oncogene 27:1045–1054
Mahajan N, Shi HY, Lukas TJ, Zhang M (2013) Tumor suppressive maspin functions as a reactive oxygen species scavenger: importance of cysteine residues. J Biol Chem 288:11611–11620
Masubuchi S et al (2010) Protein phosphatase PHLPP1 controls the light-induced resetting of the circadian clock. Proc Natl Acad Sci USA 107:1642–1647
McCormick F (2001) Cancer gene therapy: fringe or cutting edge? Nat Rev 1:130–141
Mikus P, Urano T, Liljestrom P, Ny T (1993) Plasminogen-activator inhibitor type 2 (PAI-2) is a spontaneously polymerising SERPIN. Biochemical characterisation of the recombinant intracellular and extracellular forms. Eur J Biochem 218:1071–1082
Mohsin SK, Zhang M, Clark GM, Craig Allred D (2003) Maspin expression in invasive breast cancer: association with other prognostic factors. J Pathol 199:432–435
Nakashima D et al (2006) Protein expression profiling identifies maspin and stathmin as potential biomarkers of adenoid cystic carcinoma of the salivary glands. Int J Cancer 118:704–713

Narayan M, Mirza SP, Twining SS (2011) Identification of phosphorylation sites on extracellular corneal epithelial cell maspin. Proteomics 11:1382–1390
Nawata S, Shi HY, Sugino N, Zhang M (2011) Evidence of post-translational modification of the tumor suppressor maspin under oxidative stress. Int J Mol Med 27:249–254
Nelson KJ et al (2010) Use of dimedone-based chemical probes for sulfenic acid detection methods to visualize and identify labeled proteins. Methods Enzymol 473:95–115
Ngamkitidechakul C, Burke JM, O'Brien WJ, Twining SS (2001) Maspin: synthesis by human cornea and regulation of in vitro stromal cell adhesion to extracellular matrix. Invest Ophthalmol Vis Sci 42:3135–3141
Ngamkitidechakul C, Warejcka DJ, Burke JM, O'Brien WJ, Twining SS (2003) Sufficiency of the reactive site loop of maspin for Induction of cell-matrix adhesion and inhibition of cell Invasion: conversion of ovalbumin to a maspin-like molecule. J Biol Chem 278 (34):31796–31806
Oh YL, Song SY, Ahn G (2002) Expression of maspin in pancreatic neoplasms: application of maspin immunohistochemistry to the differential diagnosis. Appl Immunohistochem Mol Morphol 10:62–66
Oshiro MM et al (2003) Mutant p53 and aberrant cytosine methylation cooperate to silence gene expression. Oncogene 22:3624–3634
Pemberton PA et al (1997) Maspin is an intracellular serpin that partitions into secretory vesicles and is present at the cell surface. J Histochem Cytochem 45:1697–1706
Poole LB, Nelson KJ (2008) Discovering mechanisms of signaling-mediated cysteine oxidation. Curr Opin Chem Biol 12:18–24
Poole LB, Karplus PA, Claiborne A (2004) Protein sulfenic acids in redox signaling. Annu Rev Pharmacol Toxicol 44:325–347
Qin L, Zhang M (2010) Maspin regulates endothelial cell adhesion and migration through an integrin signaling pathway. J Biol Chem 285:32360–32369
Ravenhill L, Wagstaff L, Edwards DR, Ellis V, Bass R (2010) G-helix of maspin mediates effects on cell migration and adhesion. J Biol Chem 285:36285–36292
Sager R et al (1994) RNA genetics of breast cancer: maspin as paradigm. Cold Spring Harb Symp Quant Biol 59:537–546
Sager R, Sheng S, Pemberton P, Hendrix MJ (1996) Maspin: a tumor suppressing serpin. Curr Top Microbiol Immunol 213:51–64
Schaefer JS et al (2010) Transcriptional regulation of p21/CIP1 cell cycle inhibitor by PDEF controls cell proliferation and mammary tumor progression. J Biol Chem 285:11258–11269
Seftor RE et al (1998) maspin suppresses the invasive phenotype of human breast carcinoma. Cancer Res 58:5681–5685
Shao ZM, Radziszewski WJ, Barsky SH (2000) Tamoxifen enhances myoepithelial cell suppression of human breast carcinoma progression in vitro by two different effector mechanisms. Cancer Lett 157:133–144
Shao LJ, Shi HY, Ayala G, Rowley D, Zhang M (2008) Haploinsufficiency of the maspin tumor suppressor gene leads to hyperplastic lesions in prostate. Cancer Res 68:5143–5151
Sheng S, Pemberton PA, Sager R (1994) Production, purification, and characterization of recombinant maspin proteins. J Biol Chem 269:30988–30993
Sheng S et al (1996) Maspin acts at the cell membrane to inhibit invasion and motility of mammary and prostatic cancer cells. Proc Natl Acad Sci USA 93:11669–11674
Sheng S et al (1998) Tissue-type plasminogen activator is a target of the tumor suppressor gene maspin. Proc Natl Acad Sci USA 95:499–504
Shi H, Liang R, Templeton N, Zhang M (2002) Inhibition of breast tumor progression by systemic delivery of maspin in a syngeneic tumor model. Mol Ther 5(6):755–761
Shi HY et al (2003) Modeling human breast cancer metastasis in mice: maspin as a paradigm. Histol Histopathol 18:201–206
Shi HY, Lydon JP, Zhang M (2004) Hormonal defect in maspin heterozygous mice reveals a role of progesterone in pubertal ductal development. Mol Endocrinol 18(9):2196–2207

Shi HY, Stafford LJ, Liu Z, Liu M, Zhang M (2007) Maspin controls mammary tumor cell migration through inhibiting Rac1 and Cdc42, but not the RhoA GTPase. Cell Motil Cytoskeleton 64:338–346

Solomon LA et al (2006) Maspin expression and localization impact on angiogenesis and prognosis in ovarian cancer. Gynecol Oncol 101:385–389

Sood AK et al (2002) The paradoxical expression of maspin in ovarian carcinoma. Clin Cancer Res 8:2924–2932

Sopel M, Kasprzyk I, Berdowska I (2005) Maspin and c-erbB-2 expression in correlation with microvessel density in invasive ductal breast cancer. Folia Histochem Cytobiol 43:109–116

Storrie H et al (2007) Supramolecular crafting of cell adhesion. Biomaterials 28:4608–4618

Sung SY, Chung LW (2002) Prostate tumor-stroma interaction: molecular mechanisms and opportunities for therapeutic targeting. Differentiation 70:506–521

Teoh SS, Whisstock JC, Bird PI (2010) Maspin (SERPINB5) is an obligate intracellular serpin. J Biol Chem 285:10862–10869

Teoh SS et al (2014) Maspin is not required for embryonic development or tumour suppression. Nat Commun 5:3164

Tsuji T et al (2007) Maspin expression is up-regulated during the progression of endometrioid endometrial carcinoma. Histopathology 51:871–874

Tzeng YJ, Guhl E, Graessmann M, Graessmann A (1993) Breast cancer formation in transgenic animals induced by the whey acidic protein SV40 T antigen (WAP-SV-T) hybrid gene. Oncogene 8:1965–1971

Umekita Y et al (2011) Maspin expression is frequent and correlates with basal markers in triple-negative breast cancer. Diagn Pathol 6:36

Velichko YS, Stupp SI, de la Cruz MO (2008) Molecular simulation study of peptide amphiphile self-assembly. J Phys Chem B 112:2326–2334

Wang DL, Wang YF, Shi GS, Huang H (2007a) Correlation of hTERT expression to maspin and bFGF expression and their significance in glioma. Ai Zheng 26:601–606

Wang SE et al (2007b) Convergence of p53 and transforming growth factor beta (TGFbeta) signaling on activating expression of the tumor suppressor gene maspin in mammary epithelial cells. J Biol Chem 282:5661–5669

Wei Y et al (1996) Regulation of integrin function by the urokinase receptor. Science 273:1551–1555

Wei Y et al (2005) Regulation of alpha5beta1 integrin conformation and function by urokinase receptor binding. J Cell Biol 168:501–511

Xia W et al (2000) High tumoral maspin expression is associated with improved survival of patients with oral squamous cell carcinoma. Oncogene 19:2398–2403

Yamada N, Tamai Y, Miyamoto H, Nozaki M (2000) Cloning and expression of the mouse Pse gene encoding a novel Ets family member. Gene 241:267–274

Yin S et al (2005) Tumor suppressive maspin regulates cell response to oxidative stress by direct interaction with glutathione S-transferase. J Biol Chem 280(41):34985–34996

Yin S et al (2006) Maspin retards cell detachment via a novel interaction with the urokinase-type plasminogen activator/urokinase-type plasminogen activator receptor system. Cancer Res 66:4173–4181

Yoshizawa K et al (2009) Loss of maspin is a negative prognostic factor for invasion and metastasis in oral squamous cell carcinoma. J Oral Pathol Med 38:535–539

Yu X, Harris SL, Levine AJ (2006) The regulation of exosome secretion: a novel function of the p53 protein. Cancer Res 66:4795–4801

Zha RH et al (2015) Supramolecular assembly of multifunctional maspin-mimetic nanostructures as a potent peptide-based angiogenesis inhibitor. Acta Biomater 12C:1–10

Zhang M (2009) PTEN in action: coordinating with p53 to regulate maspin gene expression. Cell Cycle 8:1112–1113

Zhang M, Maass N, Magit D, Sager R (1997a) Transactivation through Ets and Ap1 transcription sites determines the expression of the tumor-suppressing gene maspin. Cell Growth Differ 8:179–186

Zhang M, Magit D, Sager R (1997b) Expression of maspin in prostate cells is regulated by a positive ets element and a negative hormonal responsive element site recognized by androgen receptor. Proc Natl Acad Sci USA 94:5673–5678

Zhang M, Sheng S, Maass N, Sager R (1997c) mMaspin: the mouse homolog of a human tumor suppressor gene inhibits mammary tumor invasion and motility. Mol Med 3:49–59

Zhang M et al (1999) Maspin plays an important role in mammary gland development. Dev Biol 215:278–287

Zhang M, Volpert O, Shi YH, Bouck N (2000a) Maspin is an angiogenesis inhibitor. Nat Med 6:196–199

Zhang M, Shi Y, Magit D, Furth PA, Sager R (2000b) Reduced mammary tumor progression in WAP-TAg/WAP-maspin bitransgenic mice. Oncogene 19:6053–6058

Zhang W, Shi HY, Zhang M (2005) Maspin overexpression modulates tumor cell apoptosis through the regulation of Bcl-2 family proteins. BMC Cancer 5:50

Zheng D, Chen H, Davids J, Bryant M, Lucas A (2013) Serpins for diagnosis and therapy in cancer. Cardiovasc Hematol Disord Drug Targets 13:123–132

Zou Z et al (1994) Maspin, a serpin with tumor-suppressing activity in human mammary epithelial cells [see comments]. Science 263:526–529

Zou Z et al (2000) p53 regulates the expression of the tumor suppressor gene maspin. J Biol Chem 275:6051–6054

Zou Z et al (2002) Maspin expression profile in human prostate cancer (CaP) and in vitro induction of Maspin expression by androgen ablation. Clin Cancer Res 8:1172–1177

Protein C Inhibitor

10

Felix C. Wahlmüller

Contents

Abstract

Human protein C inhibitor (PCI, SerpinA5) was first identified in 1980 as an inhibitor of activated protein C (APC). It was soon recognized that the secreted 57 kDa glycoprotein inhibits not only APC but a variety of different proteases by forming stable 1:1 complexes. In addition to its multi-specificity, the inhibitor is

F.C. Wahlmüller (✉)
Department of Vascular Biology and Thrombosis Research, Center of Physiology and Pharmacology, Medical University of Vienna, Schwarzspanierstraße 17, 1090 Vienna, Austria
e-mail: f.wahlmueller@gmail.com

M. Geiger et al. (eds.), *The Serpin Family*, DOI 10.1007/978-3-319-22711-5_10

also present in many human tissues and secretions. Its inhibitory property toward different target proteases can be modulated by cofactors (non-protease ligands) like heparin, DNA, and phospholipids, which results in enhanced complex formation (in most cases). Furthermore PCI exhibits a variety of non-inhibitory functions. It binds to retinoic acid suggesting a role of PCI in retinoid-dependent signaling, displays broad antimicrobial activity, and acts as a lipid transferase for phosphatidylethanolamine. Interaction of PCI with phospholipids not only impacts the phagocytosis of apoptotic cells but also mediates the insertion of PCI into membranes which leads to rapid internalization of the serpin. The exact role of internalized PCI, which is detected in cytosolic and nuclear fractions, is not well understood at the moment. Since PCI interacts with intracellular phosphoinositides and alters the phosphorylation status of the PI3-kinase downstream target protein kinase B (AKT), it might influence intracellular lipid levels and contribute to tissue-specific functions. Finally PCI acts as tumor suppressor and negatively regulates tumor cell growth and metastatic potential. Recently, it was reported that higher levels of PCI in cervicovaginal fluid might be a protective factor against HIV infection. This review summarizes the current knowledge on PCI and its functions in health and disease.

10.1 Introduction

Protein C inhibitor (PCI) was first identified in 1980 by Marlar and Griffin as an inhibitor of activated protein C (APC) in human blood plasma, and the inhibition of APC by normal and various deficient plasmas was studied (Marlar and Griffin 1980). Three years later Suzuki et al. isolated PCI from human plasma and characterized some of its physicochemical properties (Suzuki et al. 1983).

10.2 Gene Structure and Tissue-Specific Expression of PCI

The gene coding for PCI is located on chromosome 14q32.1, between *SERPINA4* and α1-antichymotrypsin (AACT, *SERPINA3*) close to the genes for α1-antitrypsin (A1AT, *SERPINA1*) and corticosteroid-binding globulin (*SERPINA6*) (Billingsley et al. 1993; Meijers and Chung 1991). In 1991 the human PCI gene was isolated and its organization characterized. The gene consists of five exons and four introns and is 11.5 kb long (see Table 10.1). The overall organization of the gene closely resembles the genes coding for A1AT and AACT (Meijers and Chung 1991; Hayashi and Suzuki 1993).

Analyzing the promoter region of the PCI gene, no apparent TATA or CCAAT boxes were found, but a Sp1-binding site, an AP-2 site, and an inverted AP-2 site could be identified (Meijers and Chung 1991). Within 1.6 kb of the 5′-flanking

Table 10.1 Organization of the human PCI gene (Meijers and Chung 1991; Hayashi et al. 1998)

Exon	I		II		III		IV		V
Intron		A		B		C		D	
Length of fragment	58 bp	5.6 kb	636 bp	2.06 kb	271 bp	0.44 kb	148 bp	1.1 kb	1061 bp
Codes for amino acid	–		−19 to 187		188 to 278		278 to 327		328 to 387

bp, base pairs; kb, kilo base pairs

region, two additional AP-2-binding sites, an AP-1-binding site, an AP-3-binding site, a C/EBP-binding site, two HNF-binding sites, and a GATA-1-binding site were found, indicating a significant promoter activity in human HepG2 hepatoma cells (Hayashi and Suzuki 1993; Hayashi et al. 1998). Expression of various deletion constructs revealed that the region from −329 to −209 containing the Sp1-binding sites acts as promoter in HepG2 cells. Furthermore the upstream AP-2-binding site in the region from −452 to −330 acts as strong enhancer, and the downstream region from −176 to −147 was recognized to act as silencer (Hayashi et al. 1998). The stop codon at position 1324 divides exon V in a coding region corresponding to amino acid 328–387 and a 3′-noncoding region of 879 bp containing the polyadenylation site at position 2205 (Meijers and Chung 1991).

Northern blot analysis showed that human PCI mRNA is expressed in the liver, kidney, spleen, pancreas, skeletal muscle, heart, testis, prostate, seminal vesicles, ovary, and skin (Suzuki 2008). PCI is discussed to be a putative tumor suppressor, since loss of PCI expression is found in high-grade prostate tumors (Cao et al. 2003). Furthermore semimalignant serous ovarian borderline tumors show a 5.3-fold higher PCI mRNA expression compared to serous ovarian carcinomas, indicating a potential role for PCI as regulator for benign behavior in ovarian cancer (Sieben et al. 2005).

10.3 Characteristics of the Human PCI Protein

Human PCI cDNA was cloned and sequenced in 1987 indicating that the inhibitor consists of a 19-amino-acid signal peptide and a 387-amino-acid polypeptide chain (Suzuki et al. 1987). The molecular weight was calculated to be 43.8 kDa for the protein and 13.2 kDa for the remaining carbohydrate chains. The high homology of the amino acid sequence of PCI with A1AT (41.6 %), AACT (42.3 %), antithrombin (AT, 26.4 %), and ovalbumin (26.9 %) clearly demonstrated that PCI is a member of the serpin family (Suzuki et al. 1987; Suzuki 1993). Table 10.2 gives an overview of important properties of human PCI.

10.3.1 Glycosylation Sites

PCI is heavily glycosylated. It contains three N-glycosylation sites (see Fig. 10.1a) at positions Asn^{230}, Asn^{243}, and Asn^{319} (Sun et al. 2008). In addition two O-glycosylation sites have been predicted at positions Thr^{20} and Ser^{39} (Suzuki 1993), but further analysis revealed that only Thr^{20} is O-glycosylated and no other potential O-glycosylation sites have been confirmed (Sun et al. 2010).

Analysis of *N*-glycans derived from human urinary PCI (uPCI) revealed the structure of four major N-glycosylation types of heterogeneous, bi-antennary structure. The core unit (see Fig. 10.1b) consists of three branched mannose (Man) residues attached to two *N*-acetylglucosamines (GlcNAc). An additional fucose (Fuc) residue is found in more than 60 % of all *N*-glycans (Izutani et al. 2001). The

Table 10.2 Overview of important properties of human PCI

Property	Criterium	Refs.
Gene location	14q32.1	Billingsley et al. (1993), Meijers and Chung (1991)
Gene length/ gene structure	11.5 kb/5 exons, 4 introns	Meijers and Chung (1991)
Amino acids	406-amino-acid single polypeptide chain (19-amino-acid signal peptide and 387-amino-acid mature protein)	Suzuki et al. (1987)
Cysteins/ disulfide bonds	1/0	Suzuki (1993)
MW total	57 kDa	Suzuki et al. (1983)
MW protein/ carbohydrate	43.8 kDa/13.2 kDa	Suzuki (1993)
N-Glycosylation position	Asn^{230}, Asn^{243}, Asn^{319}	Sun et al. (2008), Suzuki (1993)
O-Glycosylation position	Thr^{20}	Sun et al. (2010), Suzuki (1993)
Reactive site peptide bond P1–P1′	Arg^{354}–Ser^{355}	Suzuki et al. (1987)
Blood plasma concentration	5 μg/mL (90 nM)	Laurell et al. (1992)
Isoelectric point (s), pI	Six pIs, 7.4, 7.6, 7.9, 8.0, 8.4, 8.6	Suzuki et al. (1983)
	4.5 – 6	Suzuki (1993)
Extinction coefficient at 280 nm	$E^{1\ \%}_{1\ \mathrm{cm}} = 14.1$	Suzuki (1993), Suzuki et al. (1983)
Half-life in the circulation:		Laurell et al. (1990)
–PCI protein	23.4 ± 5.7 h	
–APC–PCI complex	19.6 ± 3.1 min	
Optimal heparin concentration	10–100 μg/mL	Pratt and Church (1992)

trimannosyl unit seems to be important for the protease inhibitory property of uPCI as glycosidase-treated uPCI shows decreased inhibitory activity toward thrombin and plasma kallikrein (Izutani et al. 2001). PCI purified from human blood plasma exhibits a slightly different Asn-glycosylation structure compared to uPCI, as the most abundant *N*-glycan is non-fucosylated and carries sialic acid on both ends of the bi-antennary domain.

In addition to bi-antennary glycans, also tri- and tetra-antennary forms are found with molecular weights between 1.9 and 3.8 kDa (Sun et al. 2008). From glycosylation mutants of human PCI, it is known that Asn^{243}-linked glycans stimulate the

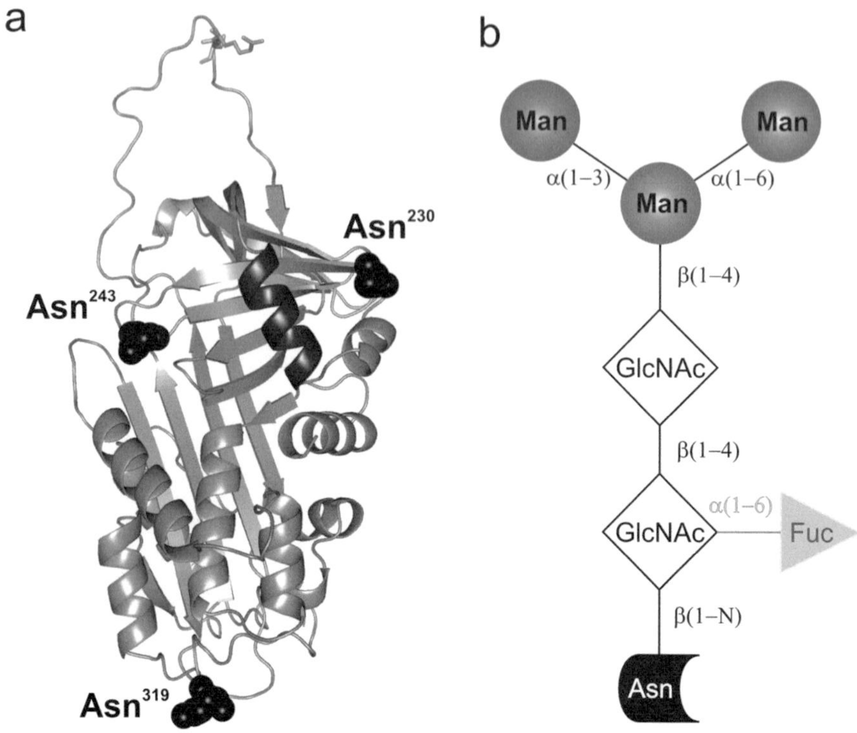

Fig. 10.1 Glycosylation sites of PCI. (**a**) Back view of the structure of native human PCI (PDB 2OL2). The three Asn-glycosylation sites Asn^{230}, Asn^{243}, and Asn^{319} and helix H are in *black*. The O-glycosylation site at position Thr20 cannot be shown, as the crystal structure lacks the amino-terminal part and starts at amino acid 28. Serine protease cleavage site P1–P1′ (Arg–Ser) of the RCL (at the top of the molecule) is shown as *sticks*. The figure was created using PYMOL (http://www.pymol.org/), (**b**) trimannosyl core unit derived from urinary PCI (uPCI). More than 60 % of all *N*-glycans derived from uPCI have a fucose residue attached to the bi-antennary core domain. Fuc, fucose; GlcNAc, *N*-acetylglucosamine; Man, mannose

inhibitory activity of PCI. On the other hand, Asn^{230}-linked glycans downregulate the protease inhibitory property of PCI, and glycans attached to Asn^{319} have no apparent effect on thrombin inhibition (Fujita et al. 2002). As differently N-glycosylated PCI mutants show only slightly increased heparin affinity but significant changes in inhibitory characteristics compared to wtPCI, it is suggested that N-glycosylation affects the conformation of PCI's reactive center, rather than the heparin-binding site (Fujita et al. 2002).

Naturally occurring size heterogeneity of blood-derived PCI can be explained by variations in the length of the N-terminus as well as variations in *N*-glycan structure and occupancy, whereas O-glycosylation does not contribute to the observed size heterogeneity (Sun et al. 2008, 2010). Therefore, purified plasma PCI from individual donors shows at least six bands when analyzed by SDS-PAGE. Conserved microheterogeneity among donors from different ethnicities supports the argument that this heterogeneity is of biological significance. Furthermore the naturally existing PCI variant lacking any glycosylation at position Asn^{243} might be functionally different from fully glycosylated PCI and other variants (Sun et al. 2010), as PCI mutants lacking any Asn^{243} glycans show reduced inhibition of target proteases (Fujita et al. 2002).

10.3.2 Polysaccharide Specificity of PCI

Negatively charged glycosaminoglycans like heparin and heparan sulfate are available at sites of serpin action (Casu and Lindahl 2001) and are known to interact with basic residues of PCI and modulate the inhibition of potential target proteases. In 1992 various polysaccharides were tested and the polyanion specificity of PCI was determined (Pratt and Church 1992). Among the glycosaminoglycans found in vertebrates, only heparin, heparan sulfate, and to a lesser extent chondroitin sulfate A stimulated PCI's inhibitory property toward APC and thrombin. Dermatan sulfate at >1000 μg/mL exhibited no modulation of thrombin inhibition, and only very weak increase in APC inhibition was observed. In addition fucoidan, a marine polysaccharide from brown algae (Li et al. 2008), and phosvitin derived from egg yolk also stimulated protease inhibition by PCI (Pratt and Church 1992).

10.3.3 Inhibitory Properties of PCI

As mentioned above PCI was originally identified as an inhibitor of APC as implicated by its name. However, it was recognized that PCI is a multi-specific inhibitor. In addition to APC (Pratt and Church 1992; Hermans and Stone 1993; Suzuki et al. 1984), human PCI inhibits many serine proteases of the vascular system involved in blood coagulation and fibrinolysis, such as thrombin (Pratt and Church 1992; Hermans and Stone 1993; Suzuki et al. 1984; Radtke et al. 1988), thrombin–thrombomodulin (TM) complex (Rezaie et al. 1995), factor (F)Xa (Pratt and Church 1992; Suzuki et al. 1984; Radtke et al. 1988), FXIa (Meijers et al. 1988), plasma kallikrein (Meijers et al. 1988), urinary plasminogen activator (urokinase, u-PA) (Stief et al. 1987; Radtke et al. 1988; Song et al. 2007; Geiger et al. 1989, 1991), and tissue-PA (t-PA) (Stump et al. 1986), but no inhibition of FXIIa or plasmin is observed (Suzuki 1984). As PCI is an inhibitor of t-PA and u-PA, it is also known as plasminogen activator inhibitor 3 (PAI-3) (Stief et al. 1987; Heeb et al. 1987). Furthermore PCI inhibits tissue kallikrein (Ecke et al. 1992; Luo and Jiang 2006), trypsin (Phillips et al. 1994), chymotrypsin (Pratt and Church 1992), the sperm protease acrosin (Hermans et al. 1994; Elisen et al. 1998a), prostate-specific antigen (PSA) (Christensson and Lilja 1994), and the FXII-like protease hepatocyte growth factor activator (HGFA) (Hayashi et al. 2007). Stable protease inhibitor complexes have also been observed with transmembrane serine proteases, such as matriptase-3 (Szabo et al. 2005), DESC1 (Hobson et al. 2004), and enteropeptidase (Prohaska et al. 2012). Interestingly PCI also inhibits the cysteine protease cathepsin L (Fortenberry et al. 2010), which makes PCI a cross-class inhibitor interacting with both chymotrypsin-like serine proteases and papain-like cysteine proteases.

In the absence of heparin, the thrombin–TM complex is inhibited the fastest with an apparent second-order rate constant of 2.4×10^6 L mol^{-1} s^{-1} (Rezaie et al. 1995). In the presence of optimal heparin concentrations, the sperm protease acrosin is inhibited the fastest with an apparent second-order rate constant of 5.6×10^7 L mol^{-1} s^{-1} (Hermans et al. 1994). Interestingly the inhibition of plasma kallikrein, FXa, and some other proteases by PCI is not enhanced by heparin or other modulators. In the case of tissue kallikrein, the addition of polysaccharides completely blocks the inhibition by PCI (Ecke et al. 1992). Table 10.3 gives an overview of the apparent second-order rate constants of the interaction of PCI with various proteases in the absence and presence of heparin.

The residues around P1–P1′ (potential cleavage site) are important determinants of serpin activity and specificity. Using recombinant PCI mutants, the effects of changing the P1, P2, and P1′ residues on the inhibition of different proteases were studied (Cooper and Church 1995). Changing the P1 residue Arg^{354} to Met results in a much poorer inhibition of thrombin, APC, trypsin, and u-PA. Changing the P2 residue Phe^{353} to Gly enhances the inhibition of thrombin or u-PA, but reduced complex formation with APC or trypsin is observed. Changing the P1′ residue Ser^{355} to Met generates a reactive site comparable to PAI-1, but did not change the inhibitory property of the mutant compared to wtPCI. Therefore, it was concluded that both P1 Arg and P2 Phe but not P1′ Ser are important for the protease specificity of PCI (Phillips et al. 1994).

10.3.4 Modulation of PCI Activity by Phospholipids

PCI is detected in human platelets and megakaryocytes at a concentration of 2.9 ng/10^9 cells and 1.5 ng/10^6 cells, respectively. Upon stimulation of resting platelets, 24–32 % of the total amount of PCI is released from the α-granules and can be detected partly within the surface-connected canalicular system and on the external side of the plasma membrane. Membrane-bound PCI is still active and forms complexes with APC (Nishioka et al. 1998). Therefore, the binding of PCI to phospholipid vesicles and the stimulation of PCI's inhibitory property toward APC, thrombin, and tissue kallikrein in the presence of different phospholipids were studied (Malleier et al. 2007; Nishioka et al. 1998). Phosphatidylcholine (PC) does not show any binding or stimulation of protease inhibition. Experiments with artificially generated phospholipid vesicles revealed that phosphatidylethanolamine (PE) is essential for the interaction with PCI (Nishioka et al. 1998).

In 2007 the interaction of PCI with specific unsaturated and oxidized phospholipids was analyzed in more detail. The used phospholipids had identical fatty acid configuration, carrying palmitic acid at position sn-1 and arachidonic acid at position sn-2 of the glycerol backbone (Malleier et al. 2007). PCI binds with high affinity to oxidized (Ox) PE, phosphatidylserine (PS), and OxPS. Binding is

Table 10.3 Apparent second-order rate constants for the interaction of PCI with various proteases

Protease	EC number	Apparent second-order rate constant L mol^{-1} s^{-1} Without heparin	With heparin	Fold enhancement by heparin	Refs.
Chymotrypsin	3.4.21.1	1.13×10^6	3.38×10^6	3.0	Pratt and Church (1992)
Trypsin	3.4.21.4	n.d.	1.70×10^4	n.a.	Phillips et al. (1994)
Thrombin	3.4.21.5	3.30×10^4	2.00×10^5	6.0	Hermans and Stone (1993)
Thrombin–TM	3.4.21.5	2.40×10^6	n.d.	n.a.	Rezaie et al. (1995)
FXa	3.4.21.6	3.18×10^3	3.50×10^3	1.1	Pratt and Church (1992)
Enteropeptidase	3.4.21.9	4.48×10^4	8.83×10^3	0.2	Prohaska et al. (2012)
Acrosin	3.4.21.10	2.41×10^5	5.60×10^7	232.0	Hermans et al. (1994)
FXIa	3.4.21.27	9.40×10^3	9.10×10^4	10.0	Meijers et al. (1988)
Plasma kallikrein	3.4.21.34	1.10×10^5	1.00×10^5	n.e.	Meijers et al. (1988)
Tissue kallikrein	3.4.21.35	2.30×10^4	No inhibition	n.a.	Ecke et al. (1992)
t-PA	3.4.21.68	Below 10^3	n.d.	n.a.	Stump et al. (1986)
APC	3.4.21.69	8.00×10^2	1.47×10^4	18.0	Hermans and Stone (1993)
u-PA	3.4.21.73	8.00×10^3	9.00×10^4	11.0	Stump et al. (1986)
PSA, KLK3	3.4.21.77	n.d.	n.d.	up to 6.0	Christensson and Lilja (1994)
HGFA	3.4.21.–	7.17×10^4	6.67×10^4	0.9	Hayashi et al. (2007)
Cathepsin L	3.4.22.15	3.00×10^5	1.90×10^5	0.6	Fortenberry et al. (2010)

n.a., not applicable; n.d., not determined; n.e., not enhanced

saturable, and stimulation of APC inhibition by PCI is observed with apparent second-order rate constants of 2.45×10^5 L mol^{-1} s^{-1} for OxPE, 3.10×10^5 L mol^{-1} s^{-1} for PS, and 2.12×10^5 L mol^{-1} s^{-1} for OxPS compared to buffer at 1.60×10^3 L mol^{-1} s^{-1}. The effect of those lipids on PCI's inhibitory property seems to be specific for PCI since thrombin inhibition by AT is not influences at all, compared to a stimulatory effect of OxPE, PS, and OxPS on thrombin inhibition by PCI (Malleier et al. 2007).

10.4 Detection of PCI Antigen in Body Fluids and Tissue Sections

In addition to its expression in various human tissues (e.g., liver, kidney, spleen, pancreas, skeletal muscle, heart, skin, reproductive organs), human PCI is present in many body fluids (e.g., plasma, urine, cerebrospinal fluid, saliva, tears, synovial fluid, graaf follicle fluid, amniotic fluid, breast milk, seminal plasma) and shows a very high concentration (~3–4 μM) in seminal plasma (Laurell et al. 1992).

To examine the pattern of PCI distribution in more detail, the protein was visualized in different human tissues. Analysis of immunostained human artery sections showed diffuse staining of PCI within the intima. Interestingly the signal is rather weak on epithelial cells and primarily localized to the endothelial basement membrane. In addition faint local staining of smooth muscle cells can be observed in the media (Cooper et al. 1996). In normal human adult skin, PCI is mainly expressed in keratinocytes of the epidermis where it might be an important inhibitor of tissue kallikrein (Krebs et al. 1999; Zhang et al. 2007).

As very high PCI levels are found in seminal plasma, a closer investigation of PCI distribution in the male reproductive organs was performed (Laurell et al. 1992). The germinal cell layer in the human testis is evenly stained. Leydig cells also stain positively in contrast to Sertoli cells, where no PCI protein can be detected. In epididymal gland and seminal vesicles, the secretory epithelium shows strong positive staining. In addition the basal cell layer of the prostate next to the stroma cells gives a positive signal, whereas no intraluminal PCI can be detected as compared with sections of the epididymal gland (Laurell et al. 1992).

In the cortex of normal kidney, PCI was found in ~25 % of tubular cross sections. Some differences between the sections have been noted, but usually the epithelial cells showed a strong but diffuse distribution pattern together with small dot-shaped accentuations (Radtke et al. 1994).

In 2010 the embryonic expression of PCI was studied using the mouse model system (Wagenaar et al. 2010). Although in adult rodents, PCI is only present in the reproductive organs, the serpin can be detected during mouse development in different tissues derived from all three germ layers. From embryonic day (ED) 9.5 onward, PCI is detected in the developing myocardium and at later stages in epithelial lineages of the lung, esophagus, nasal and oral cavities, skin, and kidney, in the developing reproductive organs, in the brain, in intervertebral disks of the vertebral column, in interdigital webs and cartilages of the paw, and in the receding notochord (Wagenaar et al. 2010). PCI expression in mouse development is summarized in Table 10.4.

The finding that PCI is present in interdigital webs of the paw as well as in the receding notochord indicates that it might play a role in cell death and apoptosis. Since PCI interacts with PS, it can accumulate on the surface of apoptotic cells at

Table 10.4 Overview of PCI expression pattern in mouse embryonic development (Wagenaar et al. 2010)

Region	ED 9.5–12.5	ED 14.5–16.5	Postnatal	No detection
Cranial region	Developing brain (epithelial cells of the choroid plexus), upper and lower jaws (skin)	Brain (choroid plexus) Nasal and oral cavities (epithelial lining)	–	–
Thoracic region	Myocardium of the heart, esophagus (epithelial lining)	Esophagus (epithelial lining), muscle cells (thorax, intercostal muscles, and diaphragm)	–	PCI is absent from the trachea
Respiratory tract	Embryonic lung (epithelial cells)	Lung (distally localized cuboidal bronchial epithelium)	Late fetal and neonatal lung (terminal bronchial epithelium)	No PCI expression in adult lung
Abdominal region	Reproductive organs, urogenital sinus	Urethra, reproductive organs, skin of urogenital region, kidney (weak in tubular epithelium)	Reproductive organs (Leydig cells)	PCI is not detected in the liver and pancreas
Vertebral column, limbs, and notochord	Developing vertebral column (weak expression), developing notochord	Paw (cartilages, interdigital webs), vertebral column (intervertebral disks), receding notochord	–	–

ED, embryonic day

sites of PS exposure (Malleier et al. 2007). Interestingly in the mouse system PCI is not detected in the pancreas or trachea and is completely absent in the liver. Compared to the human system, where PCI is mainly produced in the liver and secreted into the blood stream, mouse PCI is not found in blood plasma (Uhrin et al. 2000). Although it is obvious that regulation of extracellular matrix proteolysis is an important requirement during morphogenesis, the role of PCI in embryonic development is still unclear.

In postnatal testicular development PCI is absent from Sertoli cells and spermatogonia and restricted to early spermatids from stages V to VIII, elongated spermatids at stages VII–VIII, and Leydig cells (Odet et al. 2004). Male PCI knockout mice showed impaired spermatogenesis and a disturbed blood–testis barrier (Uhrin et al. 2000) and are sterile, but the development of PCI-deficient animals is not affected. This suggests that the functional role of PCI in development of rodents is rather redundant, and its possible functions might be replaced by other factors (Wagenaar et al. 2010; Uhrin et al. 2000).

10.5 Non-inhibitory Functions of PCI

Serpins lacking any inhibitory function are termed non-inhibitory. This means that the serpin is either poorly characterized and studies failed to demonstrate the inhibitory property due to an inappropriate choice of potential proteases or the RCL does not possess the structural requirements to insert into β-sheet A, which is necessary for efficient inhibition (Gettins et al. 1996). Based on the fact that non-inhibitory serpins fulfill important biological functions as hormone carriers (Hammond et al. 1987; Pemberton et al. 1988), storage proteins (e.g., ovalbumin), chaperones (Nagata 1996), or tumor suppressors (Zou et al. 1994; Law et al. 2005), it is assumed that inhibitory serpins might also exhibit non-inhibitory functions.

10.5.1 Interaction of PCI with Retinoic Acid

In 2001 a screening experiment was performed to identify hormones as possible new interaction partners of PCI. Aldosterone, cortisol, estradiol, progesterone, and testosterone belonging to the class of steroid hormones do not associate with PCI. Surprisingly all-*trans*-retinoic acid binds specifically to PCI, and binding to 3H-labeled retinoic acid can be competed by 9-cis-retinoic acid and retinol indicating that PCI also interacts with other members of the retinoid family (Jerabek et al. 2001). The interaction of PCI with retinoic acid can be observed in the purified system as well as in seminal plasma. The binding was dose dependent and reveals an apparent K_D of 2.43 μM with 0.8 binding sites per PCI molecule, suggesting a molar ratio of 1:1. Interestingly the inhibitory function of PCI is not altered in the presence of retinoids, and proteolytic cleavage of the RCL has no effect on retinoic acid binding, indicating that the RCL is not involved in binding of PCI to retinoic acid (Jerabek et al. 2001). However, the role of PCI on retinoid-dependent signaling is still unclear, and further experiments are needed.

10.5.2 Antimicrobial Activity of PCI

It has been shown previously that heparin-binding peptides derived from human plasma proteins show antimicrobial activity against bacteria and seem to play a role in early immune response (Andersson et al. 2004). The interaction of the antimicrobial peptide with the pathogen is mediated via the heparin-binding site, as addition of heparin blocks the binding to bacteria and abolishes the antibacterial effects (Rydengard et al. 2007). Binding of antimicrobial peptides to the membrane of bacteria often causes membrane disruption leading to cytosolic leakage and cell death (Schultz and Weiss 2007). The ability of PCI to interact with phospholipids and membranes seems to account for its antimicrobial activity toward bacteria. In 2009 the antibacterial effects of PCI and a 20-amino-acid-derived peptide corresponding to the heparin-binding H-helix region (SEKTLRKWLKMFKKRQLELY) of PCI were investigated in more detail

(Malmström et al. 2009). It could be shown that circulating PCI displays broad antimicrobial activity. This is independent from RCL cleavage or complex formation, since the treatment of PCI with a variety of proteases does not alter its antibacterial property. Interestingly, PCI is able to perforate the cell membrane of pathogens mediating cell death. Furthermore, during bacterial infection, accumulation of PCI on the surface of platelets can be observed. Hence, this might be a mechanism to locally increase PCI concentrations at sites of bacterial invasion (Malmström et al. 2009).

10.5.3 Internalization of Cell-Penetrating PCI by Cells

As shown with the leukocytic HL-60 cell line, PCI is able to enter into life cells. 10 min after incubation of cells with PCI, the serpin can be detected in the submembranous region. Time-dependent accumulation in the perinuclear space and the intranuclear compartments as well as in isolated cytosolic and nuclear fractions can be observed (Baumgärtner et al. 2007). Furthermore 30 min after the injection of PCI into the tail vain of mice, the protein is detected in the nuclei of granulocyte fractions prepared from the murine blood. PCI mutants lacking the basic residues of the H-helix cannot enter the nucleus, suggesting an important role of the H-helix region for the translocation of PCI across cellular membranes. Furthermore, radioactively labeled PCI is taken up efficiently both at 37 °C and at 6 °C by neutrophils, indicating an endocytosis-independent uptake mechanism (Baumgärtner et al. 2007). In addition PCI acts as a specific lipid transferase for phosphatidylethanolamine (PE) confirming the essential role of PE for the cellular uptake of PCI. The mechanism used by PCI to enter cells seems to be dependent on the contribution of PE, since duramycin (a specific interaction partner of PE) abolishes the cell entrance of PCI at 37 °C as well as at 6 °C (Baumgärtner et al. 2007).

In addition it was shown that the A^{+}-helix of PCI (His^{1}–Arg^{11}) exhibits cell-penetrating properties, and removal of the N-terminal A^{+}-helix prevents internalization of PCI into Jurkat T cells. This suggests that internalization and subsequent intracellular functions of PCI can be regulated by proteases (e.g., testisin) via removal of PCI's N-terminus (Yang et al. 2015a).

Till now, several proteins and peptides are known that show the ability to translocate through cellular membranes. These cell-penetrating peptides (CPPs) or protein transduction domains (PTDs) are rich in basic amino acids. Trans-activating transcriptional activator (TAT) from HIV-1 (Frankel and Pabo 1988), homeodomain of antennapedia (a transcription factor from *Drosophila*) (Joliot et al. 1991), and the herpes virus protein VP22 (Elliott and O'Hare 1997) are only some examples of proteins containing such sequences. Today CPPs are already used to deliver cargo (e.g., proteins, DNA, antibodies, nanoparticles, liposomes) into cells via endocytosis-independent pathways (Madani et al. 2011).

10.5.4 Impact of PCI on Phospholipid-Dependent Processes

As mentioned above, certain phospholipids show the ability to modulate the inhibitory property of PCI and/or are important mediators for the internalization of the serpin by cells. In addition it was shown that the direct interaction of PCI with phosphoinositides modulates the activity of the phosphoinositide-specific 5-phosphatase SHIP2 (Wahlmüller et al. 2014). The potential ability of PCI to influence intracellular lipid levels might be an underestimated mechanism how PCI could directly alter intracellular signaling cascades. This is supported by the finding that intra- and extracellular PCI, in addition to kallistatin (SerpinA4) (Shen et al. 2010) and vaspin (SerpinA12) (Jung et al. 2011), is able to activate the PI3-kinase-dependent protein kinase AKT (Wahlmüller et al. 2014).

In addition it was shown that the ability of PCI to bind to phosphatidylserine (PS) leads to a decrease of phagocytosis of PS-exposing cells (Rieger et al. 2014). Since human PCI is present in a variety of tissues and secretions, it might act as protective factor for cells exposing PS.

10.6 Animal Models

PCI shows different expression patterns in different species. In humans (Suzuki et al. 1987), monkeys (Radtke et al. 1995), and cattle (Yuasa et al. 2000), PCI mRNA is at least found in the liver, kidney, and testis. In rabbits, PCI mRNA is only found in the liver and testis. Interestingly, in chicken, PCI is expressed exclusively in the liver, and in rodents (Zechmeister-Machhart et al. 1996, 1997; Wakita et al. 1998), PCI is only expressed in the reproductive organs, suggesting that PCI plays distinct roles in different animals.

In general the preferred animal model system seems to be the mouse, because of cost efficiencies, availability of reagents, mutants, and fast reproduction rates. The organization of the mouse PCI (mPCI) gene resembles the human one, as it consists of five exons and four introns. However, the total gene length is much smaller in mice (~5 kb) due to differences in intron size. Comparison of deduced amino acid sequences of human and mPCI reveals that RCL, hinge regions, and the heparin-binding site are highly conserved (Zechmeister-Machhart et al. 1997).

As already mentioned above, mPCI expression seems to be restricted exclusively to the reproductive organs and is detected in the testis, epididymis, seminal vesicle, prostate, and ovary (Zechmeister-Machhart et al. 1996). In some animals, trace amounts of mPCI mRNA can also be discovered in the liver (e.g., 30 % of investigated mice), kidney, lung, adipose tissue, and brain (Uhrin et al. 2000). These findings might be explained due to sex-specific regulation of PCI expression in the mouse system (Wong et al. 2008). Localization of PCI in the testis of adult mice shows signals in pachytene spermatocytes and elongated spermatids. Interestingly Sertoli cells as well as spermatogonia do not show any PCI expression in adult mice, and PCI expression in Leydig cells may be age related (Uhrin et al. 2007). Another publication states that 3-week-old mice show PCI expression in Leydig cells and in the immortalized Leydig cell line mLTC-1, as well as in Sertoli cells. Furthermore it is suggested that chorionic gonadotropin acts as a negative regulator for mPCI expression in mLTC-1 cells (Odet et al. 2006). Knockout of the Sertoli cell-specific androgen receptor leads to a downregulation of testicular mPCI expression of more than twofold. Furthermore mPCI expression is significantly decreased in wild-type mice treated with testosterone or the antiandrogen flutamide (Denolet et al. 2006). In summary, testis-specific expression of PCI in mice is still controversially discussed, and the potential regulation by gonadal hormones has to be clarified.

Between 2000 and 2004, three different mouse models have been established. To study the effect of circulating PCI in the mouse, a transgenic strain has been created expressing human PCI mRNA in mouse hepatocytes using the liver-specific albumin promoter region. The plasma concentration in heterozygous animals is 3–5 μg/mL and doubles in homozygous mice, where it reaches 10 μg/mL (Wagenaar et al. 2000). Uhrin et al. generated homozygous PCI-deficient ($PCI^{-/-}$) mice by homologous recombination (Uhrin et al. 2000). These mice show no abnormalities in birth, growth, or survival, but male $mPCI^{-/-}$ mice are infertile. The blood–testis barrier is disrupted, the seminiferous tubules are stuffed with apoptotic cells, and malformation of sperm is observed (Uhrin et al. 2000). Interestingly the inactivation of t-PA, u-PA, or acrosin in $PCI^{-/-}$ mice does not restore fertility (Uhrin et al. 2007). Therefore, it can be assumed that testicular PCI might inactivate yet unidentified proteases or contribute to the generation of bioactive peptides or proteins. The third and latest transgenic mouse model was created in Japan introducing the human PCI gene into the mouse system. In addition to mPCI expression, these mice exhibit a similar expression pattern of human PCI as observed in man, although the plasma levels of human PCI in these transgenic mice are about four times higher compared to the human system and reach ~18 μg/mL (Hayashi et al. 2004). Interestingly the levels of human PCI in male animals are slightly higher than in females, which is in line with the observation that the expression of coagulation factors and inhibitors is dependent on the secretion of sex-specific growth hormone (Wong et al. 2008).

Rat PCI protein shows 85.7 % homology with mouse and 62.2 % with human PCI, and strong expression in seminal vesicle and moderate expression in testis are observed (Wakita et al. 1998). In contrast to the mouse system, rat PCI mRNA is absent from Leydig cells and also not present in pachytene, round, and elongated spermatids. In the rat testis, PCI seems to be predominantly expressed by Sertoli cells (Anway et al. 2005). The heparin-binding site and the RCL region show high homology compared to mouse and human PCI. Interestingly only one N-glycosylation site (Asn^{248}) and two O-glycosylation sites (Ser^{57}, Ser^{373}) are proposed. During development the levels of rat PCI mRNA in seminal vesicles correspond to plasma testosterone levels, and castration of rats leads to a decreased mRNA expression (Wakita et al. 1998).

10.7 PCI in Health and Disease

As mentioned above it was soon recognized that PCI is a multi-specific serpin and inhibits a variety of other serine proteases. Furthermore it can be assumed that PCI exhibits also non-inhibitory functions. However, the exact physiological role of PCI remains to be investigated. Up to now, not enough information is available concerning PCI mutations in man or PCI deficiency-related diseases of patients, making it hard to speculate on the potential physiological and pathophysiological aspects of that serpin.

10.7.1 PCI and the Clotting System

PCI directly inhibits procoagulant proteases like thrombin (Suzuki et al. 1984), FXa (Pratt and Church 1992), and FXIa (Meijers et al. 1988) as well as anticoagulant proteases like APC (Hermans and Stone 1993) and the thrombin–thrombomodulin complex (Elisen et al. 1998b). Based on second-order rate constants and physiological concentrations, PCI is considered to be the main inhibitor of APC in the vascular system (Heeb et al. 1989). Interestingly, plasma PCI levels can be used as a risk marker for certain vascular diseases, as significantly elevated levels of active PCI are found more often in male survivors of myocardial infarction (Carroll et al. 1997).

In a case–control study of venous thrombosis (Leiden Thrombophilia Study) with more than 470 patients as well as age- and sex-matched controls, increasing PCI levels are identified to be a mild risk factor for the occurrence of deep venous thrombosis (Meijers et al. 2002). Furthermore the levels of PCI in complex with proteases can be an interesting and sensitive marker for the activation of the clotting system or the fibrinolytic pathway. For example, in disseminated intravascular coagulation, high levels of the APC–PCI complex have been detected (Espana et al. 1990). Patients with aortic aneurysms also show elevated APC–PCI complex levels as compared to the control cohort. Therefore, it is possible to use the measurement of circulating APC–PCI complex levels as screening tool to predict the occurrence of aortic aneurysms (Kölbel et al. 2006; Stenflo et al. 2006).

10.7.2 Cancer and PCI

To date only a few examples of PCI expressing cell lines are known, including the hepatocellular carcinoma (HepG2) cell line (Fair and Marlar 1986; Morito et al. 1985), the epithelial kidney tumor cell line TCL-598 (Priglinger et al. 1994), the human epidermoid cell line A431 (Krebs et al. 1999), as well as the androgen-dependent malignant prostatic cell line LNCaP and the androgen-independent cell lines PC-3 and DU145 (Cao et al. 2003). A mouse xenograft model of prostate cancer confirms those findings, as PCI was expressed in androgen-dependent tumor tissue, as well as 6 days after castration in androgen-independent recurrent tumors (Glasscock et al. 2005). Interestingly, the immunostaining pattern for PCI in human prostate tumor cells differed from that of comparable serpins like A1AT and AT. For patients who underwent radical prostatectomy, the PCI level in serum is identified (among others) as biomarker to predict cancer recurrence (Rosenzweig et al. 2009; Razavi et al. 2013). Furthermore it can be shown that in higher Gleason grade tumors of the prostate, the number of PCI positive cells decreases, which might promote tumor cell invasiveness (Cao et al. 2003). In ovarian tumors, similar results have been obtained, as PCI is differently expressed in serous borderline tumors compared to carcinomas. Therefore, PCI expression might be a key regulator for the benign behavior of ovarian cancer (Sieben et al. 2005; Bijsmans et al. 2010). PCI levels in extracts from renal cell carcinoma tissues have also been significantly lower as compared to extracts from normal renal tissue, and it is shown that u-PA-dependent tumor invasiveness can be regulated by PCI in vitro (Wakita et al. 2004).

The influence of wtPCI, the inactive N-terminal fragment of cleaved PCI, as well as a reactive site mutant form of PCI (P1 Arg^{354} replaced by Ala) on tumor growth, metastasis, and angiogenesis using a breast cancer model was studied in 2007. VEGF-dependent angiogenesis, growth, and metastatic potential of human breast cancer cells are significantly inhibited by wild-type PCI as well as inactive mutants. Interestingly the invasiveness, which is probably protease dependent, is only inhibited by intact PCI (Asanuma et al. 2007). This is in line with previously published data, as the anti-angiogenic activity of AT and kallistatin is found to be rather dependent on the heparin-binding property than on the inhibitory potential (Zhang et al. 2005).

The mechanism for the regulation of tumor cell growth, invasion, and metastasis by PCI is still not fully understood. One possible explanation might be the inhibition of tumor-associated proteases like u-PA (Castelló et al. 2007), PSA (Christensson and Lilja 1994; Espana et al. 1991), prostatic kallikrein (Deperthes et al. 1995), or cathepsin L (Fortenberry et al. 2010). Recently it was reported that PCI directly binds to fibronectin and influences fibronectin–integrin signaling in hepatocellular carcinoma cell lines which could negatively influence tumor cell migration and metastatic potential (Jing et al. 2014). To complete our understanding, more data are needed concerning the influence of non-inhibitory functions of PCI on the regulation of tumor cell growth and cancer development.

10.7.3 Influence of PCI on Fertilization

PCI is detected in the reproductive organs, and very high levels of PCI (~3–4 μM) are found in human seminal plasma (Laurell et al. 1992). The secretory epithelium of the seminal vesicles seems to be the main source of human seminal plasma PCI. Interestingly the activity of PCI in seminal plasma incubated at 37 °C directly after ejaculation decreases rapidly to ~50 % after 20 min (Espana et al. 1991). The moiety of PCI in semen is found in complex with PSA (Christensson and Lilja 1994; Espana et al. 1991, 1993a). Therefore, PCI might be a protective factor concerning the degradation of semenogelin-1 and semenogelin-2 in human semen (Kise et al. 1996). Many other proteases that are present in seminal plasma like u-PA, t-PA (Espana et al. 1993b; He et al. 1999), tissue kallikrein (Espana et al. 1995), human glandular kallikrein (Deperthes et al. 1995), and acrosin (Hermans et al. 1994; Zheng et al. 1994) are complexed by PCI. Interestingly PCI levels in azoospermic seminal fluids from patients with dysfunctional seminal vesicles are ~ tenfold lower compared to the average level from samples of healthy males (Laurell et al. 1992). In the mouse system, the knockout of the PCI gene leads to infertility, a disturbed blood–testis barrier, and malformed sperm (Uhrin et al. 2000). The additional knockout of either u-PA or t-PA in PCI-deficient mice does not affect abnormal spermatogenesis, and those mice are still infertile (Uhrin et al. 2007). Furthermore cleaved and intact human PCI interferes dose dependently with binding and penetration of human sperm to zona-free hamster oocytes. The observed effect of PCI is specific as control experiments with two other serpins (A1AT and AT) do not influence the ability of sperm to penetrate a zona-free oocyte (Espana et al. 1999; Moore et al. 1993).

In 2010 fifteen different human *SERPINA5* gene single-nucleotide polymorphism (SNP) variants were identified among couples suffering from total fertilization failure after in vitro fertilization. Most of these variants are already available at the SNP database of the NCBI. Interestingly a heterozygous A/G transition in the 3′-noncoding region of the PCI gene at position 1389 (rs2069990) is detected significantly more often (5 of 46 vs. 0 of 51) in men with fertilization failure compared to controls (Bungum et al. 2010). These data suggest that PCI influences mammalian reproduction at several steps, which might be explained by its inhibitory property. In addition non-inhibitory functions in the regulation of fertilization can be assumed (Espana et al. 2007).

Testis lysates from wild-type and $PCI^{-/-}$ mice have been compared using a 2D-DIGE approach. Loss of PCI apparently leads to a tenfold upregulation of prostaglandin reductase 1 (PTGR1). In addition also changes in the isoform pattern of serpinA3K and serpinA1C are observed. This indicates that PCI strongly influences the prostaglandin metabolism in the testes which might provide new aspects on the role of PCI in male reproduction (Yang et al. 2015b).

10.7.4 PCI as Protective Factor Against HIV Infection

Recently it was reported that higher PCI levels as compared to control groups are detected in the cervicovaginal fluid of individuals which are frequently exposed to HIV but remain seronegative (Van Raemdonck et al. 2014). At the same time proteases like myeloblastin are downregulated. Therefore, the authors speculate that a reduction in total protease activity (e.g., by upregulation of serpins and other protease inhibitors and/or downregulation of proteases) is associated with resistance to HIV. However, the exact mechanism how this might be achieved is still unknown.

References

Andersson E, Rydengard V, Sonesson A, Morgelin M, Bjorck L, Schmidtchen A (2004) Antimicrobial activities of heparin-binding peptides. Eur J Biochem/FEBS 271(6):1219–1226

Anway MD, Show MD, Zirkin BR (2005) Protein C inhibitor expression by adult rat Sertoli cells: effects of testosterone withdrawal and replacement. J Androl 26(5):578–585

Asanuma K, Yoshikawa T, Hayashi T, Akita N, Nakagawa N, Hamada Y, Nishioka J, Kamada H, Gabazza EC, Ido M, Uchida A, Suzuki K (2007) Protein C inhibitor inhibits breast cancer cell growth, metastasis and angiogenesis independently of its protease inhibitory activity. Int J Cancer 121(5):955–965

Baumgärtner P, Geiger M, Zieseniss S, Malleier J, Huntington JA, Hochrainer K, Bielek E, Stoeckelhuber M, Lauber K, Scherfeld D, Schwille P, Waldele K, Beyer K, Engelmann B (2007) Phosphatidylethanolamine critically supports internalization of cell-penetrating protein C inhibitor. J Cell Biol 179(4):793–804

Bijsmans IT, Smits KM, de Graeff P, Wisman GB, van der Zee AG, Slangen BF, de Bruine AP, van Engeland M, Sieben NL, Van de Vijver KK (2010) Loss of SerpinA5 protein expression is associated with advanced-stage serous ovarian tumors. Mod Pathol 24(3):463–470

Billingsley GD, Walter MA, Hammond GL, Cox DW (1993) Physical mapping of four serpin genes: alpha 1-antitrypsin, alpha 1-antichymotrypsin, corticosteroid-binding globulin, and protein C inhibitor, within a 280-kb region on chromosome I4q32.1. Am J Hum Genet 52 (2):343–353

Bungum M, Giwercman A, Bungum L, Humaidan P, Rastkhani H, Giwercman YL (2010) Polymorphisms in the protein C inhibitor gene in in vitro fertilization failure. Fertil Steril 93 (1):277–279

Cao Y, Becker C, Lundwall A, Christensson A, Gadaleanu V, Lilja H, Bjartell A (2003) Expression of protein C inhibitor (PCI) in benign and malignant prostatic tissues. Prostate 57 (3):196–204

Carroll VA, Griffiths MR, Geiger M, Merlo C, Furlan M, Lammle B, Binder BR (1997) Plasma protein C inhibitor is elevated in survivors of myocardial infarction. Arterioscler Thromb Vasc Biol 17(1):114–118

Castelló R, Landete JM, España F, Vázquez C, Fuster C, Almenar SM, Ramón LA, Radtke KP, Estellés A (2007) Expression of plasminogen activator inhibitors type 1 and type 3 and urokinase plasminogen activator protein and mRNA in breast cancer. Thromb Res 120 (5):753–762

Casu B, Lindahl U (2001) Structure and biological interactions of heparin and heparan sulfate. Adv Carbohydr Chem Biochem 57:159–206

Christensson A, Lilja H (1994) Complex formation between protein C inhibitor and prostate-specific antigen in vitro and in human semen. Eur J Biochem/FEBS 220(1):45–53

Cooper ST, Church FC (1995) Reactive site mutants of recombinant protein C inhibitor. Biochim Biophys Acta 1246(1):29–33

Cooper ST, Neese LL, DiCuccio MN, Liles DK, Hoffman M, Church FC (1996) Vascular localization of the heparin-binding serpins antithrombin, heparin cofactor II, and protein C inhibitor. Clin Appl Thromb Hemost 2(3):185–191. doi:10.1177/107602969600200307

Denolet E, De Gendt K, Allemeersch J, Engelen K, Marchal K, Van Hummelen P, Tan KA, Sharpe RM, Saunders PT, Swinnen JV, Verhoeven G (2006) The effect of a sertoli cell-selective knockout of the androgen receptor on testicular gene expression in prepubertal mice. Mol Endocrinol 20(2):321–334

Deperthes D, Chapdelaine P, Tremblay RR, Brunet C, Berton J, Hebert J, Lazure C, Dube JY (1995) Isolation of prostatic kallikrein hK2, also known as hGK-1, in human seminal plasma. Biochim Biophys Acta 1245(3):311–316

Ecke S, Geiger M, Resch I, Jerabek I, Sting L, Maier M, Binder BR (1992) Inhibition of tissue kallikrein by protein C inhibitor. Evidence for identity of protein C inhibitor with the kallikrein binding protein. J Biol Chem 267(10):7048–7052

Elisen MG, van Kooij RJ, Nolte MA, Marquart JA, Lock TM, Bouma BN, Meijers JC (1998a) Protein C inhibitor may modulate human sperm-oocyte interactions. Biol Reprod 58 (3):670–677
Elisen MG, von dem Borne PA, Bouma BN, Meijers JC (1998b) Protein C inhibitor acts as a procoagulant by inhibiting the thrombomodulin-induced activation of protein C in human plasma. Blood 91(5):1542–1547
Elliott G, O'Hare P (1997) Intercellular trafficking and protein delivery by a herpesvirus structural protein. Cell 88(2):223–233
Espana F, Vicente V, Tabernero D, Scharrer I, Griffin JH (1990) Determination of plasma protein C inhibitor and of two activated protein C-inhibitor complexes in normals and in patients with intravascular coagulation and thrombotic disease. Thromb Res 59(3):593–608
Espana F, Gilabert J, Estelles A, Romeu A, Aznar J, Cabo A (1991) Functionally active protein C inhibitor/plasminogen activator inhibitor-3 (PCI/PAI-3) is secreted in seminal vesicles, occurs at high concentrations in human seminal plasma and complexes with prostate-specific antigen. Thromb Res 64(3):309–320
Espana F, Sanchez-Cuenca J, Vera CD, Estelles A, Gilabert J (1993a) A quantitative ELISA for the measurement of complexes of prostate-specific antigen with protein C inhibitor when using a purified standard. J Lab Clin Med 122(6):711–719
Espana F, Estelles A, Fernandez PJ, Gilabert J, Sanchez-Cuenca J, Griffin JH (1993b) Evidence for the regulation of urokinase and tissue type plasminogen activators by the serpin, protein C inhibitor, in semen and blood plasma. Thromb Haemost 70(6):989–994
Espana F, Fink E, Sanchez-Cuenca J, Gilabert J, Estelles A, Witzgall K (1995) Complexes of tissue kallikrein with protein C inhibitor in human semen and urine. Eur J Biochem/FEBS 234 (2):641–649
Espana F, Sanchez-Cuenca J, Fernandez PJ, Gilabert J, Romeu A, Estelles A, Royo M, Muller CH (1999) Inhibition of human sperm-zona-free hamster oocyte binding and penetration by protein C inhibitor. Andrologia 31(4):217–223
Espana F, Navarro S, Medina P, Zorio E, Estelles A (2007) The role of protein C inhibitor in human reproduction. Semin Thromb Hemost 33(1):41–45
Fair DS, Marlar RA (1986) Biosynthesis and secretion of factor VII, protein C, protein S, and the Protein C inhibitor from a human hepatoma cell line. Blood 67(1):64–70
Fortenberry YM, Brandal S, Bialas RC, Church FC (2010) Protein C inhibitor regulates both cathepsin L activity and cell-mediated tumor cell migration. Biochim Biophys Acta 1800 (6):580–590
Frankel AD, Pabo CO (1988) Cellular uptake of the tat protein from human immunodeficiency virus. Cell 55(6):1189–1193
Fujita M, Izutani W, Takahashi K, Nishizawa K, Shirono H, Koga J (2002) Role of each Asn-linked glycan in the anticoagulant activity of human protein C inhibitor. Thromb Res 105(1):95–102
Geiger M, Huber K, Wojta J, Stingl L, Espana F, Griffin JH, Binder BR (1989) Complex formation between urokinase and plasma protein C inhibitor in vitro and in vivo. Blood 74(2):722–728
Geiger M, Priglinger U, Griffin JH, Binder BR (1991) Urinary protein C inhibitor. Glycosaminoclycans synthesized by the epithelial kidney cell line TCL-598 enhance its interaction with urokinase. J Biol Chem 266(18):11851–11857
Gettins PGW, Patston PA, Olson ST (1996) Serpins: structure, function and biology. Springer, New York, NY
Glasscock LN, Rehault SM, Gregory CW, Cooper ST, Jackson TP, Hoffman M, Church FC (2005) Protein C inhibitor (plasminogen activator inhibitor-3) expression in the CWR22 prostate cancer xenograft. Exp Mol Pathol 79(1):23–32
Hammond GL, Smith CL, Goping IS, Underhill DA, Harley MJ, Reventos J, Musto NA, Gunsalus GL, Bardin CW (1987) Primary structure of human corticosteroid binding globulin, deduced from hepatic and pulmonary cDNAs, exhibits homology with serine protease inhibitors. Proc Natl Acad Sci USA 84(15):5153–5157

Hayashi T, Suzuki K (1993) Gene organization of human protein C inhibitor, a member of SERPIN family proteins encoded in five exons. Int J Hematol 58(3):213–224
Hayashi T, Usui M, Nishioka J, Zhang ZX, Suzuki K (1998) Regulation of the human protein C inhibitor gene expression in HepG2 cells: role of Sp1 and AP2. Biochem J 332(Pt 2):573–582
Hayashi T, Nishioka J, Kamada H, Asanuma K, Kondo H, Gabazza EC, Ido M, Suzuki K (2004) Characterization of a novel human protein C inhibitor (PCI) gene transgenic mouse useful for studying the role of PCI in physiological and pathological conditions. J Thromb Haemost 2 (6):949–961
Hayashi T, Nishioka J, Nakagawa N, Kamada H, Gabazza EC, Kobayashi T, Hattori A, Suzuki K (2007) Protein C inhibitor directly and potently inhibits activated hepatocyte growth factor activator. J Thromb Haemost 5(7):1477–1485
He S, Lin YL, Liu YX (1999) Functionally inactive protein C inhibitor in seminal plasma may be associated with infertility. Mol Hum Reprod 5(6):513–519
Heeb MJ, Espana F, Geiger M, Collen D, Stump DC, Griffin JH (1987) Immunological identity of heparin-dependent plasma and urinary protein C inhibitor and plasminogen activator inhibitor-3. J Biol Chem 262(33):15813–15816
Heeb MJ, Espana F, Griffin JH (1989) Inhibition and complexation of activated protein C by two major inhibitors in plasma. Blood 73(2):446–454
Hermans JM, Stone SR (1993) Interaction of activated protein C with serpins. Biochem J 295 (Pt 1):239–245
Hermans JM, Jones R, Stone SR (1994) Rapid inhibition of the sperm protease acrosin by protein C inhibitor. Biochemistry 33(18):5440–5444
Hobson JP, Netzel-Arnett S, Szabo R, Rehault SM, Church FC, Strickland DK, Lawrence DA, Antalis TM, Bugge TH (2004) Mouse DESC1 is located within a cluster of seven DESC1-like genes and encodes a type II transmembrane serine protease that forms serpin inhibitory complexes. J Biol Chem 279(45):46981–46994
Izutani W, Fujita M, Nishizawa K, Koga J (2001) The trimannosyl cores of N-glycans are important for the procoagulant protease-inhibitory activity of urinary protein C inhibitor. Thromb Res 104(1):65–74
Jerabek I, Zechmeister-Machhart M, Binder BR, Geiger M (2001) Binding of retinoic acid by the inhibitory serpin protein C inhibitor. Eur J Biochem/FEBS 268(22):5989–5996
Jing Y, Jia D, Wong CM, Oi-Lin Ng I, Zhang Z, Liu L, Wang Q, Zhao F, Li J, Yao M, Wu X, He X (2014) SERPINA5 inhibits tumor cell migration by modulating the fibronectin-integrin beta1 signaling pathway in hepatocellular carcinoma. Mol Oncol 8(2):366–377
Joliot A, Pernelle C, Deagostini-Bazin H, Prochiantz A (1991) Antennapedia homeobox peptide regulates neural morphogenesis. Proc Natl Acad Sci USA 88(5):1864–1868
Jung CH, Lee WJ, Hwang JY, Seol SM, Kim YM, Lee YL, Park JY (2011) Vaspin protects vascular endothelial cells against free fatty acid-induced apoptosis through a phosphatidylinositol 3-kinase/Akt pathway. Biochem Biophys Res Commun 413(2):264–269
Kise H, Nishioka J, Kawamura J, Suzuki K (1996) Characterization of semenogelin II and its molecular interaction with prostate-specific antigen and protein C inhibitor. Eur J Biochem/FEBS 238(1):88–96
Kölbel T, Strandberg K, Mattiasson I, Stenflo J, Lindblad B (2006) Activated protein C-protein C inhibitor complex: a new biological marker for aortic aneurysms. J Vasc Surg 43(5):935–939
Krebs M, Uhrin P, Vales A, Prendes-Garcia MJ, Wojta J, Geiger M, Binder BR (1999) Protein C inhibitor is expressed in keratinocytes of human skin. J Invest Dermatol 113(1):32–37
Laurell M, Stenflo J, Carlson TH (1990) Turnover of *I-protein C inhibitor and *I-alpha 1-antitrypsin and their complexes with activated protein C. Blood 76(11):2290–2295
Laurell M, Christensson A, Abrahamsson PA, Stenflo J, Lilja H (1992) Protein C inhibitor in human body fluids. Seminal plasma is rich in inhibitor antigen deriving from cells throughout the male reproductive system. J Clin Invest 89(4):1094–1101
Law RH, Irving JA, Buckle AM, Ruzyla K, Buzza M, Bashtannyk-Puhalovich TA, Beddoe TC, Nguyen K, Worrall DM, Bottomley SP, Bird PI, Rossjohn J, Whisstock JC (2005) The high

resolution crystal structure of the human tumor suppressor maspin reveals a novel conformational switch in the G-helix. J Biol Chem 280(23):22356–22364

Li B, Lu F, Wei X, Zhao R (2008) Fucoidan: structure and bioactivity. Molecules 13 (8):1671–1695

Luo LY, Jiang W (2006) Inhibition profiles of human tissue kallikreins by serine protease inhibitors. Biol Chem 387(6):813–816

Madani F, Lindberg S, Langel U, Futaki S, Graslund A (2011) Mechanisms of cellular uptake of cell-penetrating peptides. J Biophys 2011:414729

Malleier JM, Oskolkova O, Bochkov V, Jerabek I, Sokolikova B, Perkmann T, Breuss J, Binder BR, Geiger M (2007) Regulation of protein C inhibitor (PCI) activity by specific oxidized and negatively charged phospholipids. Blood 109(11):4769–4776

Malmström E, Mörgelin M, Malmsten M, Johansson L, Norrby-Teglund A, Shannon O, Schmidtchen A, Meijers JC, Herwald H (2009) Protein C inhibitor--a novel antimicrobial agent. PLoS Pathog 5(12), e1000698

Marlar RA, Griffin JH (1980) Deficiency of protein C inhibitor in combined factor V/VIII deficiency disease. J Clin Invest 66(5):1186–1189

Meijers JC, Chung DW (1991) Organization of the gene coding for human protein C inhibitor (plasminogen activator inhibitor-3). Assignment of the gene to chromosome 14. J Biol Chem 266(23):15028–15034

Meijers JC, Kanters DH, Vlooswijk RA, van Erp HE, Hessing M, Bouma BN (1988) Inactivation of human plasma kallikrein and factor XIa by protein C inhibitor. Biochemistry 27 (12):4231–4237

Meijers JC, Marquart JA, Bertina RM, Bouma BN, Rosendaal FR (2002) Protein C inhibitor (plasminogen activator inhibitor-3) and the risk of venous thrombosis. Br J Haematol 118 (2):604–609

Moore A, Penfold LM, Johnson JL, Latchman DS, Moore HD (1993) Human sperm-egg binding is inhibited by peptides corresponding to core region of an acrosomal serine protease inhibitor. Mol Reprod Dev 34(3):280–291

Morito F, Saito H, Suzuki K, Hashimoto S (1985) Synthesis and secretion of protein C inhibitor by the human hepatoma-derived cell line, Hep G2. Biochim Biophys Acta 844(2):209–215

Nagata K (1996) Hsp47: a collagen-specific molecular chaperone. Trends Biochem Sci 21 (1):22–26

Nishioka J, Ning M, Hayashi T, Suzuki K (1998) Protein C inhibitor secreted from activated platelets efficiently inhibits activated protein C on phosphatidylethanolamine of platelet membrane and microvesicles. J Biol Chem 273(18):11281–11287

Odet F, Guyot R, Leduque P, Le Magueresse-Battistoni B (2004) Evidence for similar expression of protein C inhibitor and the urokinase-type plasminogen activator system during mouse testis development. Endocrinology 145(3):1481–1489

Odet F, Verot A, Le Magueresse-Battistoni B (2006) The mouse testis is the source of various serine proteases and serine proteinase inhibitors (SERPINs): Serine proteases and SERPINs identified in Leydig cells are under gonadotropin regulation. Endocrinology 147(9):4374–4383

Pemberton PA, Stein PE, Pepys MB, Potter JM, Carrell RW (1988) Hormone binding globulins undergo serpin conformational change in inflammation. Nature 336(6196):257–258

Phillips JE, Cooper ST, Potter EE, Church FC (1994) Mutagenesis of recombinant protein C inhibitor reactive site residues alters target proteinase specificity. J Biol Chem 269 (24):16696–16700

Pratt CW, Church FC (1992) Heparin binding to protein C inhibitor. J Biol Chem 267 (13):8789–8794

Priglinger U, Geiger M, Bielek E, Vanyek E, Binder BR (1994) Binding of urinary protein C inhibitor to cultured human epithelial kidney tumor cells (TCL-598). The role of glycosaminoglycans present on the luminal cell surface. J Biol Chem 269(20):14705–14710

Prohaska TA, Wahlmüller FC, Furtmüller M, Geiger M (2012) Interaction of protein C inhibitor with the type II transmembrane serine protease enteropeptidase. PLoS One 7(6), e39262

Radtke KP, Stief TW, Heimburger N (1988) A new and simple isolation procedure for human protein C inhibitor. Evidence for a second inhibitor for activated protein C present in human plasma. Biol Chem Hoppe Seyler 369(9):965–974

Radtke KP, Fernandez JA, Greengard JS, Tang WW, Wilson CB, Loskutoff DJ, Scharrer I, Griffin JH (1994) Protein C inhibitor is expressed in tubular cells of human kidney. J Clin Invest 94 (5):2117–2124

Radtke KP, Fernandez JA, Villoutreix BO, Greengard JS, Griffin JH (1995) Characterization of a cDNA for rhesus monkey protein C inhibitor--evidence for N-terminal involvement in heparin stimulation. Thromb Haemost 74(4):1079–1087

Razavi M, Johnson LD, Lum JJ, Kruppa G, Anderson NL, Pearson TW (2013) Quantification of a proteotypic peptide from protein C inhibitor by liquid chromatography-free SISCAPA-MALDI mass spectrometry: application to identification of recurrence of prostate cancer. Clin Chem 59(10):1514–1522

Rezaie AR, Cooper ST, Church FC, Esmon CT (1995) Protein C inhibitor is a potent inhibitor of the thrombin-thrombomodulin complex. J Biol Chem 270(43):25336–25339

Rieger D, Assinger A, Einfinger K, Sokolikova B, Geiger M (2014) Protein C inhibitor (PCI) binds to phosphatidylserine exposing cells with implications in the phagocytosis of apoptotic cells and activated platelets. PLoS One 9(7), e101794

Rosenzweig CN, Zhang Z, Sun X, Sokoll LJ, Osborne K, Partin AW, Chan DW (2009) Predicting prostate cancer biochemical recurrence using a panel of serum proteomic biomarkers. J Urol 181(3):1407–1414

Rydengard V, Olsson AK, Morgelin M, Schmidtchen A (2007) Histidine-rich glycoprotein exerts antibacterial activity. FEBS J 274(2):377–389

Schultz H, Weiss JP (2007) The bactericidal/permeability-increasing protein (BPI) in infection and inflammatory disease. Clin Chim Acta 384(1–2):12–23

Shen B, Gao L, Hsu YT, Bledsoe G, Hagiwara M, Chao L, Chao J (2010) Kallistatin attenuates endothelial apoptosis through inhibition of oxidative stress and activation of Akt-eNOS signaling. Am J Physiol Heart Circ Physiol 299(5):H1419–H1427

Sieben NL, Oosting J, Flanagan AM, Prat J, Roemen GM, Kolkman-Uljee SM, van Eijk R, Cornelisse CJ, Fleuren GJ, van Engeland M (2005) Differential gene expression in ovarian tumors reveals Dusp 4 and Serpina 5 as key regulators for benign behavior of serous borderline tumors. J Clin Oncol 23(29):7257–7264

Song Z, Ma N, Hayashi T, Gabazza EC, Sugimura Y, Suzuki K (2007) Intracellular localization of protein C inhibitor (PCI) and urinary plasminogen activator in renal tubular epithelial cells from humans and human PCI gene transgenic mice. Histochem Cell Biol 128(4):293–300

Stenflo J, Kjellberg M, Strandberg K, Svensson PJ, Kolbel T, Lindblad B (2006) The complex between activated protein C and protein C inhibitor: a clinically useful indicator of aortic aneurysms? Blood Cells Mol Dis 36(2):118–121

Stief TW, Radtke KP, Heimburger N (1987) Inhibition of urokinase by protein C-inhibitor (PCI). Evidence for identity of PCI and plasminogen activator inhibitor 3. Biol Chem Hoppe Seyler 368(10):1427–1433

Stump DC, Thienpont M, Collen D (1986) Purification and characterization of a novel inhibitor of urokinase from human urine. Quantitation and preliminary characterization in plasma. J Biol Chem 261(27):12759–12766

Sun W, Parry S, Panico M, Morris HR, Kjellberg M, Engstrom A, Dell A, Schedin-Weiss S (2008) N-glycans and the N terminus of protein C inhibitor affect the cofactor-enhanced rates of thrombin inhibition. J Biol Chem 283(27):18601–18611

Sun W, Parry S, Ubhayasekera W, Engstrom A, Dell A, Schedin-Weiss S (2010) Further insight into the roles of the glycans attached to human blood protein C inhibitor. Biochem Biophys Res Commun 403(2):198–202

Suzuki K (1984) Activated protein C inhibitor. Semin Thromb Hemost 10(2):154–161

Suzuki K (1993) Protein C inhibitor. Methods Enzymol 222:385–399

Suzuki K (2008) The multi-functional serpin, protein C inhibitor: beyond thrombosis and hemostasis. J Thromb Haemost 6(12):2017–2026
Suzuki K, Nishioka J, Hashimoto S (1983) Protein C inhibitor. Purification from human plasma and characterization. J Biol Chem 258(1):163–168
Suzuki K, Nishioka J, Kusumoto H, Hashimoto S (1984) Mechanism of inhibition of activated protein C by protein C inhibitor. J Biochem 95(1):187–195
Suzuki K, Deyashiki Y, Nishioka J, Kurachi K, Akira M, Yamamoto S, Hashimoto S (1987) Characterization of a cDNA for human protein C inhibitor. A new member of the plasma serine protease inhibitor superfamily. J Biol Chem 262(2):611–616
Szabo R, Netzel-Arnett S, Hobson JP, Antalis TM, Bugge TH (2005) Matriptase-3 is a novel phylogenetically preserved membrane-anchored serine protease with broad serpin reactivity. Biochem J 390(Pt 1):231–242
Uhrin P, Dewerchin M, Hilpert M, Chrenek P, Schofer C, Zechmeister-Machhart M, Kronke G, Vales A, Carmeliet P, Binder BR, Geiger M (2000) Disruption of the protein C inhibitor gene results in impaired spermatogenesis and male infertility. J Clin Invest 106(12):1531–1539
Uhrin P, Schofer C, Zaujec J, Ryban L, Hilpert M, Weipoltshammer K, Jerabek I, Pirtzkall I, Furtmuller M, Dewerchin M, Binder BR, Geiger M (2007) Male fertility and protein C inhibitor/plasminogen activator inhibitor-3 (PCI): localization of PCI in mouse testis and failure of single plasminogen activator knockout to restore spermatogenesis in PCI-deficient mice. Fertil Steril 88(4 Suppl):1049–1057
Van Raemdonck G, Zegels G, Coen E, Vuylsteke B, Jennes W, Van Ostade X (2014) Increased Serpin A5 levels in the cervicovaginal fluid of HIV-1 exposed seronegatives suggest that a subtle balance between serine proteases and their inhibitors may determine susceptibility to HIV-1 infection. Virology 458–459:11–21
Wagenaar GT, van Vuuren AJ, Girma M, Tiekstra MJ, Kwast L, Koster JG, Rijneveld AW, Elisen MG, van der Poll T, Meijers JC (2000) Characterization of transgenic mice that secrete functional human protein C inhibitor into the circulation. Thromb Haemost 83(1):93–101
Wagenaar GT, Uhrin P, Weipoltshammer K, Almeder M, Hiemstra PS, Geiger M, Meijers JC, Schofer C (2010) Expression patterns of protein C inhibitor in mouse development. J Mol Histol 41(1):27–37
Wahlmüller FC, Sokolikova B, Rieger D, Geiger M (2014) New lipid interaction partners stimulate the inhibition of activated protein C by cell-penetrating protein C inhibitor. Thromb Haemost 111(1):41–52
Wakita T, Hayashi T, Yuasa H, Nishioka J, Kawamura J, Suzuki K (1998) Molecular cloning, tissue distribution and androgen regulation of rat protein C inhibitor. FEBS Lett 429 (3):263–268
Wakita T, Hayashi T, Nishioka J, Tamaru H, Akita N, Asanuma K, Kamada H, Gabazza EC, Ido M, Kawamura J, Suzuki K (2004) Regulation of carcinoma cell invasion by protein C inhibitor whose expression is decreased in renal cell carcinoma. Int J Cancer 108(4):516–523
Wong JH, Dukes J, Levy RE, Sos B, Mason SE, Fong TS, Weiss EJ (2008) Sex differences in thrombosis in mice are mediated by sex-specific growth hormone secretion patterns. J Clin Invest 118(8):2969–2978
Yang H, Wahlmüller FC, Sarg B, Furtmüller M, Geiger M (2015a) A + -helix of protein C inhibitor (PCI) is a cell-penetrating peptide that mediates cell membrane permeation of PCI. J Biol Chem 290(5):3081–3091
Yang H, Wahlmüller FC, Uhrin P, Baumgartner R, Mitulovic G, Sarg B, Geiger M, Zellner M (2015b) Proteome analysis of testis from infertile protein Cinhibitor deficient mice reveals predominant changes in Serpin processing and Prostaglandin metabolism. Electrophoresis. doi: 10.1002/elps.201500218
Yuasa H, Tanaka H, Hayashi T, Wakita T, Nakamura H, Nishioka J, Kawarada Y, Suzuki K (2000) Bovine protein C inhibitor has a unique reactive site and can transiently inhibit plasmin. Thromb Haemost 83(2):262–267

Zechmeister-Machhart M, Hufnagl P, Uhrin P, Xu J, Geiger M, Binder BR (1996) Molecular cloning and tissue distribution of mouse protein C inhibitor (PCI). Immunopharmacology 32 (1–3):96–98

Zechmeister-Machhart M, Hufnagl P, Uhrin P, Korschineck I, Binder BR, Geiger M (1997) Molecular cloning and sequence analysis of the mouse protein C inhibitor gene. Gene 186 (1):61–66

Zhang W, Swanson R, Izaguirre G, Xiong Y, Lau LF, Olson ST (2005) The heparin-binding site of antithrombin is crucial for antiangiogenic activity. Blood 106(5):1621–1628

Zhang C, Li X, Lian X, Wang Y, Zeng Y, Yang K, Yu J, Gao Q, Yang T (2007) Immunolocalization of protein C inhibitor in differentiation of human epidermal keratinocytes. Acta Histochem 109(6):461–467

Zheng X, Geiger M, Ecke S, Bielek E, Donner P, Eberspacher U, Schleuning WD, Binder BR (1994) Inhibition of acrosin by protein C inhibitor and localization of protein C inhibitor to spermatozoa. Am J Physiol 267(2 Pt 1):C466–C472

Zou Z, Anisowicz A, Hendrix MJ, Thor A, Neveu M, Sheng S, Rafidi K, Seftor E, Sager R (1994) Maspin, a serpin with tumor-suppressing activity in human mammary epithelial cells. Science 263(5146):526–529

Protease Nexin-1: A Serpin Involved in Pathophysiology

11

Véronique Arocas and Marie-Christine Bouton

Contents

Abstract

Protease nexin-1 (PN-1), also known as SERPINE2, is the phylogenetically closest relative of plasminogen activator type 1 (PAI-1). Unlike PAI-1, antithrombin (AT), protein C inhibitor (PCI) or heparin cofactor II (HCII), other members of the serpin family that share its ability to inhibit thrombin, PN-1 is not detectable in the circulating blood. In contrast, PN-1 is found within diverse organs, including the brain, the blood vessel wall and the lungs, among many others. PN-1 expression is tightly and finely regulated in different organs at distinct stages of development, suggesting a dynamic role in vivo depending on the cell type, the tissue and the stage of development. PN-1 is now recognized as a central regulator of vascular thrombosis and appears to play a protective role in the different tissues where it is expressed, by preventing pathological states induced by excessive proteolytic activities. In this chapter, we examine current

V. Arocas • M.-C. Bouton (✉)
LVTS, INSERM, U1148, 75018 Paris, France

Université Paris Diderot, Sorbonne Paris Cité, Paris, France
e-mail: marie-christine.bouton@inserm.fr

M. Geiger et al. (eds.), *The Serpin Family*, DOI 10.1007/978-3-319-22711-5_11

knowledge of PN-1 in the pathophysiology of the nervous, vascular, respiratory, renal and reproductive systems, as well as in the development of tumours.

11.1 Properties of PN-1

PN-1 is a 45–50 kDa glycoprotein that is encoded by the *SERPINE2* gene on human chromosome 2, region q33–q35 (Carter et al. 1995). It shares 41 % and 32 % homology with human PAI-1 and antithrombin, respectively (McGrogan et al. 1988; Sommer et al. 1987). The PN-1 gene has nine exons. Two isoforms of human PN-1, named α- and βPN-1, arising from alternative splicing have been identified. αPN-1 contains an Arg residue at position 310 where βPN-1 contains Thr–Gly residues. Both forms inhibit serine proteases (McGrogan et al. 1988), but their respective roles are still unknown.

βPN-1 consists of a single chain of 398 amino acid residues, including the signal peptide of 19 residues. The 379 amino acids of the mature βPN-1 include three cysteine residues (at positions 117, 131 and 209) that do not form disulphide bonds within the protein core of the molecule (Evans et al. 1991; Li and Huntington 2012) and three potential *N*-linked glycosylation sites (at positions 99, 140 and 365). The protease recognition region (reactive centre loop, RCL) of PN-1 is located near the carboxy-terminal end of the serpin domain, with Arg346–Ser347 as the reactive site (P1–P′1) cleaved by the target serine protease. As for many other serpins, PN-1 affinity for its target proteases can be enhanced by heparin. Indeed, PN-1 has a high affinity for heparin in the low nM range and retains this high affinity for heparin even after complex formation with the protease (Evans et al. 1991). The localization of the heparin-binding site of PN-1 has been determined by site-directed mutagenesis and contains seven lysine residues that are located between residues 70 and 87 in the amino-terminal region of the protein (Stone et al. 1994). The low-density lipoprotein receptor-related protein 1 (LRP1) of the scavenger receptor family has been shown to be involved in the endocytosis of protease–PN-1 complexes (Conese et al. 1994). The site in PN-1 that binds to LRP1 has been identified as the sequence between Pro28 and Ile39 in the amino-terminal region of the protein (Knauer et al. 1997b). A variant of PN-1 mutated at His29 and Asp30 confirmed that this region of PN-1 mediates clearance by LRP (Knauer et al. 1999).

In vitro, PN-1 exerts a broad spectrum of inhibitory actions: it inhibits trypsin and thrombin (Scott et al. 1985), tissue and urokinase plasminogen activators (tPA and uPA) (Eaton et al. 1984; Scott et al. 1985), plasmin, factor Xa (Evans et al. 1991), factor XIa (Knauer et al. 2000) and activated protein C (Hermans and Stone 1993). PN-1 can also form complexes with prostasin (Chen et al. 2004), matriptase (Myerburg et al. 2008) and factor VII-activating protease (FSAP) (Muhl et al. 2007). In contrast, the two main proteases found in leukocytes, elastase and cathepsin G, are not inhibited by PN-1 (Scott et al. 1985). Interestingly, in vitro studies demonstrated that PN-1 inhibits thrombin more effectively than antithrombin (Wallace et al. 1989), even in the presence of heparin (Evans et al. 1991; Rovelli et al. 1992; Cunningham et al. 1992). Heparin also strongly accelerates FXIa inhibition by PN-1, making PN-1 a far better inhibitor of FXIa than C1 inhibitor (Knauer et al. 2000). In contrast, heparin induces no significant change in plasminogen activators (personal data) or plasmin (Evans et al. 1991) inhibition by PN-1.

The crystal structure of the complex between thrombin and PN-1 has recently been elucidated (Li and Huntington 2012). Surprisingly, two structures with thrombin in two very different orientations have been observed. This suggests a two-step mechanism for the inhibition of thrombin by PN-1 in the presence of heparin, whereby a trimolecular complex would form first, in which heparin forms a bridge between thrombin and PN-1 (bridging mechanism), but where thrombin cannot cleave the P1–P′1 site of PN-1. This step would then be followed by a rotation of thrombin that dissociates it from heparin and would lead to the full engagement of its active site with the RCL of PN-1.

PN-1 is secreted differently according to the type of cultured cells. In glioma cells, PN-1 secretion is regulated by a heterotrimeric G protein (Lagriffoul et al. 1996) and is controlled by reorganization of the F-actin cytoskeleton (Giau et al. 2005). PN-1 is constitutively secreted in human foreskin fibroblast cultures where it represents roughly 5 % of the total cell-secreted protein (Howard and Knauer 1986). In contrast, PN-1 is not detected in the conditioned medium of cultured endothelial cells (Leroy-Viard et al. 1989) or smooth muscle cells (Richard et al. 2004), because of its high affinity for glycosaminoglycans present on the surface of these cells.

Binding of PN-1–protease complexes via a cell surface receptor was first observed in fibroblasts (Howard and Knauer 1987). LRP1 has been shown to be responsible for the clearance of protease–PN-1 complexes in fibroblasts (Knauer et al. 1997a; Muhl et al. 2007). PN-1–protease complexes are actively internalized from the pericellular environment by LRP1 via an interaction with cell surface heparins or heparin sulphate proteoglycans in fibroblasts (Knauer et al. 1997b) and astrocytoma cells (Mentz et al. 1999). The catabolism of uPA–PN-1 complexes is also mediated by LRP1 but proceeds via a pathway that utilizes the uPA receptor (uPAR) in fibroblasts (Crisp et al. 2000) and in monocytic U937 cells (Conese et al. 1994). Free PN-1 is able to bind to LRP1 (Li et al. 2006) although the formation of a complex with a protease enhances its affinity for LRP1 (Jensen et al. 2009).

A few studies have reported that PN-1 may induce signalling by a LRP1-dependent pathway in mouse embryonic fibroblasts (Li et al. 2006), murine mammary cells (Fayard et al. 2009) or cerebellar granular neuron precursors (Vaillant et al. 2007) and by a LRP1-independent pathway in mouse embryonic fibroblasts (Li et al. 2006).

PN-1 is involved in the biosynthesis of dense-core secretory granules, which are key organelles for the secretion of hormones and neuropeptides in endocrine cells and neurons, in response to stimulation. Increased PN-1 expression has been shown to stabilize and prevent granule protein degradation in the Golgi of neuroendocrine cells. Therefore, by inhibiting as yet unidentified serine proteases present in the Golgi, PN-1 leads to the promotion of granule formation (Kim and Loh 2006).

11.2 PN-1 in the Nervous System

PN-1 is present in large quantities in the brain where it is produced by many different cell types. In culture, PN-1 is expressed by diverse cells of the nervous system, including astrocytes and glioma and neuroblastoma cells. In vivo, both neuronal and glial cells express PN-1. Because its expression is tightly regulated during postnatal development, PN-1 is thought to play a role in the maturation of the central nervous system (Mansuy et al. 1993). In adulthood, PN-1 expression persists at high levels in the olfactory system where constant neuronal degeneration and regeneration occur throughout life (Hoffmann et al. 1992; Reinhard et al. 1988). In the central and peripheral nervous systems, PN-1 is upregulated in response to several types of injury. It is markedly increased after damage to the sciatic nerve (Meier et al. 1989) or after injection of excitotoxic agents in the brain (Scotti et al. 1994). PN-1 expression has been shown to be regulated by many different growth factors, cytokines and neuropeptides in cultured brain cells. Its expression is upregulated by TNFα and TGFβ in the neuroblastoma cell line SK-N-SH (Vaughan and Cunningham 1993) and in human astrocytes (Hultman et al. 2010); by IL-1 in human U373-MG astrocytoma cells (Kasza et al. 2001), SK-N-SH cells (Vaughan and Cunningham 1993) and human astrocytes (Hultman et al. 2010); by IL-6 and IL-10 in human astrocytes (Hultman et al. 2010); by vasoactive intestinal peptide in Schwann cells (Bleuel et al. 1995) and astrocytes (Festoff et al. 1996); and by IFNγ in human U373-MG astrocytoma cells (Kasza et al. 2001). Conversely, PN-1 expression is downregulated by angiotensin II and calcitonin gene-related peptide in Schwann cells (Bleuel et al. 1995).

PN-1 plays a protective role against the proteolytic activity of serine proteases that may be extravasated after impairment of the blood–brain barrier in the lesioned brain. This is illustrated in the brain of patients with Alzheimer's disease (AD) where a large decrease in free PN-1 associated with an increase in PN-1–thrombin complexes was found (Wagner et al. 1989). The neuroprotective role of PN-1 is also observed during oedema in the brain, where the PN-1–thrombin system has been shown to modulate cerebral oedema after intracerebral haemorrhage (Wu et al. 2008). By inhibiting thrombin, PN-1 promotes neurite outgrowth from cultured neuronal cells (Cunningham and Gurwitz 1989), favours stellation of astrocytes and regulates survival or death of injured neurons (Houenou et al. 1995; Smith-Swintosky et al. 1995). PN-1 also inhibits thrombin-induced neuronal damage after cerebral ischemia modelled in vitro in organotypic hippocampal slice cultures (Mirante et al. 2013). Moreover, by regulating sonic hedgehog (SHH) signalling and the *N*-methyl-D-aspartate receptor (NMDAR) availability and activity, PN-1 controls progenitor cell proliferation in the hippocampus (Lino et al. 2010) and in the cerebellum (Vaillant et al. 2007). Recently, PN-1 has been shown to affect glioma cell migration and invasiveness via the regulation of uPA and MMP9/MMP2 expression levels (Pagliara et al. 2014).

PN-1 knockout (PN-1-KO) mice did not show gross brain or behavioural defects. However, they display impaired vibrissae sensory processing (Kvajo et al. 2004) and have an altered activity of the NMDAR in the hippocampus, revealing epileptic activity and reduced long-term potentiation, a process that reflects naturally occurring neuronal plasticity (Luthi et al. 1997). Moreover, PN-1-KO mice showed delayed structural and functional recovery after sciatic nerve crush (Lino et al. 2007) and had marked impaired fear extinction, the learning process that suppresses fear when past threats no longer yield aversive outcomes (Meins et al. 2010). Mice overexpressing PN-1 specifically in the brain were found to have also altered synaptic transmission and disturbances in motor behaviour (Meins et al. 2001).

Taken together these data reinforce the idea that PN-1 plays a key role in the regulation of cerebral proteolytic activity and that its induction after injury or in some pathological situations may preserve brain tissues from neuronal death. Most of the neuroprotective role of PN-1 is attributed to its potent ability to inhibit thrombin activity; however, this does not preclude the possibility that PN-1 plays other roles in the nervous system whose mechanisms remain to be determined.

11.3 PN-1 in the Vascular System

The first report of the presence of PN-1 in the blood vessel consisted of immunohistochemical studies demonstrating an abundance of PN-1 around cerebral blood vessels (Choi et al. 1990). Direct evidence of PN-1 expression in the arterial wall was also provided in rat aortas (Bouton et al. 2003) and human aortic and mammary arteries (Mansilla et al. 2008) where it is expressed by endothelial cells (Leroy-Viard et al. 1989) and smooth muscle cells (Bouton et al. 2003). Because PN-1 has a high affinity for glycosaminoglycans, it is retained at the cell surface of vascular cells and within the extracellular matrix (ECM) of the vessel wall. Thereby, PN-1 protects vascular smooth muscle cells against thrombin-induced apoptosis (Rossignol et al. 2004). High levels of PN-1 are expressed in atherosclerotic plaques, associated with macrophages and platelets (Kanse et al. 2004; Mansilla et al. 2008), and in vascular smooth muscle cells from thoracic aneurysms of the ascending aorta, leading to the protection of cells from plasmin-induced detachment and death (Gomez et al. 2013). The protective role of PN-1 in the vessel wall is also illustrated by its interaction with thrombomodulin, a transmembrane glycoprotein specifically expressed on endothelial cells that transforms thrombin from a pro- to an anticoagulant protein. This interaction leads to an accentuation in the inhibition of fibrin formation and a limitation of the generation of activated protein C (aPC) and thrombin activatable fibrinolysis inhibitor (TAFI) (Bouton et al. 2007). Endothelial PN-1 was also shown to protect the endothelial protein C receptor from endogenous shedding, thereby favouring the cytoprotective effects of the activated protein C (Boulaftali et al. 2013).

Moreover, PN-1 is present in blood cells and in particular in monocytes and platelets (Gronke et al. 1989; Mansilla et al. 2008). Active PN-1 is secreted during platelet activation and efficiently inhibits thrombin, thereby inhibiting fibrin formation, platelet activation and amplification of thrombin generation in vitro (Boulaftali et al. 2010). Indeed, PN-1 imposes a threshold for thrombin-induced platelet activation. Complete deficiency of PN-1 in mice does not result in an alteration of the vascular phenotype. However, in vivo studies showed that thrombus formation induced after vascular injury was significantly accelerated in PN-1-deficient mice compared to wild-type mice. Thus, PN-1 accumulates at the site of the platelet thrombus where it displays its antithrombotic activity via its ability to block both the activity and the generation of thrombin.

Platelet-rich clots are known to be resistant to thrombolysis. Until recently, PAI-1 released locally by activated platelets was assumed to be the major contributor to thrombus stabilization by inhibiting endogenous fibrinolysis (Levi et al. 1992; Potter van Loon et al. 1992). However, platelet PN-1 appears to prevent premature clot dissolution by tPA. This was illustrated in vivo with PN-1-deficient mice that display accelerated and enhanced thrombolysis following treatment with tPA (Boulaftali et al. 2011). The antithrombolytic activity of platelet PN-1 is mediated by its ability to inhibit plasmin generation and activity within the clot. Interestingly, both PAI-1 and PN-1 may play complementary roles in maintaining the fibrin clot.

Local variations of PN-1 expression are likely to be a determinant for the regulation of thrombin- or plasmin-induced vascular responses. PN-1 can thus influence vascular remodelling or the development of vascular lesions. Although it remains to be established whether or not PN-1 expression is variable in different vascular beds, PN-1 may nevertheless be an attractive target to prevent or limit thrombotic disorders. However, in the present state of knowledge, the dual anticoagulant and antifibrinolytic role of PN-1 hampers the clinical use of a nonselective PN-1 inhibitor.

11.4 PN-1 in the Respiratory System

PN-1 has been detected in normal adult mouse and human lung tissue (Mansuy et al. 1993), in particular in airway epithelial cells, in the vascular adventitia ECM (Demeo et al. 2006) and in fibroblasts (Francois et al. 2014). Overexpression of PN-1 has been described in lung fibroblasts from patients with idiopathic pulmonary fibrosis (Francois et al. 2014). Up to now, no human pathology has been linked to a PN-1 mutation or polymorphism, except in lung pathology. Chronic obstructive pulmonary disease (COPD) is a complex multifactorial human disease especially influenced by environmental factors like cigarette smoking. The only proven genetic risk factor for COPD is a severe deficiency of α1-antitrypsin (Laurell and Eriksson 1963), which, however, concerns only 1 % of patients. Demeo et al. were the first to identify *SERPINE2* as a novel candidate gene for COPD susceptibility in members of the Boston Early-Onset COPD Study families (Demeo et al. 2006). Another group confirmed these data in a Norwegian population (Zhu et al. 2007). However, the link between PN-1 polymorphisms and COPD is still a matter of debate. Indeed, Chappell et al. were unable to replicate the association of *SERPINE2* with COPD in a large case–control study from the United Kingdom (Chappell et al. 2006). Moreover, no association of SNPs in the *SERPINE2* gene with the risk of COPD has been obtained in the Chinese population (Zhong et al. 2009), whereas a significant association was observed in a Korean population (Cha et al. 2009). *SERPINE2* polymorphism was also associated with emphysema in Japanese smokers (Fujimoto et al. 2010), in a North American population (Kim et al. 2011) and in Finnish construction workers (Kukkonen et al. 2011). However, to date no functional variant of the *SERPINE2* gene has been identified, and its role in the lung pathophysiology remains to be clarified. However, PN-1 has been shown to regulate the Na^+ absorption controlled by Na^+ channels in human airway epithelial cells (ENaC), through its interaction with prostasin, a serine protease-activating ENaC, and with matriptase, an upstream prostasin-activating protease (Myerburg et al. 2008). Altered PN-1 activity or expression could therefore lead to pulmonary pathologies via dysregulation of proteolytic cascades operating in the airways and lungs.

11.5 PN-1 in the Renal System

In contrast to PAI-1, PN-1 has been little studied in the kidney. Significant levels of PN-1 mRNA and protein are not only expressed in adult mouse lungs but also in the kidneys, primarily in the glomeruli (Moll et al. 1996) and in particular in mesangial cells (Taneda et al. 2008). Glomerular expression of PN-1 was markedly increased in different mouse models of glomerulonephritis (Moll et al. 1996; Taneda et al. 2008). Thus, PN-1 expressed in the glomeruli could be involved in the control of intraglomerular coagulation and therefore of fibrin deposition occurring during glomerulosclerosis.

The coordinated regulation of serine proteases and serpins plays also an important role in the sodium handling in the kidney. Indeed, ENaC is essential not only for Na^+ absorption in the lung but also in the kidney. Moreover, the expression of PN-1 in mouse renal epithelial cells has been shown to be up- and downregulated by TGFβ and aldosterone, respectively (Wakida et al. 2006). Together, these data suggest a potential role of PN-1 in salt retention or natriuresis.

11.6 PN-1 in the Reproductive System

The first report of PN-1 expression in the reproductive system described PN-1 expression in human foreskin fibroblasts (Eaton and Baker 1983). The highest level of PN-1 expression in the body was found in the male reproductive organs (Lu et al. 2011), especially in the seminal vesicles where it is regulated by testosterone. PN-1 expression is high in adult mice, low in young mice and strongly reduced in castrated animals. Testosterone treatment of castrated males rapidly restores PN-1 mRNA levels, indicating that PN-1 gene expression is under androgen control, although no response element to testosterone has been identified so far in the promoter of PN-1 (Vassalli et al. 1993). An important role for PN-1 has been suggested in the early events of testis determination (Grimmond et al. 2000). Indeed, a male-specific expression of PN-1 was observed in the developing mouse gonad before overt testis differentiation (Grimmond et al. 2000). Although it was not possible to determine the cell type expressing PN-1 at this early stage, PN-1 was detected in the somatic component of the testis cord.

Invalidation of the gene-encoding PN-1 in mice leads to a sharp decline in male fertility. PN-1-deficient male mice produce normal spermatozoa, but the composition of the seminal fluid displays an impaired antiprotease activity due to the PN-1 deficiency, resulting in an excessive proteolytic activity and in the degradation of semenoclotin, an essential protein for vaginal plug formation (Murer et al. 2001). In humans, low PN-1 levels were found in the semen of fertile men, whereas abnormally high levels were found in the semen of men with abnormal seminal vesicle secretion (Murer et al. 2001).

Studies of Lu et al. suggested that PN-1 present in the murine seminal fluid could be a sperm decapacitation factor leading to the blocking of sperm–oocyte interactions, since capacitation is an essential process for sperm fertilization of the egg in mammalian species (Lu et al. 2011). Moreover, Lu et al. showed that BSA (bovine serum albumin)-induced capacitation of sperm was reversed by the addition of PN-1, via the inhibition of cholesterol efflux, one of the initiating events of capacitation. In addition, a modification of matrix metalloproteinase 9 (MMP9) levels in the seminal fluid is known to be negatively associated with reproductive capacities (Thompson et al. 2008) and to regulate the expression levels of other serpins able to modulate capacitation (Lin et al. 2008; Ou et al. 2012). Interestingly, MMP9 cleavage of PN-1 was evidenced in seminal fluid (McKee et al. 2013). The reduction of PN-1 expression results in an increased MMP9 activity in the seminal fluid that affects the mouse reproductive system (McKee et al. 2013), via the decreased inhibition of uPA by PN-1, which in turn allows a greater activation of the zymogen form of MMP9.

A number of studies also suggested a role of PN-1 in the female reproductive tissues of mice, in particular in the ovarian follicle during ovulation, the placenta during implantation and the mammary gland during pregnancy and lactation. In cattle, PN-1 expression is regulated in a spatiotemporal pattern by FSH (follicle-stimulating hormone) and hCG (human chorionic gonadotrophin), and it has been suggested that a high expression of PN-1 may contribute to follicular growth and atresia, whereas a decrease may contribute to ovulation (Bedard et al. 2003; Cao et al. 2004, 2006). In macaques, PN-1 expression increases in the corpus luteum during early pregnancy and varies during the menstrual cycle. PN-1 was shown to regulate the activity of prostasin, a serine protease involved in the turnover of the matrix and in uterine trophoblastic invasion (Lin et al. 2006; Zhang et al. 2007).

In rats, PN-1 is also expressed in the reproductive tract by the uterus and the ovaries and is highly regulated, with an increase during implantation and a reversion to normal levels during early gestation (Kim et al. 2001). Similar results were observed in mice, in which PN-1 expression was evidenced in the placenta and uterus during the oestrous cycle, pregnancy and lactation (Chern et al. 2010).

In both mice and humans, a higher expression level of PN-1 is detected in cumulus cells surrounding immature oocytes than in those associated with mature ones (Lu et al. 2013). In mice, overexpression of PN-1 impaired oocyte maturation, probably by decreasing cumulus matrix gene expression, reducing cumulus hyaluronan contents and inhibiting uPA activity. As in rodents, PN-1 is highly expressed in the human endometrium during the secretory phase compared to the proliferative phase, suggesting a role for PN-1 in regulating tissue remodelling during implantation (Lee et al. 2011).

Taken together, these data point to a regulatory role for PN-1 in reproductive biology. However, its role and mechanism of action in reproductive physiology remain unclear.

11.7 PN-1 and Cancer

The role of PN-1 in cancer has been recently reviewed (Kousted et al. 2012). An increasing number of studies suggest that PN-1 may be important in tumour development. Indeed, its target proteases and in particular uPA are involved in proteolysis of the ECM and known to play a key role in angiogenesis and metastasis.

PN-1 is overexpressed in many human cancers and cancer cell lines, including laryngeal squamous cell carcinomas (Gao et al. 2008), pancreatic and colorectal cancers (Bergeron et al. 2010; Selzer-Plon et al. 2009; Buchholz et al. 2003), gastric carcinomas (Buchholz et al. 2003; Wang et al. 2014), breast cancer (Candia et al. 2006), testicular cancer (Yang et al. 2009; Nagahara et al. 2010) and human myxoid liposarcoma (Thelin-Jarnum et al. 1999). In human pancreatic tumours, PN-1 was found exclusively in tumour cells (Buchholz et al. 2003). However, the type of cells producing PN-1 in most human cancer remains to be determined. So far, only one malignant breast cancer cell line isolated from a patient has been reported to express reduced levels of PN-1 compared to normal cells (Liang et al. 2009). In most cases, high levels of PN-1 are correlated with poor prognosis. In comparative studies of cell line variants differing in their invasiveness and/or their metastatic mechanisms, overexpression of PN-1 was almost always found in the most aggressive variants and has thus been correlated with the process of metastasis. In breast cancer, patients with breast tumours that had elevated PN-1 levels exhibited a significantly higher probability of developing lung metastasis. Higher PN-1 mRNA levels correlated with increased expression of two other adverse prognostic markers, PAI-1 and uPA (Candia et al. 2006). PN-1 levels increased with tumour grade in breast cancer (Fayard et al. 2009) and with progression of colorectal cancers (Selzer-Plon et al. 2009), although another study found these levels increased, but independently of tumour grade (Bergeron et al. 2010).

Studies in colorectal cancer revealed that PN-1 overexpression in adenomas was a direct effect of the increased signalling of the MAPK/ERK pathway (Bergeron et al. 2010). Another study has shown that blocking the signalling of the receptor tyrosine kinase MET by HGF, including ERK, reduced both the expression of PN-1 and tumour growth (van der Horst et al. 2009).

The mechanism by which PN-1 influences tumour progression is not yet understood and may be multifactorial, probably depending on both the tumour type and the environment. PN-1 overexpression is accompanied by a massive increase in ECM deposition, as observed in chronic pancreatitis and pancreatic cancer (Buchholz et al. 2003). By interacting with the surrounding stromal cells and by acting on ECM production and degradation, PN-1 may thus regulate the development of a favourable microenvironment for tumours. Studies of Wagenblast et al. demonstrated, via a mouse model of breast cancer heterogeneity that PN-1 drives metastasis by enabling breast cancer cells to form vascular-like networks. The authors suggest that the anti-thrombin activity of PN-1 is involved in this increased capacity for metastatic progression (Wagenblast et al. 2015). MMP9 and uPA both play important roles in the progression of several types of cancer by increasing tumour growth, migration, invasion and metastasis and are associated with poor disease prognosis. Several studies indicate that the MMP9/uPA axis contributes to the role of PN-1 in cancer. Indeed, PN-1 is not only an inhibitor of uPA, but has also been shown to stimulate MMP9 expression in murine mammary carcinoma cells, via LRP-1A-dependent activation of the MAPK/ERK signalling pathway, leading to an increase of their metastatic potential (Fayard et al. 2009). In contrast, in prostate carcinoma, downregulation of MMP9 is associated with increased PN-1 levels and reduced invasiveness (Xu et al. 2010). By inhibiting uPA, PN-1 impairs MMP9 activation and thus plays a protective role against tumour invasion. However, MMP9 can degrade PN-1 (Xu et al. 2008) and thus amplify its own activation by uPA and increase tumour cell invasion. A feedback loop would be set up, in which PN-1 and MMP9 exert opposite regulatory mechanisms, the resulting effect on metastasis depending on their relative abundance. Such feedback loop has also been reported in the seminal fluid (McKee et al. 2013). Moreover, PN-1 has been shown to inhibit sonic hedgehog (SHH) signalling in prostate, a pathway largely implicated in prostate adenocarcinoma progression (McKee et al. 2012). The authors demonstrated a competing regulation of SHH signalling by PN-1 and MMP9. PN-1 inhibited tumour growth by decreasing the protein expression of SHH and its downstream effectors, and this effect could be related to PN-1 internalization via LRP and uPAR.

In addition, tumours grown in the presence of recombinant PN-1 had fewer blood vessels, although with larger diameters compared with controls (McKee et al. 2012). In agreement with these results, anti-angiogenic properties of PN-1 have been shown in vitro on HUVECs (human umbilical vein endothelial cells), ex vivo on the sprouting from aortic rings of PN-1-deficient mice compared to wild-type mice and in vivo in the Matrigel plug assay (Selbonne et al. 2012). Furthermore, our group observed that PN-1 is expressed in the mouse retina vasculature and that the postnatal angiogenesis in the retina of PN-1-deficient mice is enhanced compared to wild-type mice (Selbonne et al. 2015). This important anti-angiogenic potential of PN-1 could thus be crucial in tumour development and needs to be further explored.

In prostate and colorectal cancers, the GPI (Glycosylphosphatidylinositol)-anchored serine protease prostasin, a tumour suppressor, is downregulated relative to the corresponding normal tissues (Takahashi et al. 2003; Selzer-Plon et al. 2009). As PN-1 was identified as a major inhibitor of prostasin, it was suggested that it influenced the progression of these cancers via prostasin inhibition (Chen et al. 2004; Selzer-Plon et al. 2009). Indeed, in colorectal cancer, PN-1 expression increases with tumour grade, while levels of prostasin and of its inhibitor HAI-1 (hepatocyte growth factor activator inhibitor-1) are stable or decreased during carcinogenesis (Selzer-Plon et al. 2009). PN-1–prostasin complexes were also observed in prostate epithelial cells and prostate cancer cells, but an antitumoural effect remains to be clearly identified.

Overall, these data strongly support a significant role of PN-1 in cancer invasion and metastasis. Whether PN-1 could be a therapeutically target or a prognostic factor remains to be determined. Most of the studies so far explored the role of the antiprotease activity, but it is now known that PN-1 can act directly by binding to the cell surface or to matrix proteins, leading probably to cell signalling.

11.8 Conclusion

PN-1 has a wide tissue distribution and a broad antiprotease activity. These properties make PN-1 an important actor playing a role in many biological processes as varied as neuroprotection, thrombosis and haemostasis or tumour growth. Its overexpression is often observed in response to injury, suggesting a role in protecting tissues from dysregulated proteolytic pathways leading to disease. Improved understanding of the mechanisms involved in PN-1 cellular actions will allow the development of novel therapeutic strategies for various diseases such as ischemic and fibrotic diseases or cancer.

References

Bedard J, Brule S, Price CA, Silversides DW, Lussier JG (2003) Serine protease inhibitor-E2 (SERPINE2) is differentially expressed in granulosa cells of dominant follicle in cattle. Mol Reprod Dev 64(2):152–165

Bergeron S, Lemieux E, Durand V, Cagnol S, Carrier JC, Lussier JG, Boucher MJ, Rivard N (2010) The serine protease inhibitor serpinE2 is a novel target of ERK signaling involved in human colorectal tumorigenesis. Mol Cancer 9:271

Bleuel A, de Gasparo M, Whitebread S, Puttner I, Monard D (1995) Regulation of protease nexin-1 expression in cultured Schwann cells is mediated by angiotensin II receptors. J Neurosci 15 (1 Pt 2):750–761

Boulaftali Y, Adam F, Venisse L, Ollivier V, Richard B, Taieb S, Monard D, Favier R, Alessi MC, Bryckaert M, Arocas V, Jandrot-Perrus M, Bouton MC (2010) Anticoagulant and antithrombotic properties of platelet protease nexin-1. Blood 115(1):97–106

Boulaftali Y, Ho-Tin-Noe B, Pena A, Loyau S, Venisse L, Francois D, Richard B, Arocas V, Collet JP, Jandrot-Perrus M, Bouton MC (2011) Platelet protease nexin-1, a serpin that strongly influences fibrinolysis and thrombolysis. Circulation 123(12):1326–1334

Boulaftali Y, Francois D, Venisse L, Jandrot-Perrus M, Arocas V, Bouton MC (2013) Endothelial protease nexin-1 is a novel regulator of A disintegrin and metalloproteinase 17 maturation and endothelial protein C receptor shedding via furin inhibition. Arterioscler Thromb Vasc Biol 33 (7):1647–1654

Bouton MC, Richard B, Rossignol P, Philippe M, Guillin MC, Michel JB, Jandrot-Perrus M (2003) The serpin protease-nexin 1 is present in rat aortic smooth muscle cells and is upregulated in L-NAME hypertensive rats. Arterioscler Thromb Vasc Biol 23(1):142–147

Bouton MC, Venisse L, Richard B, Pouzet C, Arocas V, Jandrot-Perrus M (2007) Protease nexin-1 interacts with thrombomodulin and modulates its anticoagulant effect. Circ Res 100 (8):1174–1181

Buchholz M, Biebl A, Neesse A, Wagner M, Iwamura T, Leder G, Adler G, Gress TM (2003) SERPINE2 (protease nexin I) promotes extracellular matrix production and local invasion of pancreatic tumors in vivo. Cancer Res 63(16):4945–4951

Candia BJ, Hines WC, Heaphy CM, Griffith JK, Orlando RA (2006) Protease nexin-1 expression is altered in human breast cancer. Cancer Cell Int 6:16

Cao M, Sahmi M, Lussier JG, Price CA (2004) Plasminogen activator and serine protease inhibitor-E2 (protease nexin-1) expression by bovine granulosa cells in vitro. Biol Reprod 71 (3):887–893

Cao M, Nicola E, Portela VM, Price CA (2006) Regulation of serine protease inhibitor-E2 and plasminogen activator expression and secretion by follicle stimulating hormone and growth factors in non-luteinizing bovine granulosa cells in vitro. Matrix Biol 25(6):342–354

Carter RE, Cerosaletti KM, Burkin DJ, Fournier RE, Jones C, Greenberg BD, Citron BA, Festoff BW (1995) The gene for the serpin thrombin inhibitor (PI7), protease nexin I, is located on human chromosome 2q33-q35 and on syntenic regions in the mouse and sheep genomes. Genomics 27(1):196–199

Cha SI, Kang HG, Choi JE, Kim MJ, Park J, Lee WK, Kim CH, Jung TH, Park JY (2009) SERPINE2 polymorphisms and chronic obstructive pulmonary disease. J Korean Med Sci 24 (6):1119–1125

Chappell S, Daly L, Morgan K, Baranes TG, Roca J, Rabinovich R, Millar A, Donnelly SC, Keatings V, MacNee W, Stolk J, Hiemstra PS, Miniati M, Monti S, O'Connor CM, Kalsheker N (2006) The SERPINE2 gene and chronic obstructive pulmonary disease. Am J Human Genet 79(1):184–186, author reply 186–187

Chen LM, Zhang X, Chai KX (2004) Regulation of prostasin expression and function in the prostate. Prostate 59(1):1–12

Chern SR, Li SH, Lu CH, Chen EI (2010) Spatiotemporal expression of the serine protease inhibitor, SERPINE2, in the mouse placenta and uterus during the estrous cycle, pregnancy, and lactation. Reprod Biol Endocrinol 8:127

Choi BH, Suzuki M, Kim T, Wagner SL, Cunningham DD (1990) Protease nexin-1. Localization in the human brain suggests a protective role against extravasated serine proteases. Am J Pathol 137(4):741–747

Conese M, Olson D, Blasi F (1994) Protease nexin-1-urokinase complexes are internalized and degraded through a mechanism that requires both urokinase receptor and alpha 2-macroglobulin receptor. J Biol Chem 269(27):17886–17892

Crisp RJ, Knauer DJ, Knauer MF (2000) Roles of the heparin and low density lipid receptor-related protein-binding sites of protease nexin 1 (PN1) in urokinase-PN1 complex catabolism. The PN1 heparin-binding site mediates complex retention and degradation but not cell surface binding or internalization. J Biol Chem 275(26):19628–19637

Cunningham DD, Gurwitz D (1989) Proteolytic regulation of neurite outgrowth from neuroblastoma cells by thrombin and protease nexin-1. J Cell Biochem 39(1):55–64

Cunningham DD, Wagner SL, Farrell DH (1992) Regulation of protease nexin-1 activity by heparin and heparan sulfate. Adv Exp Med Biol 313:297–306

Demeo DL, Mariani TJ, Lange C, Srisuma S, Litonjua AA, Celedon JC, Lake SL, Reilly JJ, Chapman HA, Mecham BH, Haley KJ, Sylvia JS, Sparrow D, Spira AE, Beane J, Pinto-Plata V, Speizer FE, Shapiro SD, Weiss ST, Silverman EK (2006) The SERPINE2 gene is associated with chronic obstructive pulmonary disease. Am J Human Genet 78(2):253–264

Eaton DL, Baker JB (1983) Phorbol ester and mitogens stimulate human fibroblast secretions of plasmin-activatable plasminogen activator and protease nexin, an antiactivator/antiplasmin. J Cell Biol 97(2):323–328

Eaton DL, Scott RW, Baker JB (1984) Purification of human fibroblast urokinase proenzyme and analysis of its regulation by proteases and protease nexin. J Biol Chem 259(10):6241–6247

Evans DL, McGrogan M, Scott RW, Carrell RW (1991) Protease specificity and heparin binding and activation of recombinant protease nexin I. J Biol Chem 266(33):22307–22312

Fayard B, Bianchi F, Dey J, Moreno E, Djaffer S, Hynes NE, Monard D (2009) The serine protease inhibitor protease nexin-1 controls mammary cancer metastasis through LRP-1-mediated MMP-9 expression. Cancer Res 69(14):5690–5698

Festoff BW, Nelson PG, Brenneman DE (1996) Prevention of activity-dependent neuronal death: vasoactive intestinal polypeptide stimulates astrocytes to secrete the thrombin-inhibiting neurotrophic serpin, protease nexin I. J Neurobiol 30(2):255–266

Francois D, Venisse L, Marchal-Somme J, Jandrot-Perrus M, Crestani B, Arocas V, Bouton MC (2014) Increased expression of protease nexin-1 in fibroblasts during idiopathic pulmonary fibrosis regulates thrombin activity and fibronectin expression. Lab Invest 94(11):1237–1246

Fujimoto K, Ikeda S, Arai T, Tanaka N, Kumasaka T, Ishii T, Kida K, Muramatsu M, Sawabe M (2010) Polymorphism of SERPINE2 gene is associated with pulmonary emphysema in consecutive autopsy cases. BMC Med Genet 11:159

Gao S, Krogdahl A, Sorensen JA, Kousted TM, Dabelsteen E, Andreasen PA (2008) Overexpression of protease nexin-1 mRNA and protein in oral squamous cell carcinomas. Oral Oncol 44(3):309–313

Giau R, Carrette J, Bockaert J, Homburger V (2005) Constitutive secretion of protease nexin-1 by glial cells and its regulation by G-protein-coupled receptors. J Neurosci 25(39):8995–9004

Gomez D, Kessler K, Borges LF, Richard B, Touat Z, Ollivier V, Mansilla S, Bouton MC, Alkoder S, Nataf P, Jandrot-Perrus M, Jondeau G, Vranckx R, Michel JB (2013) Smad2-dependent protease nexin-1 overexpression differentiates chronic aneurysms from acute dissections of human ascending aorta. Arterioscler Thromb Vasc Biol 33(9):2222–2232

Grimmond S, Van Hateren N, Siggers P, Arkell R, Larder R, Soares MB, de Fatima Bonaldo M, Smith L, Tymowska-Lalanne Z, Wells C, Greenfield A (2000) Sexually dimorphic expression of protease nexin-1 and vanin-1 in the developing mouse gonad prior to overt differentiation suggests a role in mammalian sexual development. Hum Mol Genet 9(10):1553–1560

Gronke RS, Knauer DJ, Veeraraghavan S, Baker JB (1989) A form of protease nexin I is expressed on the platelet surface during platelet activation. Blood 73(2):472–478

Hermans JM, Stone SR (1993) Interaction of activated protein C with serpins. Biochem J 295 (Pt 1):239–245

Hoffmann MC, Nitsch C, Scotti AL, Reinhard E, Monard D (1992) The prolonged presence of glia-derived nexin, an endogenous protease inhibitor, in the hippocampus after ischemia-induced delayed neuronal death. Neuroscience 49(2):397–408

Houenou LJ, Turner PL, Li L, Oppenheim RW, Festoff BW (1995) A serine protease inhibitor, protease nexin I, rescues motoneurons from naturally occurring and axotomy-induced cell death. Proc Natl Acad Sci USA 92(3):895–899

Howard EW, Knauer DJ (1986) Biosynthesis of protease nexin-I. J Biol Chem 261 (30):14184–14190
Howard EW, Knauer DJ (1987) Characterization of the receptor for protease nexin-I:protease complexes on human fibroblasts. J Cell Physiol 131(2):276–283
Hultman K, Blomstrand F, Nilsson M, Wilhelmsson U, Malmgren K, Pekny M, Kousted T, Jern C, Tjarnlund-Wolf A (2010) Expression of plasminogen activator inhibitor-1 and protease nexin-1 in human astrocytes: response to injury-related factors. J Neurosci Res 88(11):2441–2449
Jensen JK, Dolmer K, Gettins PG (2009) Specificity of binding of the low density lipoprotein receptor-related protein to different conformational states of the clade E serpins plasminogen activator inhibitor-1 and proteinase nexin-1. J Biol Chem 284(27):17989–17997
Kanse SM, Chavakis T, Al-Fakhri N, Hersemeyer K, Monard D, Preissner KT (2004) Reciprocal regulation of urokinase receptor (CD87)-mediated cell adhesion by plasminogen activator inhibitor-1 and protease nexin-1. J Cell Sci 117(Pt 3):477–485
Kasza A, Kowanetz M, Poslednik K, Witek B, Kordula T, Koj A (2001) Epidermal growth factor and pro-inflammatory cytokines regulate the expression of components of plasminogen activation system in U373-MG astrocytoma cells. Cytokine 16(5):187–190
Kim T, Loh YP (2006) Protease nexin-1 promotes secretory granule biogenesis by preventing granule protein degradation. Mol Biolo Cell 17(2):789–798
Kim NK, Choi MJ, Chung HM, Sohn TJ, Hwang SG, Oh D, Lee HH, Lee YH, Ko JJ (2001) Increased expression and localization of a serine protease inhibitor, protease nexin-1 (PN-1), in the ovary and uterus during implantation in rat. Thromb Res 103(2):135–142
Kim WJ, Hoffman E, Reilly J, Hersh C, Demeo D, Washko G, Silverman EK (2011) Association of COPD candidate genes with computed tomography emphysema and airway phenotypes in severe COPD. Eur Respir J 37(1):39–43
Knauer MF, Hawley SB, Knauer DJ (1997a) Identification of a binding site in protease nexin I (PN1) required for the receptor mediated internalization of PN1-thrombin complexes. J Biol Chem 272(19):12261–12264
Knauer MF, Kridel SJ, Hawley SB, Knauer DJ (1997b) The efficient catabolism of thrombin-protease nexin 1 complexes is a synergistic mechanism that requires both the LDL receptor-related protein and cell surface heparins. J Biol Chem 272(46):29039–29045
Knauer MF, Crisp RJ, Kridel SJ, Knauer DJ (1999) Analysis of a structural determinant in thrombin-protease nexin 1 complexes that mediates clearance by the low density lipoprotein receptor-related protein. J Biol Chem 274(1):275–281
Knauer DJ, Majumdar D, Fong PC, Knauer MF (2000) SERPIN regulation of factor XIa. The novel observation that protease nexin 1 in the presence of heparin is a more potent inhibitor of factor XIa than C1 inhibitor. J Biol Chem 275(48):37340–37346
Kousted TM, Jensen JK, Gao S, Andreasen PA (2012) Protease nexin-1 – a serpin with a possible proinvasive role in cancer. In: Behrendt N (ed) Matrix proteases in health and disease. Wiley-VCH Verlag GmbH & Co. KGaA, Weinheim, p 251
Kukkonen MK, Tiili E, Hamalainen S, Vehmas T, Oksa P, Piirila P, Hirvonen A (2011) SERPINE2 haplotype as a risk factor for panlobular type of emphysema. BMC Med Genet 12:157
Kvajo M, Albrecht H, Meins M, Hengst U, Troncoso E, Lefort S, Kiss JZ, Petersen CC, Monard D (2004) Regulation of brain proteolytic activity is necessary for the in vivo function of NMDA receptors. J Neurosci 24(43):9734–9743
Lagriffoul A, Charpentier N, Carrette J, Tougard C, Bockaert J, Homburger V (1996) Secretion of protease nexin-1 by C6 glioma cells is under the control of a heterotrimeric G protein, Go1. J Biol Chem 271(49):31508–31516
Laurell CB, Eriksson S (1963) The electrophoretic alpha1-globulin pattern of serum in alpha1-antitrypsin deficiency. Scand J Clin Lab Invest 15:132–140
Lee RK, Fan CC, Hwu YM, Lu CH, Lin MH, Chen YJ, Li SH (2011) SERPINE2, an inhibitor of plasminogen activators, is highly expressed in the human endometrium during the secretory phase. Reprod Biol Endocrinol 9:38

Leroy-Viard K, Jandrot-Perrus M, Tobelem G, Guillin MC (1989) Covalent binding of human thrombin to a human endothelial cell-associated protein. Exp cell Res 181(1):1–10
Levi M, Biemond BJ, van Zonneveld AJ, ten Cate JW, Pannekoek H (1992) Inhibition of plasminogen activator inhibitor-1 activity results in promotion of endogenous thrombolysis and inhibition of thrombus extension in models of experimental thrombosis. Circulation 85 (1):305–312
Li W, Huntington JA (2012) Crystal structures of protease nexin-1 in complex with heparin and thrombin suggest a 2-step recognition mechanism. Blood 120(2):459–467
Li X, Herz J, Monard D (2006) Activation of ERK signaling upon alternative protease nexin-1 internalization mediated by syndecan-1. J Cell Biochem 99(3):936–951
Liang X, Huuskonen J, Hajivandi M, Manzanedo R, Predki P, Amshey JR, Pope RM (2009) Identification and quantification of proteins differentially secreted by a pair of normal and malignant breast-cancer cell lines. Proteomics 9(1):182–193
Lin HY, Zhang H, Yang Q, Wang HX, Wang HM, Chai KX, Chen LM, Zhu C (2006) Expression of prostasin and protease nexin-1 in rhesus monkey (Macaca mulatta) endometrium and placenta during early pregnancy. J Histochem Cytochem 54(10):1139–1147
Lin MH, Lee RK, Hwu YM, Lu CH, Chu SL, Chen YJ, Chang WC, Li SH (2008) SPINKL, a Kazal-type serine protease inhibitor-like protein purified from mouse seminal vesicle fluid, is able to inhibit sperm capacitation. Reproduction 136(5):559–571
Lino MM, Atanasoski S, Kvajo M, Fayard B, Moreno E, Brenner HR, Suter U, Monard D (2007) Mice lacking protease nexin-1 show delayed structural and functional recovery after sciatic nerve crush. J Neurosci 27(14):3677–3685
Lino MM, Vaillant C, Orolicki S, Sticker M, Kvajo M, Monard D (2010) Newly generated cells are increased in hippocampus of adult mice lacking a serine protease inhibitor. BMC Neurosci 11:70
Lu CH, Lee RK, Hwu YM, Chu SL, Chen YJ, Chang WC, Lin SP, Li SH (2011) SERPINE2, a serine protease inhibitor extensively expressed in adult male mouse reproductive tissues, may serve as a murine sperm decapacitation factor. Biol Reprod 84(3):514–525
Lu CH, Lee RK, Hwu YM, Lin MH, Yeh LY, Chen YJ, Lin SP, Li SH (2013) Involvement of the serine protease inhibitor, SERPINE2, and the urokinase plasminogen activator in cumulus expansion and oocyte maturation. PLoS One 8(8), e74602
Luthi A, Van der Putten H, Botteri FM, Mansuy IM, Meins M, Frey U, Sansig G, Portet C, Schmutz M, Schroder M, Nitsch C, Laurent JP, Monard D (1997) Endogenous serine protease inhibitor modulates epileptic activity and hippocampal long-term potentiation. J Neurosci 17 (12):4688–4699
Mansilla S, Boulaftali Y, Venisse L, Arocas V, Meilhac O, Michel JB, Jandrot-Perrus M, Bouton MC (2008) Macrophages and platelets are the major source of protease nexin-1 in human atherosclerotic plaque. Arterioscler Thromb Vasc Biol 28(10):1844–1850
Mansuy IM, van der Putten H, Schmid P, Meins M, Botteri FM, Monard D (1993) Variable and multiple expression of Protease Nexin-1 during mouse organogenesis and nervous system development. Development 119(4):1119–1134
McGrogan MKJ, Ping Li M, Hsu C, Scott RW, Simonsen CC, Baker JB (1988) Molecular cloning and expression of two forms of human protease nexin I. Biotechnology 6:172–177
McKee CM, Xu D, Cao Y, Kabraji S, Allen D, Kersemans V, Beech J, Smart S, Hamdy F, Ishkanian A, Sykes J, Pintile M, Milosevic M, van der Kwast T, Zafarana G, Ramnarine VR, Jurisica I, Mallof C, Lam W, Bristow RG, Muschel RJ (2012) Protease nexin 1 inhibits hedgehog signaling in prostate adenocarcinoma. J Clin Invest 122(11):4025–4036
McKee CM, Xu D, Kessler BM, Muschel RJ (2013) Proteomic analysis reveals a proteolytic feedback loop in murine seminal fluid. Prostate 73(13):1427–1440
Meier R, Spreyer P, Ortmann R, Harel A, Monard D (1989) Induction of glia-derived nexin after lesion of a peripheral nerve. Nature 342(6249):548–550
Meins M, Piosik P, Schaeren-Wiemers N, Franzoni S, Troncoso E, Kiss JZ, Brosamle C, Schwab ME, Molnar Z, Monard D (2001) Progressive neuronal and motor dysfunction in mice

overexpressing the serine protease inhibitor protease nexin-1 in postmitotic neurons. J Neurosci 21(22):8830–8841
Meins M, Herry C, Muller C, Ciocchi S, Moreno E, Luthi A, Monard D (2010) Impaired fear extinction in mice lacking protease nexin-1. Eur J Neurosci 31(11):2033–2042
Mentz S, de Lacalle S, Baerga-Ortiz A, Knauer MF, Knauer DJ, Komives EA (1999) Mechanism of thrombin clearance by human astrocytoma cells. J Neurochem 72(3):980–987
Mirante O, Price M, Puentes W, Castillo X, Benakis C, Thevenet J, Monard D, Hirt L (2013) Endogenous protease nexin-1 protects against cerebral ischemia. Int J Mol Sci 14 (8):16719–16731
Moll S, Schaeren-Wiemers N, Wohlwend A, Pastore Y, Fulpius T, Monard D, Sappino AP, Schifferli JA, Vassalli JD, Izui S (1996) Protease nexin 1 in the murine kidney: glomerular localization and up-regulation in glomerulopathies. Kidney Int 50(6):1936–1945
Muhl L, Nykjaer A, Wygrecka M, Monard D, Preissner KT, Kanse SM (2007) Inhibition of PDGF-BB by Factor VII-activating protease (FSAP) is neutralized by protease nexin-1, and the FSAP-inhibitor complexes are internalized via LRP. Biochem J 404(2):191–196
Murer V, Spetz JF, Hengst U, Altrogge LM, de Agostini A, Monard D (2001) Male fertility defects in mice lacking the serine protease inhibitor protease nexin-1. Proc Natl Acad Sci USA 98 (6):3029–3033
Myerburg MM, McKenna EE, Luke CJ, Frizzell RA, Kleyman TR, Pilewski JM (2008) Prostasin expression is regulated by airway surface liquid volume and is increased in cystic fibrosis. Am J Physiol Lung Cell Mol Physiol 294(5):L932–941
Nagahara A, Nakayama M, Oka D, Tsuchiya M, Kawashima A, Mukai M, Nakai Y, Takayama H, Nishimura K, Jo Y, Nagai A, Okuyama A, Nonomura N (2010) SERPINE2 is a possible candidate promotor for lymph node metastasis in testicular cancer. Biochem Biophys Res Commun 391(4):1641–1646
Ou CM, Tang JB, Huang MS, Sudhakar Gandhi PS, Geetha S, Li SH, Chen YH (2012) The mode of reproductive-derived Spink (serine protease inhibitor Kazal-type) action in the modulation of mammalian sperm activity. Int J Androl 35(1):52–62
Pagliara V, Adornetto A, Mammi M, Masullo M, Sarnataro D, Pietropaolo C, Arcone R (2014) Protease Nexin-1 affects the migration and invasion of C6 glioma cells through the regulation of urokinase Plasminogen Activator and Matrix Metalloproteinase-9/2. Biochim Biophys Acta 1843(11):2631–2644
Potter van Loon BJ, Rijken DC, Brommer EJ, van der Maas AP (1992) The amount of plasminogen, tissue-type plasminogen activator and plasminogen activator inhibitor type 1 in human thrombi and the relation to ex-vivo lysibility. Thromb Haemost 67(1):101–105
Reinhard E, Meier R, Halfter W, Rovelli G, Monard D (1988) Detection of glia-derived nexin in the olfactory system of the rat. Neuron 1:387–394
Richard B, Arocas V, Guillin MC, Michel JB, Jandrot-Perrus M, Bouton MC (2004) Protease nexin-1: a cellular serpin down-regulated by thrombin in rat aortic smooth muscle cells. J Cell Physiol 201(1):138–145
Rossignol P, Ho-Tin-Noe B, Vranckx R, Bouton MC, Meilhac O, Lijnen HR, Guillin MC, Michel JB, Angles-Cano E (2004) Protease nexin-1 inhibits plasminogen activation-induced apoptosis of adherent cells. J Biol Chem 279(11):10346–10356
Rovelli G, Stone SR, Guidolin A, Sommer J, Monard D (1992) Characterization of the heparin-binding site of glia-derived nexin/protease nexin-1. Biochemistry 31(13):3542–3549
Scott RW, Bergman BL, Bajpai A, Hersh RT, Rodriguez H, Jones BN, Barreda C, Watts S, Baker JB (1985) Protease nexin. Properties and a modified purification procedure. J Biol Chem 260 (11):7029–7034
Scotti AL, Monard D, Nitsch C (1994) Re-expression of glia-derived nexin/protease nexin 1 depends on mode of lesion-induction or terminal degeneration: observations after excitotoxin or 6-hydroxydopamine lesions of rat substantia nigra. J Neurosci Res 37(2):155–168
Selbonne S, Azibani F, Iatmanen S, Boulaftali Y, Richard B, Jandrot-Perrus M, Bouton MC, Arocas V (2012) In vitro and in vivo antiangiogenic properties of the serpin protease nexin-1. Mol Cell Biol 32(8):1496–1505

Selbonne S, Francois D, Raoul W, Boulaftali Y, Sennlaub F, Jandrot-Perrus M, Bouton MC, Arocas V (2015) Protease nexin-1 regulates retinal vascular development. Cell Mol Life Sci. Jun 25

Selzer-Plon J, Bornholdt J, Friis S, Bisgaard HC, Lothe IM, Tveit KM, Kure EH, Vogel U, Vogel LK (2009) Expression of prostasin and its inhibitors during colorectal cancer carcinogenesis. BMC Cancer 9:201

Smith-Swintosky VL, Zimmer S, Fenton JW II, Mattson MP (1995) Protease nexin-1 and thrombin modulate neuronal Ca2+ homeostasis and sensitivity to glucose deprivation-induced injury. J Neurosci 15(8):5840–5850

Sommer J, Gloor SM, Rovelli GF, Hofsteenge J, Nick H, Meier R, Monard D (1987) cDNA sequence coding for a rat glia-derived nexin and its homology to members of the serpin superfamily. Biochemistry 26(20):6407–6410

Stone SR, Brown-Luedi ML, Rovelli G, Guidolin A, McGlynn E, Monard D (1994) Localization of the heparin-binding site of glia-derived nexin/protease nexin-1 by site-directed mutagenesis. Biochemistry 33(24):7731–7735

Takahashi S, Suzuki S, Inaguma S, Ikeda Y, Cho YM, Hayashi N, Inoue T, Sugimura Y, Nishiyama N, Fujita T, Chao J, Ushijima T, Shirai T (2003) Down-regulated expression of prostasin in high-grade or hormone-refractory human prostate cancers. Prostate 54(3):187–193

Taneda S, Hudkins KL, Muhlfeld AS, Kowalewska J, Pippin JW, Shankland SJ, Alpers CE (2008) Protease nexin-1, tPA, and PAI-1 are upregulated in cryoglobulinemic membranoproliferative glomerulonephritis. J Am Soc Nephrol 19(2):243–251

Thelin-Jarnum S, Lassen C, Panagopoulos I, Mandahl N, Aman P (1999) Identification of genes differentially expressed in TLS-CHOP carrying myxoid liposarcomas. Int J Cancer 83 (1):30–33

Thompson AR, Drenos F, Hafez H, Humphries SE (2008) Candidate gene association studies in abdominal aortic aneurysm disease: a review and meta-analysis. Eur J Vasc Endovasc Surg 35 (1):19–30

Vaillant C, Michos O, Orolicki S, Brellier F, Taieb S, Moreno E, Te H, Zeller R, Monard D (2007) Protease nexin 1 and its receptor LRP modulate SHH signalling during cerebellar development. Development 134(9):1745–1754

van der Horst EH, Chinn L, Wang M, Velilla T, Tran H, Madrona Y, Lam A, Ji M, Hoey TC, Sato AK (2009) Discovery of fully human anti-MET monoclonal antibodies with antitumor activity against colon cancer tumor models in vivo. Neoplasia 11(4):355–364

Vassalli JD, Huarte J, Bosco D, Sappino AP, Sappino N, Velardi A, Wohlwend A, Erno H, Monard D, Belin D (1993) Protease-nexin I as an androgen-dependent secretory product of the murine seminal vesicle. EMBO J 12(5):1871–1878

Vaughan PJ, Cunningham DD (1993) Regulation of protease nexin-1 synthesis and secretion in cultured brain cells by injury-related factors. J Biol Chem 268(5):3720–3727

Wagenblast E, Soto M, Gutiérrez-Ángel S, Hartl CA, Gable AL, Maceli AR, Maceli AR, Erard N, Williams AM, Kim SY, Dickopf S, Harrell JC, Smith AD, Perou CM, Wilkinson JE, Hannon GJ, Knott SR (2015) A model of breast cancer heterogeneity reveals vascular mimicry as a driver of metastasis. Nature 520(7547):358–362

Wagner SL, Geddes JW, Cotman CW, Lau AL, Gurwitz D, Isackson PJ, Cunningham DD (1989) Protease nexin-1, an antithrombin with neurite outgrowth activity, is reduced in Alzheimer disease. Proc Natl Acad Sci USA 86(21):8284–8288

Wakida N, Kitamura K, Tuyen DG, Maekawa A, Miyoshi T, Adachi M, Shiraishi N, Ko T, Ha V, Nonoguchi H, Tomita K (2006) Inhibition of prostasin-induced ENaC activities by PN-1 and regulation of PN-1 expression by TGF-beta1 and aldosterone. Kidney Int 70(8):1432–1438

Wallace A, Rovelli G, Hofsteenge J, Stone SR (1989) Effect of heparin on the glia-derived-nexin-thrombin interaction. Biochem J 257(1):191–196

Wang K, Wang B, Xing AY, Xu KS, Li GX, Yu ZH (2014) Prognostic significance of SERPINE2 in gastric cancer and its biological function in SGC7901 cells. J Cancer Res Clin Oncol 141 (5):805–812

Wu H, Zhao R, Qi J, Cong Y, Wang D, Liu T, Gu Y, Ban X, Huang Q (2008) The expression and the role of protease nexin-1 on brain edema after intracerebral hemorrhage. J Neurol Sci 270 (1–2):172–183

Xu D, Suenaga N, Edelmann MJ, Fridman R, Muschel RJ, Kessler BM (2008) Novel MMP-9 substrates in cancer cells revealed by a label-free quantitative proteomics approach. Mol Cell Proteomics 7(11):2215–2228

Xu D, McKee CM, Cao Y, Ding Y, Kessler BM, Muschel RJ (2010) Matrix metalloproteinase-9 regulates tumor cell invasion through cleavage of protease nexin-1. Cancer Res 70 (17):6988–6998

Yang S, Dong Q, Yao M, Shi M, Ye J, Zhao L, Su J, Gu W, Xie W, Wang K, Du Y, Li Y, Huang Y (2009) Establishment of an experimental human lung adenocarcinoma cell line SPC-A-1BM with high bone metastases potency by (99m)Tc-MDP bone scintigraphy. Nucl Med Biol 36 (3):313–321

Zhang H, Lin HY, Yang Q, Wang HX, Chai KX, Chen LM, Zhu C (2007) Expression of prostasin serine protease and protease nexin-1 (PN-1) in rhesus monkey ovary during menstrual cycle and early pregnancy. J Histochem Cytochem 55(12):1237–1244

Zhong L, Fu WP, Sun C, Dai LM, Zhang YP (2009) Absence of association between SERPINE2 genetic polymorphisms and chronic obstructive pulmonary disease in Han Chinese: a case-control cohort study. BMC Med Genet 10:66

Zhu G, Warren L, Aponte J, Gulsvik A, Bakke P, Anderson WH, Lomas DA, Silverman EK, Pillai SG (2007) The SERPINE2 gene is associated with chronic obstructive pulmonary disease in two large populations. Am J Respir Crit Care Med 176(2):167–173

PEDF in the Retina

12

Jeanee Bullock and S. Patricia Becerra

Contents

Abstract

Pigment epithelium-derived factor (PEDF) is a protective protein of the eye, with neurotrophic and anti-angiogenic activities. It was discovered as a secreted factor from retinal pigment epithelium. The interphotoreceptor matrix and

J. Bullock
Laboratory of Retina Cell and Molecular Biology, Section of Protein Structure and Function, National Eye Institute, National Institutes of Health, 6 Center Drive MSC 0608, Bethesda, MD 20892-0608, USA

Biochemistry and Molecular and Cellular Biology, Georgetown University Medical Center, Washington, DC, USA
e-mail: bullockjl@nei.nih.gov

S.P. Becerra (✉)
Laboratory of Retina Cell and Molecular Biology, Section of Protein Structure and Function, National Eye Institute, National Institutes of Health, 6 Center Drive MSC 0608, Bethesda, MD 20892-0608, USA
e-mail: becerrap@nei.nih.gov

M. Geiger et al. (eds.), *The Serpin Family*, DOI 10.1007/978-3-319-22711-5_12

vitreous humor contain soluble and diffusible PEDF protein, which interacts with glycosaminoglycans, collagen, and cell-surface receptors to act on retina cells. Sequence alignment reveals that PEDF is a member of the serpin superfamily. PEDF shares hallmark features of serpins including the conserved tertiary structure with an exposed peptide loop towards its C-end. However, it is classified as a noninhibitory serpin. This review will summarize studies on the role of PEDF in the retina, in particular its anti-angiogenic, antiapoptotic, and anti-inflammatory properties. Studies with established animal models for retina diseases have demonstrated the importance of PEDF in the eye. In addition, PEDF is considered an important marker for retinal pigment epithelium derived from stem cells. The therapeutic potential of PEDF will be emphasized.

12.1 Introduction and Morphology of the Retina and Retina Disorders

The human eye functions to detect and send information to the brain, where it is processed and converted to visual images. The transparent cornea and the lens of the eye allow light to enter the eye, while the changing diameter of the pupil regulates the amount of entering light. The lens focuses light onto the retina. The retina is the innermost layer of the eye responsible for transmitting images formed by the lens to the brain by way of the optic nerve. Light must pass through several layers of cells before reaching the cells responsible for the transmission of visual information.

The retina is composed of ten layers consisting of many types of neurons connected by synapses (Fig. 12.1). The photoreceptors process the light stimulus and are composed of two types of cells: rods, which are responsible for night vision, and cones, which are responsible for color vision. The human retina contains over 6 million cone cells and over 120 million rod cells. The outer pigmented layer of the retina is composed of a monolayer of hexagonal cells called the retinal pigment epithelium (RPE). The apical, or innermost, side of the RPE layer interacts with the photoreceptor layer, while the basolateral side faces the Bruch's membrane, an inner layer of the choroid. The RPE plays many roles in the eye. The RPE is responsible for absorbing the light that is focused on the retina by the lens. The RPE is also responsible for the daily phagocytosis of the outer segments of photoreceptors, a critical immune function. It obtains nutrients from the blood and supplies these nutrients to the photoreceptors. In the visual cycle, the RPE supplies retinal, a form of vitamin A, to photoreceptors. The RPE also secretes growth factors and cytokines to maintain the homeostasis of the neural retina. One of these factors is pigment epithelium-derived factor (PEDF).

The many layers of the retina constitute a complex network of components critical for maintaining eye health and development. Disruption of the retinal layers, including the RPE layer, can lead to a host of retinal diseases and disorders such as macular degeneration, retinitis pigmentosa, and diabetic retinopathy, leading to severe loss of vision. To combat this, the retina possesses several protective

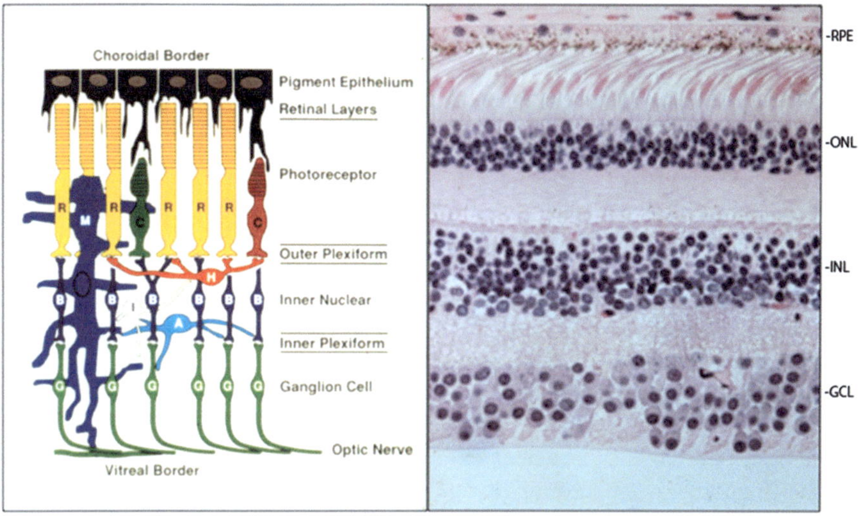

Fig. 12.1 Layers of the retina. (**a**) A scheme illustrating the layers of the retina composed of retinal ganglion cell layer (GCL), inner plexiform layer, inner nuclear layer (INL), outer plexiform layer, outer nuclear layer (ONL), external limiting layer, photoreceptor layer, retinal pigment epithelium (RPE). Labels of cell types correspond to ganglion cell (G), amacrine cell (A), bipolar cell (B), horizontal cell (H), Muller cell (M), rod photoreceptor cell (R), and cone photoreceptor cell (C). (**b**) Immunohistochemistry of a human retina

molecules and mechanisms involved in the regulation of retinal health. One such molecule is PEDF, which has been shown to be critical in eye health and possesses therapeutic value in treating diseases of the retina.

12.2 PEDF in the Retina

Although the *SERPINF1* gene that encodes PEDF is expressed by different ocular and non-ocular tissues and cells, the RPE expresses the highest levels of PEDF in the retina (Becerra et al. 2004). It was first reported that RPE-conditioned media induces neuronal differentiation in human Y79 retinoblastoma cells (Tombran Tink and Johnson 1989). PEDF was purified from RPE-conditioned media and identified as the neurite-promoting factor in human retinoblastoma Y79 cells (Steele et al. 1993). Later on, PEDF was identified in the interphotoreceptor matrix of mammalian eyes as the bona fide factor with differentiating activity on Y79 cells (Wu et al. 1995).

The human *SERPINF1* gene maps to chromosome 17p13.1-pter (Tombran-Tink et al. 1994). The *SERPINF1* gene contains eight exons and seven introns (Tombran-Tink et al. 1996). Sequencing analysis reveals that its messenger RNA is 1.5 kb (Steele et al. 1993), with an open reading frame that encodes a human polypeptide of 418 amino acids with a secretion signal peptide of 20 amino acid residues at its N-end and a glycosylation site at Asn285–Thr287. The mature product is secreted as a monomeric glycoprotein with a molecular weight of 50 kDa (Steele et al. 1993; Wu et al. 1995). The polarized RPE secretes PEDF in a preferred directional fashion towards its apical side into the interphotoreceptor matrix that bathes the outer segments of photoreceptors (Becerra et al. 2004).

The *SERPINF1* gene is expressed in fetal and adult human RPE cells (Tombran-Tink et al. 1995) as early as 17 weeks of gestation, suggesting a role in retinal development. PEDF is detected throughout the human retina (Karakousis et al. 2001). At 18 weeks of gestation, PEDF is present in horizontal cells in the outer part of the inner nuclear layer. At 21.5 weeks of gestation, PEDF is present in cytoplasmic granules in the RPE, ganglion cells of the inner nuclear layers, and differentiating cones. In the adult retina, PEDF is detected in the RPE and interphotoreceptor matrix as well as in photoreceptors and in inner retinal cell types (Karakousis et al. 2001). Aging retinal cells show a significant decrease in PEDF expression (Tombran-Tink et al. 1995). Interestingly, the PEDF distribution pattern in the developing eye is in agreement with the avascularity of the vitreous and aqueous humor and its neurotrophic role in the neural retina.

12.3 PEDF Is a Member of the Serpin Family

Amino acid sequence analysis and protein comparison data indicate that the PEDF sequence shows strong homology to members of the serine protease inhibitor (serpin) gene superfamily (Steele et al. 1993). The linear structure of PEDF possesses the accepted range of homology with other known serpins, which is between 23 and 26 % sequence identity (Steele et al. 1993) (Fig. 12.2). The overall folded structure contains the serpin-exposed peptide loop towards its carboxy end (Becerra et al. 1995). The crystal structure of PEDF further reveals that its tertiary structure is that of a serpin (Simonovic et al. 2001). Interestingly, folded PEDF has an asymmetric surface-exposed charge distribution, with a highly basic lysine-rich region on one side, and a highly acidic aspartic-acid-rich region on the opposite side, necessary for heparin and collagen binding, respectively (Fig. 12.3). When compared to other known serpins, it is clear that this charge distribution is unique to PEDF.

PEDF has been shown to behave as a substrate rather than an inhibitor of serine proteases (Becerra et al. 1993). Its P1 site is the residue leucine, specific for inhibition of leucyl-proteases, such as chymotrypsin. However, PEDF does not inhibit chymotrypsin, nor other leucyl-proteases such as cathepsin G, nor does it form an inhibitory complex with serine proteases (Becerra et al. 1993). It lacks the characteristic thermal stability of cleaved inhibitory serpins, and upon cleavage, PEDF does not undergo the conformational change from a stressed-to-relaxed heat-stable form (Becerra et al. 1995). Other noninhibitory serpins such as ovalbumin, angiotensinogen (Stein et al. 1989), and maspin (Pemberton et al. 1995) also lack this conformational change. A feature that some noninhibitory serpins share is the presence of unfavorable residues on the N-terminal side of its P1 leucine for insertion of the serpin loop into sheet A, which might be the cause of the lack of stressed-to-relax conformational change. Interestingly, cleaved PEDF at its serpin-

```
PEDF_HUMAN 1   MQALVLLLCIGALLGHSSC-----QNPASPPEEGSPDPDSTGALVEEEDPFFKVPVNKLA  55
A1AT_HUMAN 1   ---MPSSVSWGILLLAGLCCLVPVSLAEDPQGDAAQKTDT------SHHDQDHPTFNKIT  51
AACT_HUMAN 1   MERMLPLLALGL-LAAGFCPAVLC-HPNSPLDEENL-T-Q------ENQDRGTHVDLGLA  50
                  :   :. *  *  . *         .*  :             ...         ::

PEDF_HUMAN 56  AAVSNFGYDLYRVRSSTSPTTNVLLSPLSVATALSALSLGAEQRTESIIHRALYYDLI--  113
A1AT_HUMAN 52  PNLAEFAFSLYRQLAHQSNSTNIFFSPVSIATAFAMLSLGTKADTHDEILEGLNFNLTEI  111
AACT_HUMAN 51  SANVDFAFSLYKQLVLKAPDKNVIFSPLSISTALAFLSLGAHNTTLTEILKGLKFNLTET  110
                   :*.:.**:     :  .*:::**:*::**:: ****:.  *   * ..* ::*

PEDF_HUMAN 114 SSPDIHGTYKELLDTVTAPQKNLK--SASRIVFEKKLRIKSSFVAPLEKSYGTRPRVL-T  170
A1AT_HUMAN 112 PEAQIHEGFQELLRTLNQPDSQLQLTTGNGLFLSEGLKLVDKFLEDVKKLYHSEAFTVNF  171
AACT_HUMAN 111 SEAEIHQSFQHLLRTLNQSSDELQLSMGNAMFVKEQLSLLDRFTEDAKRLYGSEAFATDF  170
                . :**  ::.** *:.  ..:*:   .. :...: * : . *    :: * :.  .

PEDF_HUMAN 171 GNPRLDLQEINNWVQAQMKGKLARSTKEIPDEISILLLGVAHFKGQWVTKFDSRKTSLED  230
A1AT_HUMAN 172 GDTEEAKKQINDYVEKGTQGKIVDLVKELDRDTVFALVNYIFFKGKWERPFEVKDTEEED  231
AACT_HUMAN 171 QDSAAAKKLINDYVKNGTRGKITDLIKDLDSQTMMVLVNYIFFKAKWEMPFDPQDTHQSR  230
                :     : **::*:   :**:.   *::  :  : *:   .**.:*   *: :.*  .

PEDF_HUMAN 231 FYLDEERTVRVPMMSDPKAVLRYGLDSDLSCKIAQLPLTGSMSIIFFLPLKVTQNLTLIE  290
A1AT_HUMAN 232 FHVDQVTTVKVPMMKRLGMF-NIQHCKKLSSWVLLMKYLGNATAIFFLPDEGK--LQHLE  288
AACT_HUMAN 231 FYLSKKKWVMVPMMSLHHLTIPYFRDEELSCTVVELKYTGNASALFILPDQDK--MEEVE  288
               *::.:   * ****.           ..**. :  :   *. : :*:**  : .  : :*

PEDF_HUMAN 291 ESLTSEFIHDIDRE-LKTVQAVLTVPKLKLSYEGEVTKSLQEMKLQSLFD-SPDFSKITG  348
A1AT_HUMAN 289 NELTHDIITKFLENE-DRRSASLHLPKLSITGTYDLKSVLGQLGITKVFSNGADLSGVTE  347
AACT_HUMAN 289 AMLLPETLKRWRDSLEFREIGELYLPKFSISRDYNLNDILLQLGIEEAFTSKADLSGITG  348
                 *  : :     .      . * :**:.::   ::.. * :: : . *    *:* :*

PEDF_HUMAN 349 -KPIKLTQVEHRAGFEWNEDGAGTTPSPGLQP----AHLTFPLDYHLNQPFIFVLRDTDT  403
A1AT_HUMAN 348 EAPLKLSKAVHKAVLTIDEKGTEAAGAMFLEAI----PMSIPPEVKFNKPFVFLMIEQNT  403
AACT_HUMAN 349 ARNLAVSQVVHKAVLDVFEEGTEASAATAVKITLLSALVETRTIVRFNRPFLMIIVPTDT  408
                  : :::. *:* :   *.*: :: :  ::       :      ::*:**::::   :*

PEDF_HUMAN 404 GALLFIGKILDPRGP  418
A1AT_HUMAN 404 KSPLFMGKVVNPTQK  418
AACT_HUMAN 409 QNIFFMSKVTNPKQA  423
                  :*:.*: :*
```

Fig. 12.2 Amino acid sequence alignment of human PEDF, alpha 1-antitrypsin (A1AT_HUMAN) and antichymotrypsin (AACT_HUMAN). Human PEDF shares 27 % sequence identity and 42 % homology with antitrypsin. Human PEDF shares 27 % sequence identity and 44 % homology with antichymotrypsin

exposed loop is still able to retain its neurotrophic activity, confirming that its homologous reactive loop is dispensable for its neurotrophic activity. Thus, PEDF is classified as a noninhibitory serpin, and its biological activity is independent of its serine inhibition potential. It is believed that during evolution, PEDF lost its inhibitory function and gained other functions specific for PEDF (Table 12.1).

12.4 Biological Effects of PEDF in the Retina

12.4.1 Neuronal Differentiating Activity

PEDF is capable of morphologically differentiating Y79 retinoblastoma cells to a neuronal phenotype (Steele et al. 1993). Differentiated Y79 cells express the neuronal markers neuron-specific enolase and neurofilament proteins (Steele et al. 1993). PEDF promotes the increase and maturation of pigment granules in

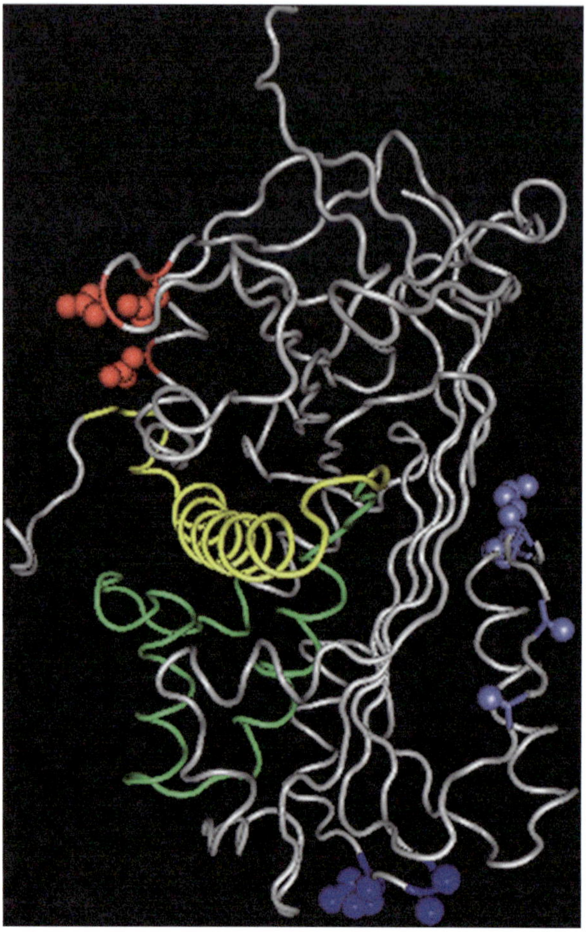

Fig. 12.3 Crystal structure of human PEDF with functional sites highlighted in color code from PDB 1IMV: neurotrophic region (*green*), anti-angiogenic region (*yellow*), collagen-binding region (*red*), heparin-binding region (*blue*), HA-binding region (*purple*)

Table 12.1 Summary of PEDF biological activity and targets in the retina

PEDF biological activity	Target
Neurotrophic	Retinoblastoma cells, Muller cells, retinal neurons
Prosurvival, antiapoptotic	Photoreceptors, retinal ganglion cells, retinal progenitor cells
Anti-angiogenic	Ocular endothelial cells
Anti-inflammatory	Muller cells and retinal ganglion cells
Stem cells, self-renewal	Retinal stem cells and human embryonic stem cells

neonatal rat pigment epithelial cells, suggesting that PEDF plays a role in RPE cell differentiation (Malchiodi-Albedi et al. 1998; Jablonski et al. 2000) reported that PEDF is morphogenetic for photoreceptors by supporting development of photoreceptor neurons during the final stages of retinal morphogenesis. The loss of differentiation of photoreceptor outer segments (Jablonski et al. 2000) and Muller cells (Jablonski et al. 2001) in *Xenopus* tadpole eyes is rescued when exogenous PEDF is introduced. PEDF promotes neurite outgrowth of retinal cells (Tanimoto et al. 2006) and can also enhance survival and axon regeneration of retinal ganglion cells (Vigneswara et al. 2013).

12.4.2 Prosurvival and Antiapoptotic Effects

PEDF is a prosurvival factor for retinal cells in vitro and in vivo. PEDF protects retinal neurons in culture against serum starvation and hydrogen peroxide-induced apoptotic cell death (Cao et al. 1999; Murakami et al. 2008; Subramanian et al. 2013). It is also protects cone photoreceptor cells from death by light damage (Rapp et al. 2014). Furthermore, it enhances survival of retinal ganglion cells and protects them against cytotoxicity (Pang et al. 2007; Vigneswara et al. 2013; Unterlauft et al. 2014).

PEDF has a protective effect in vivo in models of retinal degenerations. A group of inherited retinal diseases, collectively termed retinitis pigmentosa (RP), is characterized by the progressive and specific loss of photoreceptors, the light-transducing neurons of the retina. One advantage of retina field is the availability of animal models for retina degeneration. The *rd1/rd1* mouse is an autosomal-recessive mutant, in which a nonsense mutation in the rod photoreceptor-specific β-*phosphodiesterase* (β-PDE) gene leads to a rapid and massive death of photoreceptors in homozygous animals. In another model, a null mutation in the *peripherin/rds* gene, which encodes a structural component of photoreceptor outer segments, causes a protracted, apoptotic loss of photoreceptors. Mutations in both β-PDE and *peripherin/rds* have also been found in patients affected by dominant forms of retinal degenerations. Injection of PEDF protein into these models *rd1/rd1* and *rds/rds* can protect photoreceptors against degeneration due to death by apoptosis (Cayouette et al. 1999). PEDF also protects photoreceptors against light-induced damage. Injection of PEDF into the vitreous of rats prior to constant light exposure resulted in protection of photoreceptor function and morphology (Cao et al. 2001). A combination of PEDF and basic fibroblast growth factors improved functional protection on photoreceptors. Intraocular gene transfer of adenoviral PEDF vectors increases retinal cell survival by preventing apoptosis in pressure-induced ischemia (Takita et al. 2003).

Elucidation of the mechanism of the antiapoptotic activity of PEDF in the retina is of current interest. A receptor for PEDF in the retina with high affinity for PEDF has been identified as PEDF-R (Notari et al. 2006), which is required for the prosurvival activity of PEDF (Subramanian et al. 2013). PEDF-R is a member of the PNPLA2 family (patatin-like phospholipase A2). It is present on the surface of the retina and RPE cells. In the rat retina, PEDF-R is distributed in the inner segments of the photoreceptors and in less intensity in other neural retina cells (Notari et al. 2006). Subcellularly, the PEDF-R protein is found in plasma membranes as well as surrounding lipid droplets, where it is also known as adipose triglyceride lipase. PEDF-R is present in the plasma membrane of cone photoreceptor-derived 661W cells, which are known to respond to PEDF prosurvival effects (Rapp et al. 2014). The PEDF-R enzyme exhibits phospholipase as well as triglyceride lipase activities. It can catalyze the release of fatty acids from phospholipids, although the substrate specificity of PEDF-R is still unknown. Upon binding PEDF, the phospholipase activity of PEDF-R is enhanced (Notari et al. 2006). Therefore, fatty acids are considered putative bioactive mediators for the PEDF/PEDF-R system. In this regard, the retina contains phospholipids enriched in omega-3 fatty acids, which, in together with its derivatives like neuroprotection D1 and synaptimide, have demonstrable neuroprotective, neurite outgrowth, and anti-inflammatory properties (Bazan et al. 2013; Kim and Spector 2013). In neonatal rat R28 cells, an apoptosis-inducing factor-related pathway is an essential target of PEDF-mediated antiapoptotic activity (Murakami et al. 2008). PEDF protects R28 cells by upregulation of the antiapoptotic *B-cell lymphoma 2 (Bcl-2)* gene. Genetic silencing *Pnpla2* in R28 cells showed that PEDF-R is critical for PEDF-mediated cell survival and antiapoptosis effects (Subramanian et al. 2013). Pharmacological blocking PEDF from binding to PEDF-R with a synthetic receptor peptide leads to attenuation of the PEDF-mediated prosurvival activity. The current understanding of the mechanism of action in PEDF/PEDF-R-mediated survival activity involves the binding of PEDF to PEDF-R, which causes the release of fatty acids from plasma membranes. The release of fatty acids leads to the upregulation of *Bcl-2* ultimately leading to retinal cell survival, through a yet unknown mechanism. Further studies of the interaction between PEDF and PEDF-R will provide more insight into the mechanism of antiapoptotic activities.

12.4.3 Anti-angiogenic

A balance of proangiogenic and anti-angiogenic factors is required for ocular health. In contrast, uncontrolled angiogenesis is a common hallmark of many ocular diseases. Interestingly, the distribution of PEDF in the eye correlates with ocular avascularity. PEDF is a potent inhibitor of angiogenesis and acts as an antagonist of the proangiogenic factor vascular endothelial growth factor (VEGF) (Dawson et al. 1999). In contrast to the preferential basolateral secretion of VEGF (Blaauwgeers et al. 1999), the RPE secretes PEDF from its apicolateral side into the interphotoreceptor matrix (Becerra et al. 2004). PEDF is also present in the cornea of several mammalian species, where it prevents neovascularization (Dawson et al. 1999). In addition to VEGF, PEDF can inhibit the migration of endothelial cells in the presence of a variety of angiogenic factors, including platelet-derived growth factor, fibroblast growth factor interleukin-8, and lysophosphatidic acid. While VEGF levels increase, PEDF levels decrease in hypoxic conditions, indicating an inverse relationship of PEDF with VEGF (Notari et al. 2005).

Pathological retinal neovascularization is a feature of diabetic retinopathy and other ocular disorders characterized by retinal hypoxia. Oxygen-induced retinopathy in rodents is a model of retinal neovascularization in which vessels invade the inner retina. Choroidal neovascularization is one of the most important landmarks of neovascular age-related macular degeneration. Experimental choroidal neovascularization is performed by laser-induced breakage of the Bruch's membrane of rodents, which results in invasion of choroidal vessels into the RPE and neural retina (Campos et al. 2006). The inhibitory effects of PEDF on choroidal and retinal neovascularization in vivo have been proven by delivering purified protein by viral-mediated gene transfer into the vitreous, interphotoreceptor matrix, or subconjunctiva space (Mori et al. 2001, 2002a, b; Duh et al. 2002; Amaral and Becerra 2010). Huang et al. (2008) have shown that PEDF-deficient mice show increased retinal vasculature (Huang et al. 2008) implying that PEDF is an important regulator of the balance of retinal vascularization.

Several groups have reported on the mechanism of action of PEDF as an anti-angiogenic factor. Structure-function relationships have shown that a stretch of 34 amino acids from the amino-terminal end of the PEDF polypeptide, termed the 34-mer (positions 44-77), is responsible for the PEDF-mediated anti-angiogenic activity (Filleur et al. 2005; Amaral and Becerra 2010; Longeras et al. 2012). The pathways targeted by PEDF for inhibiting neovascularization have been summarized before (Becerra and Notario 2013). It appears that PEDF effects are cell context, and it can interact with a variety of partners and receptors including laminin receptor, cell-surface ATP synthase, gamma-secretase pathway, FAS/FASL pathway, etc.

12.4.4 Anti-inflammatory

PEDF can decrease the levels of proinflammatory cytokines. In addition to being an angiogenic factor, VEGF is also proinflammatory factor capable of upregulating expression of ICAM-1, another inflammatory factor (Lu et al. 1999). Therefore, as PEDF is a VEGF antagonist, it is considered anti-inflammatory. Zhang et al. (2006) have reported that a rat model of endotoxin-induced uveitis (EIU), characterized by retinal vascular hyperpermeability, shows decreased levels of PEDF and increased levels of proinflammatory cytokines (Zhang et al. 2006). Furthermore, PEDF injections into rat models of streptozotocin-induced diabetes and oxygen-induced retinopathy decrease vascular hyperpermeability and retinal inflammatory factors. In hypoxic conditions, PEDF reduces inflammatory factors TNF-α and ICAM-1 expression in cultured retinal capillary endothelial cells. Silencing of PEDF in retinal Muller cells results in the increase of VEGF and TNF-α secretion (Zhang et al. 2006).

Wang et al. (2013) have reported that injections of human recombinant PEDF protein into vitreous and subconjunctiva also lower the levels of anti-inflammatory molecules in the DKO rd8 mouse model of retina degeneration (Wang et al. 2013).

In addition they observed attenuation of focal lesion, less photoreceptor and RPE degeneration, and lower expression of apoptotic molecules.

12.4.5 Stem Cells and Self-Renewal

Retinal stem cell research is emerging as an attractive clinical therapy tool. There are several successful studies of differentiation of human embryonic stem cells and RPE human-induced pluripotent stem cells, which yields cells similar to those of native RPE. The ciliary body of adult mammals represents a source of quiescent retinal stem cells, which are neural progenitors with a limited self-renewal potential in vitro. De Marzo et al. (2010) have shown that a combination of PEDF and fibroblast growth factor supplemented in the media of retinal stem cells can influence positively their self-renewal (De Marzo et al. 2010). In experiments where human embryonic stem cell-derived RPE was being evaluated for its ability to promote survival of retinal progenitor cells, differentiated RPE cells were able to secrete high levels of PEDF capable of supporting retinal progenitor cell survival (De Marzo et al. 2010). The observations imply that PEDF can serve to overcome the limitation of the self-renewal potential of stem cells.

Several other research groups have used PEDF as a key marker of RPE derived from stem cells. Differentiated RPE cells derived from human embryonic and induced pluripotent stem cells have been generated and display hallmark functions of RPE (Vaajasaari et al. 2011). In addition to phagocytosis of photoreceptor outer segments, RPE derived from human embryonic and induced pluripotent stem cells secrete PEDF (Klimanskaya et al. 2004; Vaajasaari et al. 2011). Zhu et al. (2011) have reported that polarized human embryonic stem cell-derived RPE secrete large amounts of PEDF towards its apical side (Zhu et al. 2011) and that the conditioned media from the polarized human stem cell-derived RPE exhibits cells' survival effects on human fetal retinal progenitor cells, which were decreased when a blocking antibody to PEDF was added to the media. These results imply that PEDF plays an important role in retinal progenitor cell growth and protection.

12.5 Therapeutic Implications of PEDF in Retinal Diseases

The neuroprotective, anti-angiogenic, and anti-inflammatory properties of PEDF make it an attractive therapeutic agent. Decreased ocular levels of PEDF are reported for many ocular diseases including neovascularization (Duh et al. 2004), age-related macular degeneration (Holekamp et al. 2002), neuroretinal dystrophies (Ogata et al. 2004), and diabetic retinopathy (Spranger et al. 2001). So far, there are no known reports of toxicity by PEDF. For example, PEDF overexpressed in neonatal mice (Wong et al. 2004) yielding levels up to 3.5-fold than endogenous levels had no significant toxic effect. Strategies involving the delivery of PEDF and PEDF functional peptides could prove useful in treating many ocular diseases.

The photoreceptor protective properties of PEDF may serve to develop therapeutic drugs from this protein for preventing blindness originated by photoreceptor cell death. Some promising results of preclinical experiments with PEDF have been reported. As mentioned above, administration of PEDF protein into the vitreous results in lowering apoptosis of photoreceptors in retinal degeneration models *rd1* and *rds*. Intraocular gene transfer of PEDF for a mouse model of light-induced photoreceptor degeneration in Lewis rats results in photoreceptor cell rescue (Imai et al. 2005). Intravitreal injection of PEDF-impregnated nanoparticles delays photoreceptor degeneration by inhibiting apoptosis in Royal College of Surgeons (RSC) rats (Akiyama et al. 2012). Ocular therapies where patients are treated with PEDF could prove useful in delaying photoreceptor degeneration, thus preventing vision loss.

The anti-inflammatory effects of PEDF can be beneficial towards glaucoma, which is characterized by loss of retinal ganglion cells caused by increases in proinflammatory factors (Tezel et al. 2001). PEDF expression is reduced with age in the eyes of the DBA/2J glaucoma mouse model (Zhou et al. 2009). Transfection of PEDF in DBA/2J mice results in reduced retinal ganglion cells, reduced loss of the nerve fiber layer, and reduced expression of proinflammatory factors (Zhou et al. 2009).

The ability for PEDF to regulate neovascularization makes its therapeutic potential in many ocular diseases evident. It has been established that the imbalance of proangiogenic and anti-angiogenic factors plays a role in retinal neovascularization. In this regard, Gao et al. (2001) reported that there is a five-fold increase in VEGF and a two-fold decrease in PEDF in ischemia-induced rats with retinal neovascularization (Gao et al. 2001). To compensate the imbalance that is permissive to angiogenesis, several research groups have administered PEDF into animal models of ocular angiogenic diseases. For example, in transgenic mice with VEGF expression and a mouse model of choroid neovascularization, PEDF gene transfer results in regression of ocular neovascularization by promoting apoptosis of retinal cells with lesions (Mori et al. 2002b). Gene transfer of an adenoviral vector expressing PEDF results in higher intraocular levels of PEDF and suppressed choroid neovascularization (Gehlbach et al. 2003).

An imbalance of angiogenic regulators in the eye is also a hallmark of diabetic retinopathy (DR). Patients suffering from DR have an increase in retinal neovascularization caused by lack of blood flow, in which angiogenic factors are upregulated, while anti-angiogenic factors are downregulated. The anti-angiogenic properties of PEDF may constitute the base for potential therapies for DR patients. Spranger et al. (2001) have reported that the ocular fluids from patients with proliferative diabetic retinopathy show lower levels of PEDF than non-diabetic patients (Spranger et al. 2001). Patients who previously had photocoagulation, a surgery used to treat abnormal blood vessels in DR patients, showed higher levels of PEDF than those who did not. In a diabetes-like retinopathy mouse model, overexpression of PEDF via gene transfer leads to long-term prevention of neovascularization (Haurigot et al. 2012). The single injection also results in normalization of intraocular levels of VEGF and other proangiogenic molecules.

Additional complications of DR also benefit from PEDF treatment. Two PEDF derivatives of the anti-angiogenic region (positions 60–77) and neuroprotective region (positions 78–121) of PEDF were used to examine the effects of retinal complications caused by diabetes (Liu et al. 2012). In mouse retinas, the neuroprotective peptide reduces proinflammatory cytokines and prevents diabetes-induced microglia activation, retinal ganglion cell death, and inner plexiform layer thinning. Both peptides reduce vascular leakage. PEDF has therapeutic potential towards preventing patient loss for patients with DR.

The macula is a central part of the retina responsible for detailed vision. Macular degeneration results in central vision loss. Age-related macular degeneration (AMD) affects older adults. Choroidal neovascularization is a characteristic of the neovascular form of AMD called wet AMD, a severe form that can lead to retinal detachment. The levels of PEDF in patients with choroidal neovascularization due to AMD have been examined (Holekamp et al. 2002). Undiluted vitreous samples from patients with neovascular AMD have lower levels of PEDF. The vitreous of patients with AMD has lower levels of anti-angiogenic activity, probably due to lower levels of PEDF. In a study of Taiwan Chinese patients with AMD, a PEDF Met72Thr allele was found to be a possible risk factor for neovascular AMD (Lin et al. 2008). The other form, called dry AMD, is more prevalent than the wet one, in which photoreceptors undergo atrophy and RPE degeneration by yet unknown genetic and/or environmental causes. Elevation of proinflammatory cytokines is associated with these AMD forms. The mouse retina does not have a macula; therefore, mouse models for AMD mimic certain features of human AMD such as focal photoreceptor atrophy and RPE degeneration (Chu et al. 2012). The DKO rd8 mouse model retina shows some features of geographic atrophy with focal photoreceptor lesions, and its retina and RPE has decreased PEDF levels (Wang et al. 2013). Injection of PEDF into DKO rd8 mouse eyes shows attenuation of focal lesion, less photoreceptor and RPE degeneration, and lower expression of apoptotic and inflammatory molecules. The irreversible vision loss caused by AMD can be attenuated by PEDF treatment. Thus, the neurotrophic, anti-angiogenic, and anti-inflammatory properties of this interesting serpin point to promising PEDF-based therapies for AMD.

12.6 Conclusion

PEDF is a multifunctional protein and a member of the serpin family. It is ubiquitously expressed throughout the body and plays an important protective role in the retina, as demonstrated in numerous studies in vitro and in vivo. The therapeutic potential for PEDF is clear, and future preclinical studies towards the generation of PEDF-based drugs will prove beneficial towards treating retinal diseases.

Acknowledgments This research was supported by the Intramural Research Program of the National Eye Institute (NIH).

References

Akiyama G et al (2012) Photoreceptor rescue of pigment epithelium-derived factor-impregnated nanoparticles in Royal College of Surgeons rats. Mol Vis 18:3079–3086

Amaral J, Becerra SP (2010) Effects of human recombinant PEDF protein and PEDF-derived peptide 34-mer on choroidal neovascularization. Invest Ophthalmol Vis Sci 51(3):1318–1326. doi:10.1167/iovs.09-4455

Bazan NG et al (2013) Docosahexaenoic acid and its derivative neuroprotectin D1 display neuroprotective properties in the retina, brain and central nervous system. Nestle Nutr Inst Workshop Ser 77:121–131. doi:10.1159/000351395

Becerra SP, Notario V (2013) The effects of PEDF on cancer biology: mechanisms of action and therapeutic potential. Nat Rev Cancer 13(4):258–71. doi:10.1038/nrc3484

Becerra SP et al (1993) Overexpression of fetal human pigment epithelium-derived factor in Escherichia coli. A functionally active neurotrophic factor. J Biol Chem 268(31):23148–23156

Becerra SP et al (1995) Pigment epithelium-derived factor behaves like a noninhibitory serpin. Neurotrophic activity does not require the serpin reactive loop. J Biol Chem 270 (43):25992–25999. doi:10.1074/jbc.270.43.25992

Becerra SP et al (2004) Pigment epithelium-derived factor in the monkey retinal pigment epithelium and interphotoreceptor matrix: apical secretion and distribution. Exp Eye Res 78 (2):223–234. doi:10.1016/j.exer.2003.10.013

Blaauwgeers HG et al (1999) Polarized vascular endothelial growth factor secretion by human retinal pigment epithelium and localization of vascular endothelial growth factor receptors on the inner choriocapillaris. Evidence for a trophic paracrine relation. Am J Pathol 155(2):421–428. doi:10.1016/S0002-9440(10)65138-3

Campos M et al (2006) A novel imaging technique for experimental choroidal neovascularization. Invest Ophthalmol Vis Sci 47(12):5163–5170. doi:10.1167/iovs.06-0156

Cao W et al (1999) Pigment epithelium-derived factor protects cultured retinal neurons against hydrogen peroxide-induced cell death. J Neurosci Res 57(6):789–800. doi:10.1002/(SICI)1097-4547(19990915)57:6<789::AID-JNR4>3.0.CO;2-M

Cao W et al (2001) In vivo protection of photoreceptors from light damage by pigment epithelium-derived factor. Invest Ophthalmol Vis Sci 42(7):1646–1652

Cayouette M et al (1999) Pigment epithelium-derived factor delays the death of photoreceptors in mouse models of inherited retinal degenerations. Neurobiol Dis 6(6):523–532. doi:10.1006/nbdi.1999.0263

Chu XK et al (2012) Controversial view of a genetically altered mouse model of focal retinal degeneration. Bioengineered 4(3):130–135. doi:10.4161/bioe.22949

Dawson DW et al (1999) Pigment epithelium-derived factor: a potent inhibitor of angiogenesis. Science 285(5425):245–248. doi:10.1126/science.285.5425.245

De Marzo A et al (2010) PEDF promotes retinal neurosphere formation and expansion in vitro. Adv Exp Med Biol 664:621–630. doi:10.1007/978-1-4419-1399-9_71

Duh EJ et al (2002) Pigment epithelium-derived factor suppresses ischemia-induced retinal neovascularization and VEGF-induced migration and growth. Invest Ophthalmol Vis Sci 43 (3):821–829

Duh EJ et al (2004) Vitreous levels of pigment epithelium-derived factor and vascular endothelial growth factor: implications for ocular angiogenesis. Am J Ophthalmol 137(4):668–674. doi:10.1016/j.ajo.2003.11.015

Filleur S et al (2005) Two functional epitopes of pigment epithelial-derived factor block angiogenesis and induce differentiation in prostate cancer. Cancer Res 65(12):5144–5152. doi:10.1158/0008-5472.CAN-04-3744

Gao G et al (2001) Unbalanced expression of VEGF and PEDF in ischemia-induced retinal neovascularization. FEBS Lett 489:270–276

Gehlbach P et al (2003) Periocular injection of an adenoviral vector encoding pigment epithelium-derived factor inhibits choroidal neovascularization. Gene Ther 10(8):637–646. doi:10.1038/sj.gt.3301931
Haurigot V et al (2012) Long-term retinal PEDF overexpression prevents neovascularization in a murine adult model of retinopathy. PLoS One 7(7):e41511. doi:10.1371/journal.pone.0041511
Holekamp NM et al (2002) Pigment epithelium-derived factor is deficient in the vitreous of patients with choroidal neovascularization due to age-related macular degeneration. Am J Ophthalmol 134(2):220–227. doi:10.1016/S0002-9394(02)01549-0
Huang Q et al (2008) PEDF-deficient mice exhibit an enhanced rate of retinal vascular expansion and are more sensitive to hyperoxia-mediated vessel obliteration. Exp Eye Res 87(3):226–241. doi:10.1016/j.exer.2008.06.003
Imai D et al (2005) Intraocular gene transfer of pigment epithelium-derived factor rescues photoreceptors from light-induced cell death. J Cell Physiol 202(2):570–578. doi:10.1002/jcp.20155
Jablonski MM et al (2000) Pigment epithelium-derived factor supports normal development of photoreceptor neurons and opsin expression after retinal pigment epithelium removal. J Neurosci 20(19):7149–7157
Jablonski MM et al (2001) Pigment epithelium-derived factor supports normal Müller cell development and glutamine synthetase expression after removal of the retinal pigment epithelium. Glia 35(1):14–25. doi:10.1002/glia.1066
Karakousis PC et al (2001) Localization of pigment epithelium derived factor (PEDF) in developing and adult human ocular tissues. Mol Vis 7:154–163
Kim HY, Spector AA (2013) Synaptamide, endocannabinoid-like derivative of docosahexaenoic acid with cannabinoid-independent function. Prostaglandins Leukot Essent Fatty Acids 88 (1):121–125. doi:10.1016/j.plefa.2012.08.002
Klimanskaya I et al (2004) Derivation and comparative assessment of retinal pigment epithelium from human embryonic stem cells using transcriptomics. Cloning Stem Cells 6(3):217–245. doi:10.1089/clo.2004.6.217
Lin JM et al (2008) Pigment epithelium-derived factor gene Met72Thr polymorphism is associated with increased risk of wet age-related macular degeneration. Am J Ophthalmol 145 (4):716–721. doi:10.1016/j.ajo.2007.11.006
Liu Y et al (2012) Pigment epithelium-derived factor (PEDF) peptide eye drops reduce inflammation, cell death and vascular leakage in diabetic retinopathy in Ins2(Akita) mice. Mol Med 18:1387–1401. doi:10.2119/molmed.2012.00008
Longeras R et al (2012) A PEDF-derived peptide inhibits retinal neovascularization and blocks mobilization of bone marrow-derived endothelial progenitor cells. Exp Diabetes Res 2012:518426
Lu M et al (1999) VEGF increases retinal vascular ICAM-1 expression in vivo. Invest Opthalmol Vis Sci 40(8):1808–1812
Malchiodi-Albedi F et al (1998) PEDF (pigment epithelium-derived factor) promotes increase and maturation of pigment granules in pigment epithelial cells in neonatal albino rat retinal cultures. Int J Dev Neurosci 5:423–432. doi:10.1016/S0736-5748(98)00014-8
Mori K et al (2001) Pigment epithelium-derived factor inhibits retinal and choroidal neovascularization. J Cell Physiol 188(2):253–263. doi:10.1002/jcp.1114
Mori K et al (2002a) AAV-mediated gene transfer of pigment epithelium-derived factor inhibits choroidal neovascularization. Invest Ophthalmol Vis Sci 43(6):1994–2000
Mori K et al (2002b) Regression of ocular neovascularization in response to increased expression of pigment epithelium-derived factor. Invest Ophthalmol Vis Sci 43(7):2428–2434
Murakami Y et al (2008) Inhibition of nuclear translocation of apoptosis-inducing factor is an essential mechanism of the neuroprotective activity of pigment epithelium-derived factor in a rat model of retinal degeneration. Am J Pathol 173(5):1326–1338. doi:10.2353/ajpath.2008.080466

Notari L et al (2005) Pigment epithelium-derived factor is a substrate for matrix metalloproteinase type 2 and type 9: implications for downregulation in hypoxia. Invest Ophthalmol Vis Sci 46(8):2736–2747. doi:10.1167/iovs.04-1489

Notari L et al (2006) Identification of a lipase-linked cell membrane receptor for pigment epithelium-derived factor. J Biol Chem 281(49):38022–38037. doi:10.1074/jbc.M600353200

Ogata N et al (2004) Decreased levels of pigment epithelium-derived factor in eyes with neuroretinal dystrophic diseases. Am J Ophthalmol 137(6):1129–1130. doi:10.1016/j.ajo.2003.11.080

Pang IH et al (2007) Pigment epithelium-derived factor protects retinal ganglion cells. BMC Neurosci 8:11. doi:10.1186/1471-2202-8-11

Pemberton PA et al (1995) The tumor suppressor maspin does not undergo the stressed to relaxed transition or inhibit trypsin-like serine proteases. Evidence that maspin is not a protease inhibitory serpin. J Biol Chem 270(26):15832–15837. doi:10.1074/jbc.270.26.15832

Rapp M et al (2014) Pigment epithelium-derived factor protects cone photoreceptor-derived 661W cells from light damage through Akt activation. Adv Exp Med Biol 801:813–820. doi:10.1007/978-1-4614-3209-8_102

Simonovic M et al (2001) Crystal structure of human PEDF, a potent anti-angiogenic and neurite growth-promoting factor. Proc Natl Acad Sci U S A 98(20):11131–11135. doi:10.1073/pnas.211268598

Spranger J et al (2001) Loss of the antiangiogenic pigment epithelium-derived factor in patients with angiogenic eye disease. Diabetes 50(12):2641–2645. doi:10.2337/diabetes.50.12.2641

Steele FR et al (1993) Pigment epithelium-derived factor: neurotrophic activity and identification as a member of the serine protease inhibitor gene family. Proc Natl Acad Sci USA 90(4):1526–1530. doi:10.1073/pnas.90.4.1526

Stein PE et al (1989) Ovalbumin and angiotensinogen lack serpin S-R conformational. Biochem J 262(1):103–107

Subramanian P et al (2013) Pigment epithelium-derived factor (PEDF) prevents retinal cell death via PEDF Receptor (PEDF-R): identification of a functional ligand binding site. J Biol Chem 288(33):23928–23942. doi:10.1074/jbc.M113.487884

Takita H et al (2003) Retinal neuroprotection against ischemic injury mediated by intraocular gene transfer of pigment epithelium-derived factor. Invest Ophthalmol Vis Sci 44(10):4497–4504. doi:10.1167/iovs.03-0052

Tanimoto S et al (2006) Pigment epithelium-derived factor promotes neurite outgrowth of retinal cells. J Med Sci 55(4):109–116

Tezel G et al (2001) TNF-alpha and TNF-alpha receptor-1 in the retina of normal and glaucomatous eyes. Invest Ophthalmol Vis Sci 42(8):1787–1794

Tombran-Tink J, Johnson LV (1989) Neuronal differentiation of retinoblastoma cells induced by medium conditioned by human RPE cells. Invest Ophthalmol Vis Sci 30(8):1700–1707

Tombran-Tink J et al (1994) Localization of the gene for pigment epithelium-derived factor (PEDF) to chromosome 17p13.1 and expression in cultured human retinoblastoma cells. Genomics 19(2):266–272. doi:10.1006/geno.1994.1057

Tombran-Tink J et al (1995) Expression, secretion, and age-related downregulation of pigment epithelium-derived factor, a serpin with neurotrophic activity. J Neurosci 15(7):4992–5003

Tombran-Tink J et al (1996) Organization, evolutionary conservation, expression and unusual Alu density of the human gene for pigment epithelium-derived factor, a unique neurotrophic serpin. Mol Vis 2(11)

Unterlauft JD et al (2014) Enhanced survival of retinal ganglion cells is mediated by Müller glial cell-derived PEDF. Exp Eye Res 127:206–214. doi:10.1016/j.exer.2014.08.004

Vaajasaari H et al (2011) Toward the defined and xeno-free differentiation of functional human pluripotent stem cell-derived retinal pigment epithelial cells. Mol Vis 17:558–575

Vigneswara V et al (2013) Pigment epithelium-derived factor is retinal ganglion cell neuroprotective and axogenic after optic nerve crush injury. Invest Ophthalmol Vis Sci 54(4):2624–2633. doi:10.1167/iovs.13-11803

Wang Y et al (2013) Pigment epithelium-derived factor reduces apoptosis and pro-inflammatory cytokine gene expression in a murine model of focal retinal degeneration. ASN Neuro 5(5): e00126. doi:10.1042/AN20130028
Wong WT et al (2004) Effect of over-expression of pigment epithelium derived factor (PEDF) on developing retinal vasculature in the mouse. Mol Vis 10:837–844
Wu YQ et al (1995) Identification of pigment epithelium-derived factor in the interphotoreceptor matrix of bovine eyes. Protein Expr Purif 6(4):447–456. doi:10.1006/prep.1995.1060
Zhang SX et al (2006) Pigment epithelium-derived factor (PEDF) is an endogenous antiinflammatory factor. FASEB J 20(2):323–325. doi:10.1096/fj.05-4313fje
Zhou X et al (2009) Anti-inflammatory effect of pigment epithelium-derived factor in DBA/2J mice. Mol Vis 15:438–450
Zhu D et al (2011) Polarized secretion of PEDF from human embryonic stem cell-derived RPE promotes retinal progenitor cell survival. Invest Ophthalmol Vis Sci 52(3):1573–1585. doi:10.1167/iovs.10-6413

Defining the Proteostasis Network Responsible for Managing the Fate of Newly Synthesized Alpha1-Antitrypsin

13

Richard N. Sifers

Contents

Abstract

Alpha1-antitrypsin (AAT) is a serine proteinase inhibitor (serpin) secreted from hepatocytes. Its normal circulating concentration in the bloodstream is sufficient to prevent the destruction of lung elastin fibers by excessive elastase released from activated neutrophils, thereby maintaining the organ's elasticity and function. Many naturally occurring genetic variants of AAT, unable to acquire correct structural maturation following biosynthesis, are subjected to intracellular proteolysis in the hepatocyte secretory pathway. This, in turn, can contribute to the development of chronic obstructive pulmonary disease. In addition, the inappropriate accumulation of structurally aberrant AAT within hepatocytes can contribute to the etiology of liver disease. The mechanistic analysis of intracellular systems that manage AAT fate has led to the identification of dedicated

R.N. Sifers (✉)
Department of Pathology and Immunology, Baylor College of Medicine, Houston, TX 77030, USA
e-mail: rsifers@bcm.edu

M. Geiger et al. (eds.), *The Serpin Family*, DOI 10.1007/978-3-319-22711-5_13

systems that facilitate proper polypeptide folding and orchestrate the selective elimination of molecules that fail to acquire conformational maturation. These systems, which contribute to the diverse cellular proteostasis network, are expected to aid in the identification of disease modifiers and development of novel strategies for therapeutic intervention of the lung and liver diseases, as well as additional conformational disorders that involve the secretory pathway. This review focuses on the discovery and characterization of proteostasis systems that operate in the secretory pathway that selectively eliminate different forms of structurally aberrant AAT.

13.1 Conformational Diseases

Many inherited diseases result from mechanistic complications that take place at the level of encoded proteins, rather than the gene (Balch et al. 2008). This perspective stems from the understanding that genetic information is transformed into biological activity in response to translation of the encoded polypeptide which then undergoes conformational maturation and deployment. Therefore, a current challenge in both basic science and biomedical research is to elucidate how these disrupted processes (Fig. 13.1) contribute to, or influence, the molecular pathogenesis of numerous inherited disorders. The term "proteostasis" represents a relatively new designation for the totality of intracellular systems and molecular machinery

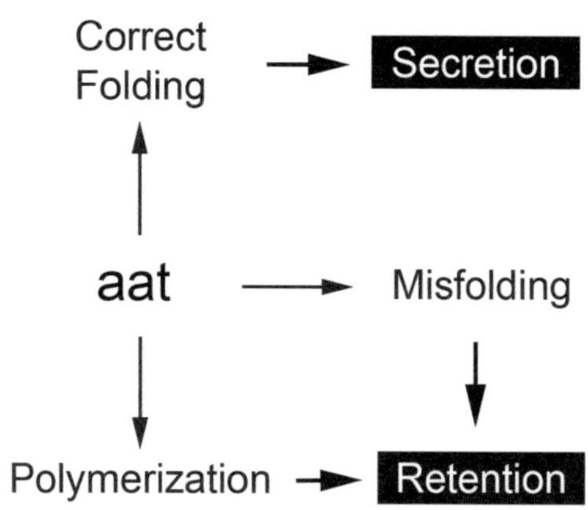

Fig. 13.1 Potential fates for the newly synthesized alpha1-antitrypsin polypeptide. The polypeptide translocated into the endoplasmic reticulum exists in a nonnative conformation (AAT). The addition of three asparagine-linked oligosaccharides (not shown) can allow lectin-like molecular chaperones to facilitate correct polypeptide folding which leads to productive intracellular transport and secretion. Alternatively, polypeptides that misfold are unable to acquire the native conformation, leading to persistent interaction with the molecular chaperones and prolonged intracellular retention. Finally, for some genetic variants, mutations in the polypeptide promote a fraction of the newly synthesized molecules to undergo spontaneous polymerization, leading to chaperone-independent retention

responsible for managing all aspects of cellular proteome homeostasis (Hutt et al. 2009). Because two of its central roles are to facilitate the correct conformational maturation of translated proteins and orchestrate the selective removal of those molecules unable to attain this structural milestone, the mechanistic characterization of these pathways is appropriate to elucidate the underlying mechanisms that might contribute to the development of numerous conformational disorders and especially those operating in the secretory pathway of eukaryotic cells (Sifers 2010).

13.1.1 Lung Disease

To this end, several naturally occurring genetic variants of AAT have been identified that coincide with a diminished circulating concentration of the proteinase inhibitor (Sifers et al. 1992) as an etiologic agent of lung disease (Lomas and Mahadeva 2002). The heritable phenomenon of severe AAT deficiency can lead to clinically relevant consequences. As the most abundant proteinase inhibitor in human plasma, a primary physiologic role for AAT is to protect lung elastin fibers from excessive degradation by neutrophil elastase (Eriksson 1965). In many individuals, a severe deficiency (10–15 % of normal) of the circulating proteinase inhibitor is an underlying cause for excessive proteolytic destruction of lung elastic tissues, resulting from the unopposed hydrolytic action of elastase released from activated neutrophils (Eriksson 1965; Lomas and Mahadeva 2002). Early-onset panlobular emphysema has been designated as a primary loss-of-function phenotype. The situation is exacerbated because the available *Z* variant is less effective than the wild-type molecule (fivefold diminished association rate between the *Z* variant and neutrophil elastase) (Carrell et al. 1997). The problem is further intensified by the inappropriate structural conversion caused by the amino acid substitution (see later) that converts AAT to a chemoattractant for human neutrophils (Parmar et al. 2002).

13.1.2 Liver Disease

In addition, the accumulation of a fraction of undegraded AAT molecules that have undergone inappropriate, but spontaneous, polymerization (Crowther et al. 2004) leads to the formation of diastase-resistant inclusion bodies derived from the hepatocyte ER (Graham et al. 1990), and these are thought to function as an etiologic agent for the development of liver disease. In most cases, the detection of periodic acid–Schiff-positive inclusion bodies in hepatocytes corresponds to distended regions of the hepatocyte endoplasmic reticulum (ER), resulting from the inappropriately accumulated *Z* variant (Carlson et al. 1989). The observation also led to the eventual identification of the liver as the primary source for AAT biosynthesis (Sifers et al. 1992). Transgenic animal models have confirmed that the inclusion bodies can contribute to chronic liver injury as a primary gain-of-toxic-function phenotype (Perlmutter 1993). In some individuals the disorder can progress to hepatitis, cirrhosis, and/or hepatocellular carcinoma (Perlmutter 1993, 2006), the latter of which has also been detected in some, but not all, transgenic mouse models. Presently, orthotopic organ transplantation is most commonly used treatment to alleviate the end-stage liver and lung diseases (Perlmutter 1993).

13.2 Proteostasis in the Secretory Pathway

These observations have made AAT deficiency a clinically relevant paradigm to initiate a mechanistic investigation of how cells facilitate the structural maturation and intracellular transport of proteins in the secretory pathway of eukaryotic cells (Sifers et al. 1989) and how these processes might contribute to the extensive phenotypic variability associated with either disorder (Volpert et al. 2000). As more fully described in the next section, intensive experimental investigation led to the discovery that correct conformational maturation of the newly synthesized AAT polypeptide is a prerequisite for its productive transport along compartments of the secretory pathway. Subsequent discoveries identified a role for the processing of asparagine-linked oligosaccharides in promoting the conformational maturation of newly synthesized glycoproteins in the early secretory pathway (Liu et al. 1997; Molinari 2007). In addition, a functionally linked system was discovered through which additional covalent modification of the carbohydrate moieties orchestrates the proteolytic elimination of numerous structurally aberrant glycoproteins retained in the endoplasmic reticulum, as well as the utilization of autophagic disposal to eliminate inappropriately polymerized AAT (Teckman and Perlmutter 2000). Finally, recent studies have uncovered the existence of an unconventional disposal branch by which the misfolded monomeric AAT polypeptide is eliminated (Pan et al. 2011; Pan et al. 2013; Iannotti et al. 2014), leading to the identification of the system's participation in accelerating onset of the end-stage liver disease phenotype (Pan et al. 2009).

The secretory pathway functions as a physical route through which newly synthesized proteins destined for secretion or membrane association are transported (Gething and Sambrook 1992). A proteostasis network manages all aspects of this multistep process (Balch et al. 2008). This particular subset of polypeptides contain a signal sequence and are targeted into the endoplasmic reticulum as they emerge from ribosomes bound to the Sec61 translocon. Well-defined systems utilize molecular chaperones to facilitate conformational maturation (Gething and Sambrook 1992) that is needed for productive transport to more distant compartments (Ellgaard and Helenius 2001). Nevertheless, the cycling of numerous chaperone-bound proteins between the ER and early regions of the Golgi complex is also common in mammalian cells (Caldwell et al. 2001). Proteins that fail to achieve conformational maturation or properly assemble into larger complexes are subjected to "ER-associated degradation" (ERAD) which culminates with their dislocation into the cytoplasm for elimination by 26S proteasomes. Some known aspects of facilitated protein folding and quality control are described below.

13.2.1 Facilitated Glycoprotein Folding

As a general rule, the final three-dimensional conformation of a protein is dictated by the primary amino acid sequence (Ellgaard and Helenius 2001). Throughout the cell, the efficiency of protein conformational maturation process is greatly enhanced in response to physical engagement with molecular chaperones, all of which hinder nonproductive folding events as a means to facilitate the attainment of native structure (Gething and Sambrook 1992). The endoplasmic reticulum (ER) is no exception as it contains a specialized "protein-folding environment" where molecular chaperones and chaperone-related proteins engage newly synthesized polypeptides as they emerge during translocation across the ER membrane (Gething and Sambrook 1992; Plemper and Wolf 1999; Fewell et al. 2001). Arguably, the intracellular machinery responsible for facilitating protein conformational maturation is best understood for asparagine-linked glycoproteins (Fig. 13.2). Numerous polypeptides translocated across the ER membrane during biosynthesis, including AAT, are subjected to asparagine-linked glycosylation at specific asparagine residues (Sifers et al. 1988). The polar nature of these complex appendages helps maintain the solubility of polypeptides and generate a scaffold through which a group of oligosaccharide-processing enzymes can mediate the glycoprotein folding pathway (Cabral et al. 2001; Molinari 2007). Each newly added appendage consists of a branched, 14-unit oligosaccharide. Cotranslational removal of the outer glucose units by glucosidases I and II generates a monoglycosylated oligosaccharide recognized by the lectin-like molecular chaperones, calnexin, and calreticulin. Release from either lectin coincides with the enzymatic removal of the remaining glucose unit by ER glucosidase II (Helenius et al. 1992; Hammond and Helenius 1995). Subsequently, UDP-glucose:glycoprotein glucosyltransferase functions as a glycoprotein-folding sensor that uses UDP-glucose as a donor to catalyze the transfer of a single glucose unit back to high mannose-type glycans attached to nonnative proteins (Ellgaard et al. 1999; Cabral et al. 2002). This conformation-driven posttranslational event regenerates the monoglycosylated oligosaccharide to promote reassembly of nonnative glycoproteins with the glycoprotein-folding machinery described above (Sousa et al. 1992). Following several rounds of interaction, successful conformational maturation releases native glycoproteins from the protein-folding machinery for productive delivery to downstream compartments (Klausner and Sitia 1990). For some proteins, the conformational maturation process is coupled to rounds of cycling between the ER and early regions of the Golgi complex (Caldwell et al. 2001), requiring that molecular chaperones are no longer bound to allow productive transport beyond early regions of the Golgi process (i.e., cis-Golgi Network).

13.2.2 Conventional Mannosidase-Mediated Glycoprotein ERAD

As mentioned above, newly synthesized proteins that either remain unfolded for long periods, eventually misfold, or fail to assemble into multimeric complexes will

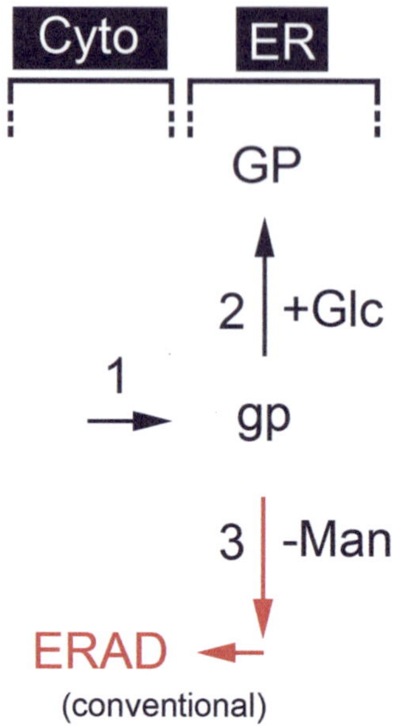

Fig. 13.2 Conventional glycoprotein folding and ERAD systems. The folding and intracellular degradation of many newly synthesized glycoproteins are facilitated by the covalent modification of asparagine-linked oligosaccharides linked to the polypeptide. Following the insertion of the ribosome-associated polypeptide from the cytoplasm (Cyto) into the endoplasmic reticulum (ER), the nonnative polypeptide is generated and glycosylated (GP) (step1). The existence of a terminal glucose unit (step 2) promotes physical interaction with the molecular chaperones calnexin and calreticulin to promote folding of the glycoprotein into its native conformation (GP). For those polypeptides unable to achieve this structural milestone in a biologically relevant time frame, two mannose units are removed from the asparagine-linked oligosaccharides to generate a degradation signal (step 3) that leads to the conventional ERAD routing system

persistently interact with molecular chaperones (Gething and Sambrook 1992). Rather than compete with other proteins for molecular chaperones, resulting in nonspecific protein aggregation and disruption of normal secretory pathway function, these conformationally impaired proteins are selectively eliminated by a complex disposal system collectively coined ERAD under basal conditions which culminates in proteolytic elimination by cytoplasmic 26S proteasomes (Bonifacino and Weissman 1998; Plemper and Wolf 1999).

The exact combination of events by which any individual protein is selectively eliminated is still under investigation. However, the intracellular machinery responsible for orchestrating the selective disposal of misfolded asparagine-linked glycoproteins retained in the ER is currently best understood (Cabral et al. 2002; Molinari 2007; Olivari and Molinari 2007). The capacity of alpha-1,2-mannosidase inhibitors to abrogate glycoprotein ERAD provided an initial clue that asparagine-linked oligosaccharides, in addition to promoting glycoprotein folding, can act as an appendage through which a degradation signal can be generated. This repeated observation from numerous labs (Cabral et al. 2002; Hosokawa et al. 2007;

Molinari 2007; Bernasconi et al. 2010) eventually led to an understanding that the opportunistic removal of 1,2-linked mannose units from asparagine-linked oligosaccharides (Fig. 13.2) in response to prolonged ER residence (Bernasconi et al. 2008; Satoh et al. 2010) designates that the associated glycoprotein has failed to acquire native structure within a biologically relevant time frame (Wu et al. 2003). The modified glycans promote the extraction of glycoproteins from the calnexin cycle that facilitates the folding of asparagine-linked glycoproteins (Sousa et al. 1992) and, in combination with nonnative protein structure, are suspected to complete the formation of a proposed bipartite glycoprotein ERAD signal (Wu et al. 2003). The temporal manner in which the bipartite signal is formed is suspected to distinguish chronically misfolded glycoproteins from nonnative wild-type folding intermediates, such that only the former population is degraded (Wu et al. 2003). Although ER mannosidase I (ERManI), which is also designated MAN1B1, was initially considered to play a major role in this process (Hosokawa et al. 2003; Wu et al. 2003; Karaveg et al. 2005), a recent publication has definitively negated this possibility (Ninagawa et al. 2014). It was determined that the removal of alpha-1,2-mannoses from misfolded proteins retained in the ER is sequentially catalyzed by ER degradation-enhancing mannosidases (EDEMs) 2 and 3 (Ninagawa et al. 2014) which are members of the class 47 glycosylhydrolase family, as is ERManI. Subsequent dislocation of the tagged proteins into the cytosol for proteasomal destruction involves the cooperation of additional events and proteins, many of which are currently under intense investigation.

13.3 Unconventional ERAD

A series of more recent studies have demonstrated that ERManI/MAN1B1, which is essential for ERAD of the misfolded AAT monomer (Liu et al. 1997; Cabral et al. 2000; Hosokawa et al. 2003), contributes to a newly discovered unconventional ERAD branch, as discussed below. For example, both newly synthesized recombinant human and endogenous mouse ERManI/MAN1B1 orthologs are subject to rapid lysosomal downregulation shortly following biosynthesis (Wu et al. 2007). Based on several characteristics of this event, it has been suggested that most of the newly synthesized ERManI/MAN1B1 molecules are subjected to ERAD tuning (Bernasconi and Molinari 2011), a lysosomal targeting process known to manage the intracellular concentration of numerous ERAD regulators under basal conditions (Cali et al. 2008; Reggiori et al. 2010). Consistent with this notion, the lysosomal downregulation of ERAD regulators is hindered in response to ER stress, and this is also observed for ERManI/MAN1B1 (Wu et al. 2007) via the inositol-responsive element 1–X box binding protein (Ire1-Xbp1) branch of the mammalian UPR (Shen et al. 2001). It should be noted, however, that unlike most ERAD regulators whose genes are transcribed more intensely during ER stress, translation of the ERManI/MAN1B1 gene (*MAN1B1*) is not transcriptionally upregulated in response to conditions of ER stress (Yoshida et al. 2003).

Collectively, the aforementioned observations began to hint about a likely paradigm shift in the manner that ERAD pathways might operate in lower and higher eukaryotic cells. This notion was greatly strengthened by the recent published observation that two known ER degradation-enhancing mannosidases (EDEMs) 2 and 3 (Ninagawa et al. 2014) contribute to the conventional ERAD of asparagine-linked glycoproteins retained in the ER, whereas ERManI/MAN1B1 is completely dispensable (Ninagawa et al. 2014). Importantly, these findings support our recent published observations that the bulk of endogenous ERManI in numerous human cell lines resides in the Golgi complex where the protein receives four O-linked oligosaccharides, each of which contains a terminal sialic acid residue (Pan et al. 2011). Golgi residence of the endogenous protein has also been reported by additional groups (Rafiq et al. 2011; Rymen et al. 2013), and organelle-specific proteomics has detected ERManI/MAN1B1 in COPI vesicles recycling from the Golgi complex (Gilchrist et al. 2006). Consistent with these observations, Kopito and colleagues (Christianson et al. 2011) were unable to detect endogenous ERManI/MAN1B1 in the protein linkage map for members of the conventional ERAD machinery.

In subsequent studies, we identified the existence of gamma-COP-binding sites in the amino-terminal cytoplasmic tail of ERManI (Pan et al. 2013), as well as the co-immunoprecipitation of ERManI/MAN1B1 with a misfolded monomeric AAT variant degraded by 26S proteasomes (Pan et al. 2013). In addition, we observed that removal of the catalytic domain does not prevent or even hinder recombinant ERManI/MAN1B1 from inducing proteolytic degradation (Iannotti et al. 2014) and then identified a conserved decapeptide sequence (conserved among vertebrate orthologs) within the remaining stem domain that is essential for both co-immunoprecipitation and proteolytic degradation of misfolded AAT (Iannotti et al. 2014). Because the elevated expression of recombinant ERManI/MAN1B1 will inappropriately target newly synthesized wild-type AAT and transferrin for proteasomal degradation (Wu et al. 2003), this combination of recent findings fits a model in which ERManI/MAN1B1 contributes to the functioning of a unique cargo capture and retrieval complex that manages proteins, like AAT, that escape the ER as part of the normal folding and trafficking itinerary (Iannotti et al. 2014). Because ERManI/MAN1B1 is not utilized in conventional ERAD, and the knockdown of EDEMs 2 and 3 does not interfere with proteasomal degradation mediated by ERManI/MAN1B1 (E. Young, unpublished observations), we conclude that ERManI/MAN1B1 contributes to the operation of a newly discovered unconventional ERAD branch in higher eukaryotes (Fig. 13.3). In our opinion, the existence of this additional unconventional ERAD branch to the preexisting conventional ERAD system (Brodsky 2012) likely represents an evolutionary expansion of the proteostasis network (Balch et al. 2008; Iannotti et al. 2014) whose operation is governed by distinct mechanistic principals, including the exclusion of oligosaccharide degradation signals on the protein substrate to promote proteasomal degradation (Iannotti et al. 2014).

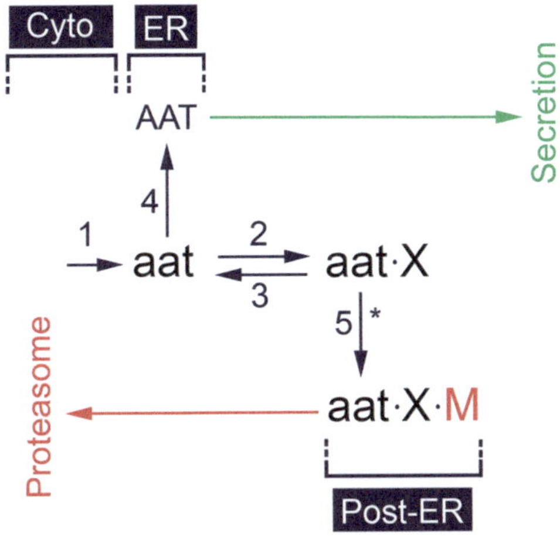

Fig. 13.3 Nonnative alpha1-antitrypsin monomers are managed by an unconventional ERAD branch. The newly synthesized alpha1-antitrypsin (AAT) polypeptide is translocated from the cytoplasm (Cyto) into the endoplasmic reticulum (ER) where it is glycosylated and bound to a variety of molecular chaperones (step1). Rather than remaining in the ER, it is eventually transported to a post-endoplasmic reticulum (post-ER) compartment (step 2) where it is captured by one or more presumptive cargo receptors (factor X). Retrieval of the chaperone-bound AAT folding intermediate back to the ER (step 3) allows the alpha1-antitrypsin polypeptide to reengage with the polypeptide folding machinery, allowing for conformational maturation (AAT) (step 4) and secretion. For AAT molecules that fail to correctly fold, ERManI/MAN1B1 (factor M) eventually associates with the AAT-cargo receptor complex (step 5). Existence of ERManI/MAN1B1, as a new member of the retrieval complex, plays a deterministic/adaptor role that promotes proteasomal elimination of associated AAT. Currently, it is suspected that the experimental elevation of ERManI/MAN1B1 enhances the rate in which step 5 occurs, thereby enhancing the elimination of misfolded AAT as well as the newly synthesized wild-type polypeptide (*asterisk*). Presently, it is assumed that polymerization of the late folding intermediate occurs following step 3 and possibly triggered by repeated introduction to the acidic environment of post-ER compartments

13.4 AAT Polymerization and Autophagic Disposal

The fate of misfolded, monomeric AAT has been discussed up to this point. However, approximately 95 % of patients that exhibit severe AAT deficiency are homozygous for the proteinase inhibitor Z (PI Z) variant (incidence of homozygosity is ~1/1800 live births) (Sifers et al. 1992; Crowther et al. 2004). A single amino acid substitution (Glu394Lys) at the base of the reactive center loop diminishes the molecule's secretion in response to hindered folding kinetics (Lomas et al. 1992), such that only a very small fraction of monomers are secreted from hepatocytes (Le et al. 1992; Qu et al. 1996). The mutation promotes the specific conformational rearrangement of a late, monomeric folding intermediate (Fig. 13.1) that favors subsequent polymerization (Yu et al. 1995). Several mechanisms of polymerization have been proposed and will be the subject of a chapter within this series. Nevertheless, intermolecular polymerization also underlies the severe deficiency associated with the rare *Siiyama* (Ser-53Phe) and *Mmalton* (deletion of residue 52) deficiency variants of AAT, plus the milder deficiency variants designated *S* (Glu264Val) and *I* (Arg39Cys) (Lomas and Mahadeva 2002). Metabolic pulse-chase radiolabeling studies have convincingly demonstrated that the rate in which the monomeric Z variant is subject to intracellular transport is greatly hindered as compared with correctly folded monomers (Le et al. 1990; Teckman and Perlmutter 1996; Teckman et al. 2001). The structural anomaly is considered pathogenic because mutations suspected to most favor polymerization also coincide with the greatest risk of liver injury and most severe plasma deficiency (Carrell et al. 1997). Moreover, several members of the serpin family exhibit mutations that result in their spontaneous polymerization, intrahepatic retention, and plasma deficiency. Identification of the common molecular mechanism has allowed for the different disorders to be designated a new class of disorders called the *serpinopathies* (Lomas and Mahadeva 2002). Finally, the recent use of a monoclonal antibody specific for the detection of AAT polymers led to the detection of Z polymers in emphysematous tissue that colocalized with neutrophils in the lung alveoli (Lomas and Mahadeva 2002). Therefore, polymerization not only limits the secretion of AAT from hepatocytes but might also directly contribute to the destruction of lung elastin fibers. The proinflammatory nature of the material was further substantiated by the influx of neutrophils into the lungs of mice instilled with polymers. Based on these observations, one must now consider whether the chemoattractant nature of Z polymers might promote the progression of severe lung damage.

Although accumulation of the Z variant induces ER stress, several reports have convincingly demonstrated that the classical unfolded protein response (UPR) pathway is not activated either in transfected cultured cell lines or in transgenic mice (Graham et al. 1990; Hidvegi et al. 2005). The exact mechanism(s) by which the accumulated polymers lead to cellular injury is currently under intense investigation and includes augmented transcription of nuclear factor-B (Hidvegi et al. 2005) and a marked autophagic response that strongly correlates with the absolute amount of the mutant protein accumulated within the individual cell (Teckman and Perlmutter 2000). Stimulation of the apoptotic cascade with specific patterns of both mitochondrial autophagy and mitochondrial injury has also been detected (Teckman and Perlmutter 2000; Teckman et al. 2004). Based on these findings, plus the successful use of autophagy enhancers as an experimental therapeutic treatment (Hidvegi et al. 2010), it is now apparent that autophagy is playing a pivotal role in endeavoring to remove the accumulated material in an attempt to protect hepatocytes from injury.

13.5 Genetic Modifier of the End-Stage Liver Disease

Importantly, accumulation of polymerized AAT is apparently necessary, but not sufficient, to promote liver disease. In fact, only a subset (10–15 %) of patients who accumulate Z polymers in the hepatocyte ER actually develop clinically relevant liver disease (Perlmutter 1993; Qu et al. 1997). Because patients with liver disease exhibit extensive phenotypic variability in terms of the time of disease onset (Perlmutter 2000), the involvement of genetic modifiers is suspected to control the rate at which the disease is manifested and possibly its severity. Notably, Z polymers accumulate in the distended hepatocyte ER (Carlson et al. 1989; Graham et al. 1990), rather than in the cytoplasm, implying that an early step in the disposal process may lack the ability to initially target the secretion-impaired molecules into the cytoplasm. In support of this notion, a delay in the degradation of the *Z* variant has been observed in transduced fibroblasts from patients who eventually underwent liver transplantation (Wu et al. 1994), implying that the intracellular degradation process likely contributes to disease pathogenesis. One suggestion is that severe liver injury results from a greater burden (Perlmutter 1991) of accumulated Z polymers in response to their impaired intracellular clearance.

Because a monomeric folding intermediate functions as the direct precursor of toxic AAT polymers, and ERManI/MAN1B1 contributes to the elimination of the monomer, we asked whether a single-nucleotide polymorphism in the human *MAN1B1* gene might contribute to phenotypic variability of the liver disease. Therefore, genomic DNA was extracted from resected livers obtained during transplantation to allow for the candidate gene approach. With the use of statistical analyses, it was demonstrated that homozygosity for a single-nucleotide polymorphism (SNP) in the 3′-untranslated region of the ERManI/MAN1B1 mRNA transcript (i.e., the "A" nucleotide allele as rs4567) coincides with onset of end-stage liver disease in Caucasian ZZ patients at 2 years of age or less, requiring orthotopic liver transplantation (Pan et al. 2009). In the inclusion of addition of quantifiable biochemical methodology in a set of functional assays, it was then concluded that the "A" allele of the SNP induces the conditional suppression of ERManI/MAN1B1 translation in response to ER stress caused by the intracellular accumulation of the Z variant. Even more recently, a combination of predictive algorithms and miRNA profiling determined that the "A" allele was responsible for generating a new microRNA (miRNA) binding site and that elevated concentrations of the corresponding miRNA are sufficient to suppress the translation of endogenous ERManI/MAN1B1 in a SNP-dependent manner in both transfected laboratory cell lines and genotyped ZZ patient fibroblasts (S. Pan, unpublished observation). Based on these findings, a model has been proposed in which the single-nucleotide polymorphism generates a conditional hypomorphic allele for ERManI/MAN1B1 that, when inherited in a homozygous fashion, can impair the liver's capacity to efficiently degrade Z polymers, thereby accelerating the rate at which toxic polymers are formed, accelerating the attainment of a toxic threshold. It should be noted that homozygosity for the SNP is not sufficient to promote the development of liver disease in most ZZ patients (Qu et al. 1997; Perlmutter 2000), indicating that it likely functions as a disease modifier and that the underlying major causal defect is still unknown. Nevertheless, these observations support the notion that proteasomal elimination of the newly synthesized Z monomeric intermediate, which is the direct precursor for the formation of polymers, is directed by the unconventional ERAD branch for which ERManI/MAN1B1 plays a pivotal role.

13.6 Conclusions and Future Directions

The analysis of AAT as a secreted serpin has played a major role in identifying the proteostasis machinery of the secretory pathway and defining the principles by which the network operates. The information led to the discovery of roles played by asparagine-linked oligosaccharides in the facilitation of glycoprotein folding and orchestration of the conventional ERAD system. More recently, the identification of a novel unconventional ERAD branch has emerged in which a putative cargo receptor must still be identified, and the deterministic role played by bound ERManI/MAN1B1 must be clearly defined at a mechanistic level. If it holds true that the catalytic activity of ERManI/MAN1B1 is dispensable for its role in unconventional ERAD, then it is likely that its role in biology has been repurposed during evolution. Additional future pursuits will be designed to understand how the unconventional ERAD pathway emerged during the evolution of cellular compartmentalization and to elucidate the underlying mechanistic principles in which the new system operates under basal and stress conditions.

The contribution of a lowered ERManI/MAN1B1 concentration in accelerating the onset of the end-stage liver disease is a clinically relevant observation that must be further analyzed and may be amenable to the use of synthetic complimentary antagomers to diminish the expression of the responsible microRNA. Therefore, the underlying biology of both ERManI/MAN1B1 and the unconventional ERAD branch will be further investigated as sources for new prognostic indicators and novel sites for therapeutic intervention of both the liver disease and lung diseases (Fig. 13.4), plus elucidation of their roles in promoting the phenotypic variability of both disorders.

Potential treatment options:

$\downarrow$ Synthesis

$\uparrow$ Conformational maturation

$\uparrow$ Unconventional ERAD

$\uparrow$ Autophagy

Fig. 13.4 Potential treatment options for associated conformational diseases. Knowing the manner in which the conformational maturation and elimination of the alpha1-antitrypsin monomer is orchestrated, plus having identified most of the corresponding molecular machinery, several potential options have conceptually emerged to treat the associated liver and lung diseases. One option is to use interfering RNAi technology to selectively hinder the synthesis of the alpha1-antitrypsin polypeptide. This is a reasonable strategy because the formation of toxic polymers should be halted, and additional genetic and/or environmental factors are apparently necessary for lowered circulating levels of the inhibitor to cause lung disease. Another proposed option is to use combinations of pharmacological chaperones to increase the efficiency of conformational maturation. This strategy is predicted to diminish both polypeptide misfolding and polymerization, allowing for more protein to be secreted. In another strategy, attempts will be made to enhance the efficiency of unconventional ERAD. This should prevent the formation of toxic polymers and should counteract the destructive role of the identified SNP rs4567 (see text). Finally, early attempts to enhance autophagy are being explored to prevent the intracellular accumulation of toxic polymer

References

Balch WE, Morimoto RI et al (2008) Adapting proteostasis for disease intervention. Science 319 (5865):916–919
Bernasconi R, Molinari M (2011) ERAD and ERAD tuning: disposal of cargo and of ERAD regulators from the mammalian ER. Curr Opin Cell Biol 23(2):176–183
Bernasconi R, Pertel T et al (2008) A dual task for the Xbp1-responsive OS-9 variants in the mammalian endoplasmic reticulum: inhibiting secretion of misfolded protein conformers and enhancing their disposal. J Biol Chem 283(24):16446–16454
Bernasconi R, Galli C et al (2010) Stringent requirement for HRD1, SEL1L, and OS-9/XTP3-B for disposal of ERAD-LS substrates. J Cell Biol 188(2):223–235
Bonifacino JS, Weissman AM (1998) Ubiquitin and the control of protein fate in the secretory and endocytic pathways. Annu Rev Cell Dev Biol 14:19–57
Brodsky JL (2012) Cleaning up: ER-associated degradation to the rescue. Cell 151(6):1163–1167
Cabral CM, Choudhury P et al (2000) Processing by endoplasmic reticulum mannosidases partitions a secretion-impaired glycoprotein into distinct disposal pathways. J Biol Chem 275(32):25015–25022
Cabral CM, Liu Y et al (2001) Dissecting glycoprotein quality control in the secretory pathway. Trends Biochem Sci 26(10):619–624
Cabral CM, Liu Y et al (2002) Organizational diversity among distinct glycoprotein endoplasmic reticulum-associated degradation programs. Mol Biol Cell 13(8):2639–2650
Caldwell SR, Hill KJ et al (2001) Degradation of endoplasmic reticulum (ER) quality control substrates requires transport between the ER and Golgi. J Biol Chem 276(26):23296–23303
Cali T, Vanoni O et al (2008) The endoplasmic reticulum crossroads for newly synthesized polypeptide chains. Prog Mol Biol Transl Sci 83:135–179
Carlson JA, Rogers BB et al (1989) Accumulation of PiZ alpha 1-antitrypsin causes liver damage in transgenic mice. J Clin Invest 83(4):1183–1190
Carrell R, Lomas D et al (1997) Dysfunctional variants and the structural biology of the serpins. Adv Exp Med Biol 425:207–222
Christianson JC, Olzmann JA et al (2011) Defining human ERAD networks through an integrative mapping strategy. Nat Cell Biol 14(1):93–105
Crowther DC, Belorgey D et al (2004) Practical genetics: alpha-1-antitrypsin deficiency and the serpinopathies. Eur J Hum Genet 12(3):167–172
Ellgaard L, Helenius A (2001) ER quality control: towards an understanding at the molecular level. Curr Opin Cell Biol 13(4):431–437
Ellgaard L, Molinari M et al (1999) Setting the standards: quality control in the secretory pathway. Science 286(5446):1882–1888
Eriksson S (1965) Studies in alpha 1-antitrypsin deficiency. Acta Med Scand Suppl 432:1–85
Fewell SW, Travers KJ et al (2001) The action of molecular chaperones in the early secretory pathway. Annu Rev Genet 35:149–191
Gething MJ, Sambrook J (1992) Protein folding in the cell. Nature 355(6355):33–45
Gilchrist A, Au CE et al (2006) Quantitative proteomics analysis of the secretory pathway. Cell 127(6):1265–1281
Graham KS, Le A et al (1990) Accumulation of the insoluble PiZ variant of human alpha 1-antitrypsin within the hepatic endoplasmic reticulum does not elevate the steady-state level of grp78/BiP. J Biol Chem 265(33):20463–20468
Hammond C, Helenius A (1995) Quality control in the secretory pathway. Curr Opin Cell Biol 7 (4):523–529
Helenius A, Marquardt T et al (1992) The endoplasmic reticulum as a protein-folding compartment. Trends Cell Biol 2(8):227–231
Hidvegi T, Schmidt BZ et al (2005) Accumulation of mutant alpha1-antitrypsin Z in the endoplasmic reticulum activates caspases-4 and -12, NFkappaB, and BAP31 but not the unfolded protein response. J Biol Chem 280(47):39002–39015

Hidvegi T, Ewing M et al (2010) An autophagy-enhancing drug promotes degradation of mutant alpha1-antitrypsin Z and reduces hepatic fibrosis. Science 329(5988):229–232

Hosokawa N, Tremblay LO et al (2003) Enhancement of endoplasmic reticulum (ER) degradation of misfolded Null Hong Kong alpha1-antitrypsin by human ER mannosidase I. J Biol Chem 278(28):26287–26294

Hosokawa N, You Z et al (2007) Stimulation of ERAD of misfolded null Hong Kong alpha1-antitrypsin by Golgi alpha1,2-mannosidases. Biochem Biophys Res Commun 362(3):626–632

Hutt DM, Powers ET et al (2009) The proteostasis boundary in misfolding diseases of membrane traffic. FEBS Lett 583(16):2639–2646

Iannotti MJ, Figard L et al (2014) A Golgi-localized mannosidase (MAN1B1) plays a non-enzymatic gatekeeper role in protein biosynthetic quality control. J Biol Chem 289 (17):11844–11858

Karaveg K, Siriwardena A et al (2005) Mechanism of class 1 (glycosylhydrolase family 47) {alpha}-mannosidases involved in N-glycan processing and endoplasmic reticulum quality control. J Biol Chem 280(16):16197–16207

Klausner RD, Sitia R (1990) Protein degradation in the endoplasmic reticulum. Cell 62 (4):611–614

Le A, Graham KS et al (1990) Intracellular degradation of the transport-impaired human PiZ alpha 1-antitrypsin variant. Biochemical mapping of the degradative event among compartments of the secretory pathway. J Biol Chem 265(23):14001–14007

Le A, Ferrell GA et al (1992) Soluble aggregates of the human PiZ alpha 1-antitrypsin variant are degraded within the endoplasmic reticulum by a mechanism sensitive to inhibitors of protein synthesis. J Biol Chem 267(2):1072–1080

Liu Y, Choudhury P et al (1997) Intracellular disposal of incompletely folded human alpha1-antitrypsin involves release from calnexin and post-translational trimming of asparagine-linked oligosaccharides. J Biol Chem 272(12):7946–7951

Lomas DA, Mahadeva R (2002) Alpha1-antitrypsin polymerization and the serpinopathies: pathobiology and prospects for therapy. J Clin Invest 110(11):1585–1590

Lomas DA, Evans DL et al (1992) The mechanism of Z alpha 1-antitrypsin accumulation in the liver. Nature 357(6379):605–607

Molinari M (2007) N-glycan structure dictates extension of protein folding or onset of disposal. Nat Chem Biol 3(6):313–320

Ninagawa S, Okada T et al (2014) EDEM2 initiates mammalian glycoprotein ERAD by catalyzing the first mannose trimming step. J Cell Biol 206(3):347–356

Olivari S, Molinari M (2007) Glycoprotein folding and the role of EDEM1, EDEM2 and EDEM3 in degradation of folding-defective glycoproteins. FEBS Lett 581(19):3658–3664

Pan S, Huang L et al (2009) Single nucleotide polymorphism-mediated translational suppression of endoplasmic reticulum mannosidase I modifies the onset of end-stage liver disease in alpha1-antitrypsin deficiency. Hepatology 50(1):275–281

Pan S, Wang S et al (2011) Golgi localization of ERManI defines spatial separation of the mammalian glycoprotein quality control system. Mol Biol Cell 22(16):2810–2822

Pan S, Cheng X et al (2013) Golgi-situated endoplasmic reticulum alpha-1, 2-mannosidase contributes to the retrieval of ERAD substrates through a direct interaction with gamma-COP. Mol Biol Cell 24(8):1111–1121

Parmar JS, Mahadeva R et al (2002) Polymers of alpha(1)-antitrypsin are chemotactic for human neutrophils: a new paradigm for the pathogenesis of emphysema. Am J Respir Cell Mol Biol 26 (6):723–730

Perlmutter DH (1991) The cellular basis for liver injury in alpha 1-antitrypsin deficiency. Hepatology 13(1):172–185

Perlmutter DH (1993) Liver disease associated with alpha 1-antitrypsin deficiency. Prog Liver Dis 11:139–165

Perlmutter DH (2000) Liver injury in alpha 1-antitrypsin deficiency. Clin Liver Dis 4(2):387–408, vi

Perlmutter DH (2006) Pathogenesis of chronic liver injury and hepatocellular carcinoma in alpha-1-antitrypsin deficiency. Pediatr Res 60(2):233–238
Plemper RK, Wolf DH (1999) Endoplasmic reticulum degradation. Reverse protein transport and its end in the proteasome. Mol Biol Rep 26(1–2):125–130
Qu D, Teckman JH et al (1996) Degradation of a mutant secretory protein, alpha1-antitrypsin Z, in the endoplasmic reticulum requires proteasome activity. J Biol Chem 271(37):22791–22795
Qu D, Teckman JH et al (1997) Review: alpha 1-antitrypsin deficiency associated liver disease. J Gastroenterol Hepatol 12(5):404–416
Rafiq MA, Kuss AW et al (2011) Mutations in the alpha 1,2-mannosidase gene, MAN1B1, cause autosomal-recessive intellectual disability. Am J Hum Genet 89(1):176–182
Reggiori F, Monastyrska I et al (2010) Coronaviruses Hijack the LC3-I-positive EDEMosomes, ER-derived vesicles exporting short-lived ERAD regulators, for replication. Cell Host Microbe 7(6):500–508
Rymen D, Peanne R et al (2013) MAN1B1 deficiency: an unexpected CDG-II. PLoS Genet 9(12), e1003989
Satoh T, Chen Y et al (2010) Structural basis for oligosaccharide recognition of misfolded glycoproteins by OS-9 in ER-associated degradation. Mol Cell 40(6):905–916
Shen X, Ellis RE et al (2001) Complementary signaling pathways regulate the unfolded protein response and are required for C. elegans development. Cell 107(7):893–903
Sifers RN (2010) Intracellular processing of alpha1-antitrypsin. Proc Am Thorac Soc 7 (6):376–380
Sifers RN, Brashears-Macatee S et al (1988) A frameshift mutation results in a truncated alpha 1-antitrypsin that is retained within the rough endoplasmic reticulum. J Biol Chem 263 (15):7330–7335
Sifers RN, Finegold MJ et al (1989) Alpha-1-antitrypsin deficiency: accumulation or degradation of mutant variants within the hepatic endoplasmic reticulum. Am J Respir Cell Mol Biol 1 (5):341–345
Sifers RN, Finegold MJ et al (1992) Molecular biology and genetics of alpha 1-antitrypsin deficiency. Semin Liver Dis 12(3):301–310
Sousa MC, Ferrero-Garcia MA et al (1992) Recognition of the oligosaccharide and protein moieties of glycoproteins by the UDP-Glc:glycoprotein glucosyltransferase. Biochemistry 31 (1):97–105
Teckman JH, Perlmutter DH (1996) The endoplasmic reticulum degradation pathway for mutant secretory proteins alpha1-antitrypsin Z and S is distinct from that for an unassembled membrane protein. J Biol Chem 271(22):13215–13220
Teckman JH, Perlmutter DH (2000) Retention of mutant alpha(1)-antitrypsin Z in endoplasmic reticulum is associated with an autophagic response. Am J Physiol Gastrointest Liver Physiol 279(5):G961–G974
Teckman JH, Burrows J et al (2001) The proteasome participates in degradation of mutant alpha 1-antitrypsin Z in the endoplasmic reticulum of hepatoma-derived hepatocytes. J Biol Chem 276(48):44865–44872
Teckman JH, An JK et al (2004) Mitochondrial autophagy and injury in the liver in alpha 1-antitrypsin deficiency. Am J Physiol Gastrointest Liver Physiol 286(5):G851–G862
Volpert D, Molleston JP et al (2000) Alpha1-antitrypsin deficiency-associated liver disease progresses slowly in some children. J Pediatr Gastroenterol Nutr 31(3):258–263
Wu Y, Whitman I et al (1994) A lag in intracellular degradation of mutant alpha 1-antitrypsin correlates with the liver disease phenotype in homozygous PiZZ alpha 1-antitrypsin deficiency. Proc Natl Acad Sci USA 91(19):9014–9018
Wu Y, Swulius MT et al (2003) Elucidation of the molecular logic by which misfolded alpha 1-antitrypsin is preferentially selected for degradation. Proc Natl Acad Sci USA 100 (14):8229–8234
Wu Y, Termine DJ et al (2007) Human endoplasmic reticulum mannosidase I is subject to regulated proteolysis. J Biol Chem 282(7):4841–4849

Yoshida H, Matsui T et al (2003) A time-dependent phase shift in the mammalian unfolded protein response. Dev Cell 4(2):265–271
Yu MH, Lee KN et al (1995) The Z type variation of human alpha 1-antitrypsin causes a protein folding defect. Nat Struct Biol 2(5):363–367

Endoplasmic Reticulum Stress and the Protein Overload Response in the Serpinopathies

14

Adriana Ordóñez and Stefan J. Marciniak

Contents

Abstract

The endoplasmic reticulum (ER) is an essential compartment for the synthesis and maturation of transmembrane and secretory proteins. Perturbations of protein folding by environmental or genetic factors result in the accumulation of misfolded proteins within the lumen of the ER. This leads to ER stress, a state defined as an imbalance between the rate of secretory protein synthesis and the capacity of the ER to fold new client proteins. The unfolded protein response (UPR) and the ER overload response (EOR) are two stress-induced signalling pathways emanating from the ER during the accumulation of either misfolded or well-folded client proteins, respectively. The serpins are unique as a class of proteins, since naturally occurring mutations appear to cause disease by triggering either the UPR or the EOR depending upon the specific nature of the mutation, such that different mutants of the same protein can preferentially

A. Ordóñez, PhD • S.J. Marciniak, MD PhD (✉)
Department of Medicine, Cambridge Institute for Medical Research (CIMR), University of Cambridge, Cambridge, UK
e-mail: aog23@cam.ac.uk; sjm20@cam.ac.uk

M. Geiger et al. (eds.), *The Serpin Family*, DOI 10.1007/978-3-319-22711-5_14

trigger one or other response. Here, we discuss these signalling responses and how study of the serpinopathies has shed light on this area of cell biology.

14.1 Endoplasmic Reticulum: Structure and Function

The endoplasmic reticulum (ER) is composed of an interconnected network of tubular membranes and flattened cisternae in continuity with the nuclear membrane that together enclose an internal space, the ER lumen, that is topologically separated from the cytosol (Voeltz et al. 2002). It is a multifunctional organelle involved in the synthesis, folding and maturation of secreted and transmembrane proteins prior to their trafficking along the secretory pathway. An appropriate balance between the load of client proteins within the ER and the folding capacity of the organelle is essential for its proper function. The ER is also the major site for the intracellular storage of calcium, which can be released to mediate intracellular signalling (Rutkowski and Kaufman 2004). Since many of the chaperones within the ER lumen have evolved to function only in a high calcium environment to facilitate productive folding of proteins, depletion of ER calcium stores can precipitate secretory protein misfolding (Lievremont et al. 1997).

ER proteins that misfold will fail to pass the ER quality control checkpoint that ensures only correctly folded proteins exit the ER (Ellgaard and Helenius 2003). Such misfolded ER client proteins are retained within the lumen of ER where repeated cycles of interaction with ER-resident chaperones continue to promote folding towards the native conformation (Fewell et al. 2001; Chambers and Marciniak 2014). However, proteins that remain misfolded are eventually degraded either by specific ER-associated protein degradation (ERAD) (McCracken and Brodsky 1996; Meusser et al. 2005; Vembar and Brodsky 2008) or by autophagy of portions of the ER and its contents (Ravikumar et al. 2010). When protein-folding homeostasis (proteostasis) fails, misfolded proteins can accumulate within the lumen and lead to 'ER stress' (Marciniak and Ron 2006; Walter and Ron 2011). Eukaryotic cells have therefore evolved a surveillance system to sense and respond to ER stress by activating a homeostatic pathway termed the 'unfolded protein response' (UPR) (Walter and Ron 2011) (Fig. 14.1). Ultimately, if the cell fails to re-establish proteostasis, ER stress leads to programmed cell death by apoptosis (Tabas and Ron 2011).

The precise mechanism by which the UPR is triggered remains a matter for debate. One model suggests that when the level of free BiP falls owing to an increased load of misfolded protein, it releases transmembrane signalling molecules PERK, IRE1 and ATF6 from an inhibitory interaction (Bertolotti et al. 2000; Shen et al. 2002). This links the activation of the UPR to BiP's known interaction with exposed hydrophobic residues of incompletely folded and misfolded proteins (Gething et al. 1986; Dorner et al. 1992). However, it has been reported that mutants of yeast Ire1p lacking the BiP-interaction site remain able to respond to ER stress (Kimata et al. 2004; Pincus et al. 2010). An alternative model suggests that these ER

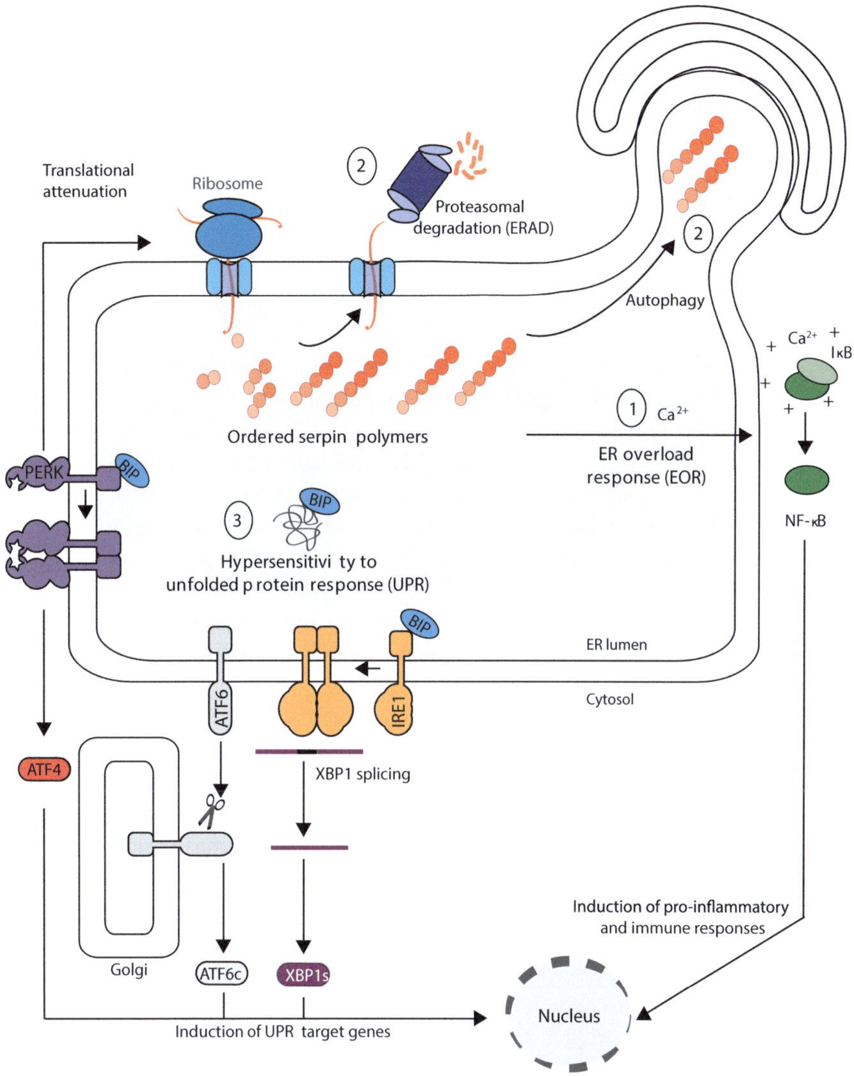

Fig. 14.1 Signalling responses from the lumen of the endoplasmic reticulum (ER). Newly synthesised serpin molecules enter the ER cotranslationally, are folded by resident chaperones and continue along the secretory pathway. (1) Polymers of mutant serpins activate NF-κB via a calcium-dependent ER overload response (EOR) triggering pro-inflammatory responses. (2) In a cell type-dependent manner, serpin mutants are eliminated either by autophagy of insoluble polymers or by ER-associated degradation (ERAD) of soluble monomers. (3) The presence of serpin polymers makes the ER hypersensitised to the activation of the unfolded protein response (UPR) in response to a second insult. PERK attenuates global translation while increasing ATF4 synthesis; IRE1 initiates splicing of XBP1 mRNA; and ATF6 is cleaved to a soluble transcription factor (ATF6c). Ultimately, these pathways enable the cell to adapt to ER stress by inducing chaperones and protein degradation machinery or else trigger cell death

stress sensors can interact directly with misfolded proteins (Credle et al. 2005; Gardner and Walter 2011). These two models are not mutually exclusive, but they both stress the importance of elevated misfolded protein as the trigger. It is therefore unsurprising that the accumulation of well-folded proteins within the ER lumen fails to activate the UPR (Hidvegi et al. 2005; Davies et al. 2009). Instead, such well-folded proteins have been shown to induce a distinct and less well-understood stress pathway called the 'ER overload response' (EOR) characterised by NF-κB activation (Pahl and Baeuerle 1995; Lawrence 2009) (Fig. 14.1).

14.2 ER Dysfunction and Serpinopathies

In this chapter, we will explore the role of ER stress pathways in the development of the serpinopathies, diseases caused by point mutants of *ser*ine *p*rotease *in*hibitors (serpins) (Lomas and Mahadeva 2002). An unusual feature of the serpinopathies is that many of their point mutants accumulate as large polymerised assemblies of relatively well-folded protein, e.g. Z (E342K) α_1-antitrypsin (Lomas et al. 1992; Yamasaki et al. 2008, 2011). Ordered polymers of serpin molecules are not trafficked but instead accumulate as inclusions within the ER (Granell et al. 2008; Miranda et al. 2008; Ordonez et al. 2013) (Fig. 14.2a). Although the mechanism of polymer toxicity is not fully understood, these mutant proteins cause disease both in the synthesising tissue by a toxic gain of function (Dycaico et al. 1988) and at distant sites through loss of function allowing uncontrolled proteolytic cascades (Eriksson 1965; Elliott et al. 1998). In addition to these polymeric mutants, serpinopathies have been associated with truncated (misfolding) variants (e.g. the null Hong Kong (Sifers et al. 1988) and Saar (Faber et al. 1994) α_1-antitrypsin) that are efficient substrates for ERAD (Brodbeck and Brown 1992; Liu et al. 1997) and potently induce ER stress if allowed to accumulate (Hidvegi et al. 2005; Ordonez et al. 2013). As such, the serpinopathies are excellent models for the study of ER dysfunction since well-characterised disease-associated and synthetic mutants of the same protein can trigger either the UPR or the EOR depending upon the consequence of each mutation for protein folding.

14.3 UPR and EOR Signalling

It appears that the accumulation of misfolded or of polymerised well-folded proteins represents two distinct classes of stress to the cell, since not only are the stress sensors distinct, but the downstream signalling differs (Ron and Walter 2007; Pahl and Baeuerle 1995) (Fig. 14.1). But, whereas the UPR has been studied extensively, the EOR remains somewhat mysterious.

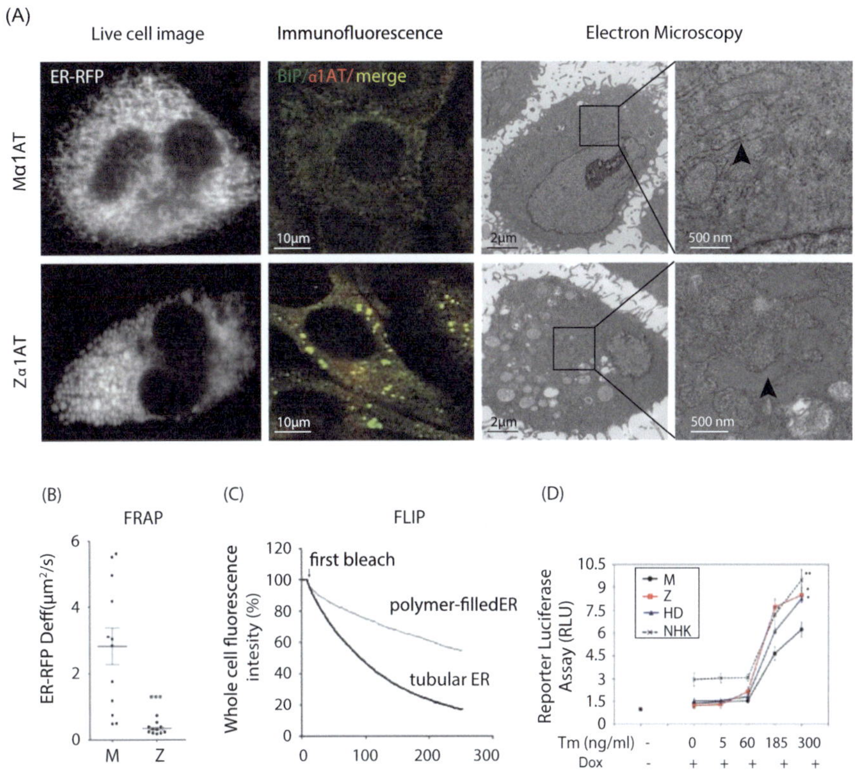

Fig. 14.2 Polymers of α_1-antitrypsin disrupt ER morphology, impair luminal protein mobility and increase sensitivity to ER stress. (**a**) Representative live cell image (*left panels*), immunofluorescence (*middle panels*) and electron micrograph (*right panels*) of CHO-Tet-On cell lines expressing either M (wild type) or Z human α_1-antitrypsin. For live cell images, cells were co-transfected with the ER fluorescent marker, KDEL-RFP (or ER-RFP). For immunofluorescence, cells were fixed and co-immunostained with an α_1-antitrypsin polyclonal antibody and ER protein marker (BiP). For electron micrographs, *left panels* represent whole cell, with higher power images shown in the *right panels*. *Black arrows* indicate the ER cisterns. Inclusion bodies were only identified in micrographs of the polymerogenic mutant (Z α_1-antitrypsin). (**b**) Fluorescence recovery after photobleaching (FRAP) analysis of M and Z α_1-antitrypsin-expressing cells demonstrating impaired ER protein mobility on the expression of Z α_1-antitrypsin. (**c**) Fluorescence loss in photobleaching (FLIP) analysis of cells expressing Z α_1-antitrypsin cells and ER-RFP within both a tubular ER network cell and an α_1-antitrypsin polymer-filled ER. (**d**) ATF6 activation during weak induction of ER stress with the indicated doses of tunicamycin (Tm) for 16 h. α_1-Antitrypsin expression was induced for 48 h. Twenty-four hours prior to lysis, cells were transfected for 6 h with an ATF6-luciferase reporter. Reproduced from Ordonez et al. (2013) with permission

14.3.1 Unfolded Protein Response

ER stress can be defined as the threat that misfolded protein will accumulate within the ER. Factors that cause ER stress do so by impairing protein maturation by a number of means: through a deficiency of energy, through impaired chaperone function (e.g. impaired N-glycosylation on treatment of tunicamycin or depletion of ER calcium on treatment with the SERCA pump inhibitor thapsigargin), through lipid deprivation, or through difficult (or impossible) to fold mutated client proteins. The UPR of mammals, mediated by IRE1, PERK and ATF6, alleviates ER stress through several mechanisms (Fig. 14.1). There is an immediate lessening of secretory protein synthesis, which reduces the influx of newly synthesised client proteins into the lumen and thus helps prevent the ER from being further overwhelmed (Harding et al. 1999). Next, folding capacity is increased through enhanced expression (Kozutsumi et al. 1988; Dorner et al. 1992) and activation of dormant ER-resident chaperones (Flynn et al. 1991; Blond-Elguindi et al. 1993; Chambers et al. 2012), while expansion of the ER membrane allows these components to be accommodated without increased molecular crowding (Bernales et al. 2006; Schuck et al. 2009). In parallel, the efficient degradation of misfolded proteins is enhanced via upregulation of ERAD components (Travers et al. 2000) (Fig. 14.1).

IRE1 is the most conserved of the ER sensors being found in all eukaryotes (Cox et al. 1993). It has a cytoplasmic domain with kinase and nuclease activities. During ER stress, oligomerisation of IRE1 leads to its *trans*-autophosphorylation, which activates the endoribonuclease domain (Sidrauski and Walter 1997). One function of this domain is to initiate the unconventional splicing of X-box binding protein-1 (*XBP1*) mRNA (Calfon et al. 2002). This is achieved by excising an intron that, following religated by RtcB, causes a frameshift in XBP1 generating the transcription factor XBP1s (Calfon et al. 2002; Lu et al. 2014b). Targets of XBP1s include many UPR genes involved in protein folding and ERAD (Yamamoto et al. 2007). In addition, IRE1 has been shown to degrade many other mRNAs encoding secretory proteins, both because of their proximity to the ER membrane and because of specific IRE1 cleavage sites (So et al. 2012). This so-called regulated IRE1-dependent mRNA decay (RIDD) contributes to the off loading of the organelle and frees ribosomes to translate mRNAs of UPR target genes (Hollien et al. 2009). IRE1 has additional signalling functions beyond its nucleolytic activity, with prolonged activation being linked to cell death through a pro-apoptotic IRE1-TRAF2-JNK signalling pathway (Urano et al. 2000). The cytoplasmic domain of IRE1 interacts with the adaptor protein TRAF2 that ultimately activates Jun amino-terminal kinases (JNKs) that interact with pro-apoptotic components such as caspase-12 (Yoneda et al. 2001). It is worth noting, however, that active caspase-12 is restricted to humans from sub-Saharan Africa, suggesting the additional ER stress death pathways must exist (Saleh et al. 2004).

PERK is a protein kinase that mediates the translational effects of ER stress by phosphorylating the translation initiation factor eIF2α (Harding et al. 1999). Once phosphorylated, phospho-eIF2α potently inhibits the activity of its own guanine nucleotide exchange factor, eIF2B, and thus global protein synthesis is inhibited. However, phosphorylation of eIF2α also promotes the translation of a subset of mRNAs, notably that encoding activating transcription factor 4 (ATF4), whose target genes include amino acid transporters and antioxidant factors that contribute to the cell's adaptation to stress (Harding et al. 2000). ATF4 also induces expression of the transcription factor CHOP, and then together ATF4 and CHOP induce GADD34 (Novoa et al. 2001, 2003). GADD34, which is also known as PPP1R15A owing to its role as a regulatory subunit of protein phosphatase 1 (PP1), selectively dephosphorylates eIF2α and so completes a negative feedback loop promoting the recovery of protein translation (Novoa et al. 2001, 2003; Marciniak et al. 2004). During chronic ER stress, the recovery of protein translation mediated by this ATF4–CHOP–GADD34 axis can contribute to cellular toxicity (Marciniak et al. 2004). Recently, it has been suggested that death receptor 5 (DR5) is induced during ER stress by CHOP and contributes to cell death (Lu et al. 2014a). During transient ER stress, IRE1 degrades DR5 mRNA via RIDD to prevent death, but during chronic stress, the eventual deactivation of IRE1 may enable DR5 to be synthesised leading to activation of apoptosis.

The third branch of the UPR is governed by ATF6. During ER stress, this protein traffics from the ER to the Golgi apparatus where it is cleaved by resident proteases that release its cytoplasmic domain (ATF6c) (Haze et al. 1999). This is a transcription factor that translocates to the nucleus to upregulate genes containing ER stress response elements (ERSE), such as BiP (Chen et al. 2002). Since ATF6 can also induce XBP1 expression, this arm of the UPR can synergise with IRE1 signalling (Yoshida et al. 2001).

14.3.2 ER Overload Response

In 1995, Pahl and colleagues reported that the accumulation of well-folded proteins within the ER initiated nuclear factor κB (NF-κB) signalling via a poorly defined mechanism that was termed the ER overload response (EOR) (Pahl and Baeuerle 1995). This was initially described in the context of viral infection, but the overexpression of other proteins within the ER, such as MHC class I or the ΔF508 mutant of CFTR, also induces NF-κB signalling indicating that the EOR is not limited to viral protein accumulation (Pahl et al. 1996; Knorre et al. 2002). Subsequently, studies linked activation of NF-κB to the accumulation of proteins within the ER lumen, including ordered polymers of α_1-antitrypsin and neuroserpin (Lawless et al. 2004; Hidvegi et al. 2005; Davies et al. 2009; van't Wout et al. 2014). Therefore, the term EOR is used for the cellular response elicited by the ER retention of proteins that have achieved their native or near-native conformations.

When inactive, NF-κB is located in the cytoplasm bound to its inhibitor, IκB. External stimuli or endogenous stimuli can induce degradation of IκB to release NF-κB which translocates into the nucleus where it upregulates pro-inflammatory and immune-modulatory factors (Fig. 14.1). The precise mechanism by which ER overload activates NF-κB remains unclear, but the use of small molecule antagonists supports a role for the release of ER calcium into the cytoplasm and subsequent generation of reactive oxygen species (ROS) (Pahl and Baeuerle 1996; Davies et al. 2009), although definitive studies are required to define the mechanism by which distention of the ER leads to calcium release. It is important to distinguish activation of NF-κB during the EOR from its activation during ER stress, since these appear to be distinct processes. While the EOR involves active degradation of IκB, the UPR can cause NF-κB signalling in a different manner. The inhibition of cap-dependent mRNA translation by PERK during ER stress can be sufficient to block synthesis of IκB if the stress is severe (Jiang et al. 2003). Because the half-life of IκB is shorter than that of NF-κB, eventually sufficient free NF-κB is available to activate its downstream transcriptional programme simply through failure to synthesise its inhibitor. More recently, it has been suggested that IRE1 can form a complex with the inhibitor κB kinase (IKK) through TRAF2, leading to phosphorylation and degradation of IκB (Tam et al. 2012).

14.4 Do Serpin Polymers Activate the UPR, EOR or Both?

The ability of ER stress sensors to initiate NF-κB signalling and the relatively opaque nature of EOR signalling have led some to argue that the EOR is little more than a weak UPR. In this regard, the serpinopathies represent a powerful tool with which to examine this question, in that they can result from the retention of ordered polymers of protein within the ER, but classical misfolding mutants for the same proteins also exist (Irving et al. 2011). Serpins whose mutants are prone to polymerisation include antithrombin (Bruce et al. 1994; Picard et al. 2003), C1-inhibitor (Aulak et al. 1993; Eldering et al. 1995) and α_1-antichymotrypsin (Faber et al. 1993; Poller et al. 1993), causing thrombosis, angioedema and emphysema, respectively. Arguably, however, the most characterised serpinopathies result from point mutations of *SERPINA1* (encoding α_1-antitrypsin), causing cirrhosis (Eriksson et al. 1986) and emphysema (Elliott et al. 1998), and of *SERPINI1* (encoding neuroserpin) in the rare dementia familial encephalopathy with neuroserpin inclusion bodies (FENIB) (Davis et al. 2002). Ordered polymers of α_1-antitrypsin are seen in tissue of patients expressing the E342K, 'Z' allele or the H334D mutant (Lomas et al. 1992; Miranda et al. 2010), while polymers of neuroserpin occur within neurons in the brains of S52R and G392E FENIB patients (Davis et al. 1999; Coutelier et al. 2008). In contrast, the naturally occurring truncated NHK mutant of α_1-antitrypsin caused by the C-terminal truncation of 61 residues and addition of 15 new amino acids (Sifers et al. 1988) or the synthetic truncation of neuroserpin (ΔNS) lacking the C-terminal 134 residues (Davies et al. 2009) are both efficiently degraded by ERAD, which prevents their accumulation (Muensch et al. 1986; Brodbeck and Brown 1992).

It is well known that the truncated mutants of α_1-antitrypsin and neuroserpin invoke the UPR, and they have been used in many studies as positive controls for ER stress (Davies et al. 2009; Kroeger et al. 2009; Ordonez et al. 2013). In contrast, the response of cells to the expression of polymerogenic serpins has been the subject of much debate with apparently conflicting reports that polymers of protein within the ER lumen do (Carroll et al. 2010; Schipanski et al. 2013) or do not trigger the UPR (Graham et al. 1990; Lawless et al. 2004; Hidvegi et al. 2005; Davies et al. 2009; Ordonez et al. 2013; van't Wout et al. 2014, 2015).

The first observation that polymers of Z α_1-antitrypsin do not activate the UPR was made in 1990 (Graham et al. 1990), when studies performed in primary hepatocytes from ZZ α_1-antitrypsin transgenic mice showed levels of BiP to be unaffected. Remarkably, they also showed that polymers triggered cells to upregulate BiP more than in wild-type α_1-antitrypsin-expressing cells upon treatment with an ER stress inducer or 'second hit'. A subsequent study using Chinese hamster ovary (CHO) and human embryonic kidney (HEK) 293 cell lines transiently expressing either wild-type or Z α_1-antitrypsin suggested that accumulation of Z α_1-antitrypsin increased BiP and GRP94 expression consistent with activation of the UPR (Lawless et al. 2004). Interestingly, this was only observed in combination with a second stress, such as tunicamycin or incubation at 42 °C. Moreover, under unstressed conditions, the expression of Z α_1-antitrypsin failed to increase the phosphorylation of eIF2α. Tet-Off HeLa cell lines (human cervical carcinoma) made to express Z α_1-antitrypsin did not increase the expression of either BiP or GRP94 acutely (2 days) or chronically (7 days) (Hidvegi et al. 2005). More recently, stable Tet-On CHO or PC12 (rat pheochromocytoma) cell lines expressing polymerogenic of either variants α_1-antitrypsin (E342K and H334D) (Ordonez et al. 2013) or neuroserpin (S53R and G392E) (Davies et al. 2009) were developed to examine UPR signalling in more depth. Both lines showed no induction of UPR target proteins (BiP, GRP94 or PDI) nor activation of proximal signalling pathways (splicing of XBP1 or activation of an ATF6 reporter), but retention of polymers of α_1-antitrypsin significantly sensitised cells to activation of the UPR by a second insult (Ordonez et al. 2013). More recently, studies with transgenic mice expressing the mutant S49P neuroserpin have reported UPR activation in neuron suggesting that tissue-specific difference may exist (Schipanski et al. 2013).

We recently observed that primary bronchial epithelial cells isolated from PiZZ patients show no evidence of constitutive ER stress when grown *in vitro* with normal levels of *XBP1s*, *CHOP* and *GADD34* mRNA and normal levels of chaperone proteins (van't Wout et al. 2014). Similarly, stable Tet-On A549 lung carcinoma cells induced to express Z α_1-antitrypsin did not activate the UPR. In both cases, however, neither cell type could accumulate detectable levels of polymers of α_1-antitrypsin. This may plausibly reflect the low expression of α_1-antitrypsin in lung epithelial cells, approximately 100-fold lower than hepatocytes, and the concentration dependence of serpin polymerisation (van't Wout et al. 2014). Of note, in these cells, expression of Z α_1-antitrypsin did not affect the sensitivity of the UPR.

Despite much evidence that ordered serpin polymers do not evoke a UPR, it was recently suggested that the accumulation of mutant α_1-antitrypsin in monocytes isolated from homozygous PiZZ α_1-antitrypsin patients was associated with increased mRNA levels of several UPR target genes including *BiP*, *GRP94*, *Derlin-1* and *p97* (Carroll et al. 2010). This might reflect a tissue-specific effect of serpin polymers, although a subsequent study failed to replicate those findings in PiZZ patient-derived monocytes or in monocyte-derived macrophages (van't Wout et al. 2015). Moreover, like the airway epithelium, monocytes and monocyte-derived macrophages appear to make insufficient α_1-antitrypsin to generate polymers.

Several groups have independently reported that polymerogenic serpins such as Z α_1-antitrypsin and S52R or G392E neuroserpin initiate NF-κB signalling independent of the UPR (Hidvegi et al. 2005; Lawless et al. 2004; Davies et al. 2009; van't Wout et al. 2014). How serpin polymers (or any trigger of the EOR) might activate NF-κB remains to be elucidated fully, but early studies implicated cytosolic calcium as a component of the signalling pathway (Carlson et al. 1989). It is plausible that the marked changes to the ER structure caused by the accumulation of polymers might disrupt ER calcium homeostasis leading to release into the cytosol (Fig. 14.2a, electron microscopy) (Ordonez et al. 2013). Our own work has shown that chelation of intracellular, but not extracellular calcium, ameliorates the activation of NF-κB in a cell culture model of FENIB (Davies et al. 2009). However, recent work has shown that in primary bronchial epithelial cells from PiZZ α_1-antitrypsin patients NF-κB signalling is active despite the absence of detectable polymers (van't Wout et al. 2014). It was shown that NF-κB activation in this case was mediated through hyperactivity of the ERK signalling pathway in a manner dependent upon ADAM17. The mechanism for hyperactivation of the ERK signalling pathway in cells deficient in α_1-antitrypsin is unclear. It has been suggested α_1-antitrypsin can inhibit ADAM17, an enzyme responsible for the generation of EGFR ligands upstream of ERK signalling (Bergin et al. 2010), but detailed analysis of ADAM17 *in vitro* recently failed to replicate this (van't Wout et al. 2014).

On balance, it appears likely that overloading the ER with polymers of mutant serpin molecules does not efficiently activate a classical UPR in most cell types (Graham et al. 1990; Lawless et al. 2004; Hidvegi et al. 2005; Davies et al. 2009; Ordonez et al. 2013). It has been proposed that serpins polymerise via the insertion of a reactive centre loop into the underlying beta sheet of an adjacent molecule (Lomas et al. 1992). This model suggests that both monomer and polymer will present few (or no) residues to the environment that are normally buried in their native structure, and so interaction with chaperones, in particular with BiP, the putative master regulator of the UPR, would appear unlikely. Recently, a competing model for polymerisation has emerged in which individual protomers exchange larger domains during polymerisation, as was shown for a stable trimer of α_1-antitrypsin formed by denaturation-renaturation (Yamasaki et al. 2011). It seems unlikely that the specific domain swap observed in that crystal structure occurs naturally since polymer-selective monoclonal antibodies that recognise polymers generated *in vivo* fail to recognise polymers formed by similar refolding of denatured α_1-antitrypsin (Ekeowa et al. 2010). However, the possibility that other domain swaps might play some role in serpin polymerisation persists (Yamasaki et al. 2011). In fact, the ER chaperones BiP, GRP94 and GRP170 have been shown to interact with mutant α_1-antitrypsin retained within the ER (Schmidt and

Perlmutter 2005), whereas other studies suggest association with BiP but not with GRP94 (Cabral et al. 2002). This therefore remains a fertile and important area for research.

We recently made use of live cell imaging techniques to examine the relative mobilities of proteins within the ER in cells expressing wild-type or polymerogenic mutants of α_1-antitrypsin, since one could imagine that polymer accumulation could impede protein movement either directly or through alterations to ER structure (Ordonez et al. 2013). Using photobleaching techniques, we demonstrated that accumulation of serpin polymers within the ER impaired the mobility of other soluble marker proteins (Fig. 14.2b, c), which correlated with their increased susceptibility of affected cells to ER stress caused by a second insult (Ordonez et al. 2013). Based on these observations, we proposed a model in which impaired mobility of ER chaperones due to the accumulation of ER polymers might increase heterogeneity within the organelle. That is to say, localised mismatches in the load of client proteins and the availability of chaperones might result from the reduced protein diffusion, either through increased viscosity of the luminal environment or because of changes to the normal ER structure. This would render the cell vulnerable to regional defects in protein homeostasis and so increase the likelihood of activating the UPR during mild ER stress (Fig. 14.2d).

Subcellular fractionation has suggested that ER inclusion bodies containing serpin polymers are physically separated from the main ER network (Granell et al. 2008). Further studies are required to determine if such inclusions might communicate with the 'healthy' ER and whether this aids the cell to withstand the ultrastructural changes that accompany polymer formation.

14.5 Consequences of Stress Signalling in the Serpinopathies

The lack of functional α_1-antitrypsin may also have additional pro-apoptotic effects. Accumulation of Z α_1-antitrypsin has been linked with the activation of ER-associated caspase-12 in mice and caspase-4 in human cells (Hidvegi et al. 2005). The precise mechanisms are unclear, although BAP31, an integral membrane protein involved in the ER retention of several proteins, has been suggested to mediate pro-apoptotic signals between the ER and mitochondria during accumulation of Z α_1-antitrypsin (Schamel et al. 2003). Indeed, mitochondrial dysfunction has been described in mouse models of α_1-antitrypsin deficiency and in the livers of PiZZ patients (Teckman et al. 2004).

It is curious that although PiZZ patients frequently show evidence of intraluminal α_1-antitrypsin accumulation, only 10 % of patients develop clinically relevant liver disease (Sveger 1976). Although this accumulation can induce cell death, it appears also to cause hepatocellular proliferation and may contribute to the development of hepatocellular carcinoma (Rudnick et al. 2004). It is tempting to speculate that this might involve a 'regenerative stimulus' from dying cells to compensate for their loss. Equally, NF-κB signalling arising as part of the EOR may also represent a cell-autonomous proliferative signal (Biswas et al. 2004). Polymers of α_1-antitrypsin are also detectable in the circulation of individuals with α_1-antitrypsin deficiency caused by polymerogenic mutations (Tan et al. 2014). Owing to the pro-inflammatory nature of extracellular polymers, these may explain the association of this disease with panniculitis and vasculitis (de Serres and Blanco 2014).

14.6 Degrading Serpins: ERAD and Autophagy

A proportion of all proteins made, even in health, will misfold. This is prevented from causing harm by the identification and degradation of the offending proteins. In the broadest terms, such proteins can be degraded by two main mechanisms. The ubiquitin proteasome pathway tags misfolded proteins for destruction by the addition of ubiquitin chains (Hiller et al. 1996; Brodsky and Wojcikiewicz 2009), while macro-autophagy (often referred to simply as 'autophagy') involves the engulfment of organelles or portions of the cytoplasm by specialised double-membrane structures that fuse with lysosomes to degrade their contents (Ravikumar et al. 2010) (Fig. 14.1).

There has been some controversy regarding the role of each of the two pathways in the degradation of serpin polymers (Qu et al. 1996; Teckman and Perlmutter 2000; Kruse et al. 2006; Kamimoto et al. 2006; Kroeger et al. 2009). Mutant α_1-antitrypsin was among the first substrates shown to be degraded by ERAD (Le et al. 1990, 1992; Werner et al. 1996; Liu et al. 1997, 1999). The NHK variant of α_1-antitrypsin caused by the C-terminal truncation of 61 residues cannot fold and is efficiently cleared by ERAD (Sifers et al. 1988; Brodbeck and Brown 1992; Liu et al. 1997). Accordingly, treatment of cells with lactacystin, an irreversible inhibitor of the proteasome, reduces NHK α_1-antitrypsin turnover by 90 % (Cabral et al. 2000). ERAD has also been shown to be important in the degradation of other variants, including polymerogenic Z α_1-antitrypsin (Wu et al. 1994; Werner et al. 1996; Qu et al. 1996; Teckman et al. 2001; Kroeger et al. 2009). Similarly, polymerogenic variants of neuroserpin can be cleared by ERAD (Ying et al. 2011; Kroeger et al. 2009; Roussel et al. 2013). Different cell types, however, differ in their reliance upon ERAD for the degradation of polymeric serpins. For example, in PC12 cells, ERAD is the dominant mechanism for the degradation of polymerising neuroserpin, while in mouse embryonic fibroblasts, autophagy appears to be more important (Kroeger et al. 2009).

Because the proteasome is located in the cytosol, if ER client proteins fail to reach their native conformation, they are retrotranslocated to the cytosol to be degraded by the proteasome (Kruse et al. 2006). Although the mechanism by which these processes occur is not entirely clear, for *N*-glycoproteins interaction with lectin-like chaperones appears to be key (Ou et al. 1993; Hebert et al. 1995; Liu et al. 1997). Serpins such as α_1-antitrypsin are therefore ideal subtracts for the study of this glycoprotein ERAD (GERAD) (Sifers 1992; Cabral et al. 2001, 2002). The enzyme ER α-1,2 mannosidase I (ERManI) trims mannose residues from the N-glycan of such proteins and in so doing generates an early signal for degradation (Cabral et al. 2000, 2001; Fagioli and Sitia 2001). Accordingly, overexpression of ERManI in cells improves the degradation of newly synthesised NHK and Z α_1-antitrypsin (Hosokawa et al. 2003), while treatment with chemical inhibitors of the ERManI, e.g. kifunensine, has been shown to retard this (Wu et al. 2003). An enzymatically inactive homologue of ERManI called EDEM also interacts selectively with terminally misfolded glycoproteins like NHK α_1-antitrypsin, to enhance their GERAD (Hosokawa et al. 2006, 2010; Molinari et al. 2003).

Remarkably, although ERMan1 was thought to reside within the ER, recent work showed that it cycles between the ER and Golgi apparatus and that its primer site of action is within the Golgi (Pan et al. 2011; Iannotti et al. 2014). This suggests that its substrates, including NHK α_1-antitrypsin, are tagged for GERAD in the Golgi before being returned to the ER (Pan et al. 2011). Indeed, ERManI has recently been shown to be able to interact simultaneously with NHK α_1-antitrypsin in the lumen of the cis-Golgi and with γ-COP, a component of the vesicle coat on the cytosolic side that mediates retrograde traffic of proteins from the Golgi to the ER (Pan et al. 2013). Since only 10 % of patients with α_1-antitrypsin deficiency develop clinically apparent liver disease, other genetic or environmental factors must contribute. Homozygosity for the minor allele of a polymorphism of *MAN1B1* (the gene encoding ERManI) has been reported to reduce expression levels and to be overrepresented in infants who had required transplantation for ZZ liver disease (Pan et al. 2009). A subsequent study failed to demonstrate an association between this polymorphism and the development of established liver disease, which argues that rather than causing liver disease, this variant may accelerate the onset of disease (Chappell et al. 2009).

Early work suggested that targets of ERAD are returned to the cytosol by being thread back through the pore composed of Sec61 complex, also known as the translocon (Zhou and Schekman 1999), but more recent studies identified an important role for derlin and suggested it instead might contribute to the structure of the retrotranslocon (Ye et al. 2004). The force to extract ERAD targets from the ER is generated by p97-ATPase (also called VCP), which resides in the cytosol, couples the hydrolysis of ATP to the movement of substrates through the retrotranslocon (Ye et al. 2005) and has been required for the degradation of mutant neuroserpin (Ying et al. 2011). Also on the cytosolic side of the ER membrane, the ubiquitin E3 ligases Hrd1 and gp78 cooperate with their E2 ligases, UBE2j1 and UBE2g2, respectively, in the degradation of many misfolded proteins (Chen et al. 2006). It has been shown that Hrd1 and gp78 are both involved in the degradation of α_1-antitrypsin (Christianson et al. 2008; Wang et al. 2011) and neuroserpin (Ying et al. 2011; Roussel et al. 2013). Hrd1 and UBE2j1 appear to show some selectivity towards polymerogenic forms of neuroserpin, whereas gp78 and UBE2g2 preferentially target truncated forms of neuroserpin (Roussel et al. 2013). Recent work has hinted at an interaction between ER membrane cholesterol metabolism and the machinery involved in the ERAD of serpins, although the details remain to be clarified (Roussel et al. 2013).

The narrowness of the retrotranslocon pore requires that proteins cross the membrane in an unfolded state (Rapoport 2007). It is therefore likely that ERAD is important for the degradation of monomeric species that have yet to be incorporated into polymer chains. This is an attractive idea, since there is compelling evidence that monomers of Z α_1-antitrypsin adopt a non-native intermediate conformation prior to polymerisation (Ekeowa et al. 2010). There is also much evidence that the rate of folding of Z α_1-antitrypsin is slowed compared to wild type, and so this variant will inevitably be more prone to identification by timer-based ERAD mechanisms (Kruse et al. 2006; Miranda et al. 2010). Existing ordered polymers of mutant serpins would require disassembly prior to degradation by ERAD, but a mechanism for such disassembly is currently lacking. Instead, established polymers would more plausibly be degraded by autophagy as occurs for the degradation of other large aggregates of mutant protein, for example, huntingtin in neurons affected by Huntington's disease (Kegel et al. 2000). Macro-autophagy can either be a non-selective bulk-turnover process, or it can be targeted towards specific targets via adaptor molecules like sequestosome-1. It has been proposed to be the primary mode of degradation of α_1-antitrypsin (Perlmutter 2006; Hidvegi et al. 2011) as suggested by the numerous autophagosomes observed in the livers of mice expressing Z α_1-antitrypsin and in liver biopsy material from homozygous PiZZ α_1-antitrypsin patients (Teckman and Perlmutter 2000). Several mouse and cell-based models have supported a role for autophagy in the degradation of Z α_1-antitrypsin and other polymerogenic mutant serpins (Hidvegi et al. 2010, 2011; Kroeger et al. 2009; Kamimoto et al. 2006; Kruse et al. 2006). Indeed, it has been claimed that autophagy selectively degrades insoluble polymers of α_1-antitrypsin (Kamimoto et al. 2006; Kruse et al. 2006), and treatment of mice carbamazepine, an antiepileptic drug known also to enhance autophagy, was found to decrease the hepatic load of Z α_1-antitrypsin and reduce liver damage (Hidvegi et al. 2010; Marciniak and Lomas 2010). But it remains unclear to what extent the autophagy of serpin polymers is a selective process and how much this simply reflects the role of autophagy in the turnover of most organelles including the ER. It is noteworthy that in the neuronal-like PC12 cell line, autophagic degradation of wild-type and polymerising variants of neuroserpin shows no selectivity towards the polymeric form (Kroeger et al. 2009). In contrast, heterologous expression in fibroblasts leads to apparent preferential degradation of polymeric neuroserpin by autophagy, but interpretation was hampered by the failure of wild-type protein to accumulate significantly in this cell type. Conceptually, for autophagy to show selectivity towards polymer-containing ER, there would be a need for a method for the autophagic machinery to recognise these structures. At present, such a mechanism has not been identified, although it is tempting to speculate that the reported induction of macro-autophagy by NF-κB signalling (Criollo et al. 2010) might plausibly link activation of the EOR to this process. Further work in this area is required.

14.7 Concluding Remarks

Over the last two decades, the study of polymerogenic variants of α_1-antitrypsin and neuroserpin has contributed to our understanding, not only of disease mechanisms, but also of the basic cell biology that protects the cell from ER dysfunction. The truncated mutants are efficient activators of the classical UPR, while ordered polymers trigger the more mysterious EOR response, which may be crucial to their cell-autonomous pro-inflammatory effects. But little is known about the molecular mechanisms by which ER overload triggers its downstream signalling or how the presence of serpin polymers within the lumen of the ER alters the cell's sensitivity to ER stress caused by a second insult. Studies that address these questions will expand our understanding of ER stress signalling and may provide novel therapeutic targets for this intriguing class of disease.

References

Aulak KS, Eldering E, Hack CE, Lubbers YP, Harrison RA, Mast A, Cicardi M, Davis AE III (1993) A hinge region mutation in C1-inhibitor (Ala436→Thr) results in nonsubstrate-like behavior and in polymerization of the molecule. J Biol Chem 268(24):18088–18094

Bergin DA, Reeves EP, Meleady P, Henry M, McElvaney OJ, Carroll TP, Condron C, Chotirmall SH, Clynes M, O'Neill SJ, McElvaney NG (2010) alpha-1 Antitrypsin regulates human neutrophil chemotaxis induced by soluble immune complexes and IL-8. J Clin Invest 120 (12):4236–4250. doi:10.1172/JCI41196

Bernales S, McDonald KL, Walter P (2006) Autophagy counterbalances endoplasmic reticulum expansion during the unfolded protein response. PLoS Biol 4(12), e423. doi:10.1371/journal.pbio.0040423

Bertolotti A, Zhang Y, Hendershot LM, Harding HP, Ron D (2000) Dynamic interaction of BiP and ER stress transducers in the unfolded-protein response. Nat Cell Biol 2(6):326–332. doi:10.1038/35014014

Biswas DK, Shi Q, Baily S, Strickland I, Ghosh S, Pardee AB, Iglehart JD (2004) NF-kappa B activation in human breast cancer specimens and its role in cell proliferation and apoptosis. Proc Natl Acad Sci U S A 101(27):10137–10142. doi:10.1073/pnas.0403621101

Blond-Elguindi S, Fourie AM, Sambrook JF, Gething MJ (1993) Peptide-dependent stimulation of the ATPase activity of the molecular chaperone BiP is the result of conversion of oligomers to active monomers. J Biol Chem 268(17):12730–12735

Brodbeck RM, Brown JL (1992) Secretion of alpha-1-proteinase inhibitor requires an almost full length molecule. J Biol Chem 267(1):294–297

Brodsky JL, Wojcikiewicz RJ (2009) Substrate-specific mediators of ER associated degradation (ERAD). Curr Opin Cell Biol 21(4):516–521. doi:10.1016/j.ceb.2009.04.006

Bruce D, Perry DJ, Borg JY, Carrell RW, Wardell MR (1994) Thromboembolic disease due to thermolabile conformational changes of antithrombin Rouen-VI (187 Asn→Asp). J Clin Invest 94(6):2265–2274. doi:10.1172/JCI117589

Cabral CM, Choudhury P, Liu Y, Sifers RN (2000) Processing by endoplasmic reticulum mannosidases partitions a secretion-impaired glycoprotein into distinct disposal pathways. J Biol Chem 275(32):25015–25022. doi:10.1074/jbc.M910172199

Cabral CM, Liu Y, Sifers RN (2001) Dissecting glycoprotein quality control in the secretory pathway. Trends Biochem Sci 26(10):619–624

Cabral CM, Liu Y, Moremen KW, Sifers RN (2002) Organizational diversity among distinct glycoprotein endoplasmic reticulum-associated degradation programs. Mol Biol Cell 13 (8):2639–2650. doi:10.1091/mbc.E02-02-0068

Calfon M, Zeng H, Urano F, Till JH, Hubbard SR, Harding HP, Clark SG, Ron D (2002) IRE1 couples endoplasmic reticulum load to secretory capacity by processing the XBP-1 mRNA. Nature 415(6867):92–96. doi:10.1038/415092a

Carlson JA, Rogers BB, Sifers RN, Finegold MJ, Clift SM, DeMayo FJ, Bullock DW, Woo SL (1989) Accumulation of PiZ alpha 1-antitrypsin causes liver damage in transgenic mice. J Clin Invest 83(4):1183–1190. doi:10.1172/JCI113999

Carroll TP, Greene CM, O'Connor CA, Nolan AM, O'Neill SJ, McElvaney NG (2010) Evidence for unfolded protein response activation in monocytes from individuals with alpha-1 antitrypsin deficiency. J Immunol 184(8):4538–4546. doi:10.4049/jimmunol.0802864, jimmunol.0802864 [pii]

Chambers JE, Marciniak SJ (2014) Protein misfolding and ER stress. Am J Physiol Cell Physiol 307(8):C657–70. doi:10.1152/ajpcell.00183.2014

Chambers JE, Petrova K, Tomba G, Vendruscolo M, Ron D (2012) ADP ribosylation adapts an ER chaperone response to short-term fluctuations in unfolded protein load. J Cell Biol 198 (3):371–385. doi:10.1083/jcb.201202005

Chappell S, Guetta-Baranes T, Hadzic N, Stockley R, Kalsheker N (2009) Polymorphism in the endoplasmic reticulum mannosidase I (MAN1B1) gene is not associated with liver disease in individuals homozygous for the Z variant of the alpha1-antitrypsin protease inhibitor (PiZZ individuals). Hepatology 50(4):1315. doi:10.1002/hep.23170, author reply 1315–1316

Chen X, Shen J, Prywes R (2002) The luminal domain of ATF6 senses endoplasmic reticulum (ER) stress and causes translocation of ATF6 from the ER to the golgi. J Biol Chem 277 (15):13045–13052

Chen B, Mariano J, Tsai YC, Chan AH, Cohen M, Weissman AM (2006) The activity of a human endoplasmic reticulum-associated degradation E3, gp78, requires its Cue domain, RING finger, and an E2-binding site. Proc Natl Acad Sci U S A 103(2):341–346. doi:10.1073/pnas.0506618103

Christianson JC, Shaler TA, Tyler RE, Kopito RR (2008) OS-9 and GRP94 deliver mutant alpha1-antitrypsin to the Hrd1-SEL1L ubiquitin ligase complex for ERAD. Nat Cell Biol 10 (3):272–282. doi:10.1038/ncb1689

Coutelier M, Andries S, Ghariani S, Dan B, Duyckaerts C, van Rijckevorsel K, Raftopoulos C, Deconinck N, Sonderegger P, Scaravilli F, Vikkula M, Godfraind C (2008) Neuroserpin mutation causes electrical status epilepticus of slow-wave sleep. Neurology 71(1):64–66. doi:10.1212/01.wnl.0000316306.08751.28

Cox JS, Shamu CE, Walter P (1993) Transcriptional induction of genes encoding endoplasmic reticulum resident proteins requires a transmembrane protein kinase. Cell 73(6):1197–1206

Credle JJ, Finer-Moore JS, Papa FR, Stroud RM, Walter P (2005) On the mechanism of sensing unfolded protein in the endoplasmic reticulum. Proc Natl Acad Sci U S A 102 (52):18773–18784. doi:10.1073/pnas.0509487102, 0509487102 [pii]

Criollo A, Senovilla L, Authier H, Maiuri MC, Morselli E, Vitale I, Kepp O, Tasdemir E, Galluzzi L, Shen S, Tailler M, Delahaye N, Tesniere A, De Stefano D, Younes AB, Harper F, Pierron G, Lavandero S, Zitvogel L, Israel A, Baud V, Kroemer G (2010) The IKK complex contributes to the induction of autophagy. EMBO J 29(3):619–631. doi:10.1038/emboj.2009.364

Davies MJ, Miranda E, Roussel BD, Kaufman RJ, Marciniak SJ, Lomas DA (2009) Neuroserpin polymers activate NF-kappaB by a calcium signaling pathway that is independent of the unfolded protein response. J Biol Chem 284(27):18202–18209. doi:10.1074/jbc.M109.010744

Davis RL, Shrimpton AE, Holohan PD, Bradshaw C, Feiglin D, Collins GH, Sonderegger P, Kinter J, Becker LM, Lacbawan F, Krasnewich D, Muenke M, Lawrence DA, Yerby MS, Shaw CM, Gooptu B, Elliott PR, Finch JT, Carrell RW, Lomas DA (1999) Familial dementia caused by polymerization of mutant neuroserpin. Nature 401(6751):376–379. doi:10.1038/43894

Davis RL, Shrimpton AE, Carrell RW, Lomas DA, Gerhard L, Baumann B, Lawrence DA, Yepes M, Kim TS, Ghetti B, Piccardo P, Takao M, Lacbawan F, Muenke M, Sifers RN, Bradshaw CB, Kent PF, Collins GH, Larocca D, Holohan PD (2002) Association between conformational mutations in neuroserpin and onset and severity of dementia. Lancet 359 (9325):2242–2247. doi:10.1016/S0140-6736(02)09293-0

de Serres F, Blanco I (2014) Role of alpha-1 antitrypsin in human health and disease. J Intern Med 276(4):311–335. doi:10.1111/joim.12239

Dorner AJ, Wasley LC, Kaufman RJ (1992) Overexpression of GRP78 mitigates stress induction of glucose regulated proteins and blocks secretion of selective proteins in Chinese hamster ovary cells. EMBO J 11(4):1563–1571

Dycaico MJ, Grant SG, Felts K, Nichols WS, Geller SA, Hager JH, Pollard AJ, Kohler SW, Short HP, Jirik FR et al (1988) Neonatal hepatitis induced by alpha 1-antitrypsin: a transgenic mouse model. Science 242(4884):1409–1412

Ekeowa UI, Freeke J, Miranda E, Gooptu B, Bush MF, Perez J, Teckman J, Robinson CV, Lomas DA (2010) Defining the mechanism of polymerization in the serpinopathies. Proc Natl Acad Sci USA 107(40):17146–17151. doi:10.1073/pnas.1004785107

Eldering E, Verpy E, Roem D, Meo T, Tosi M (1995) COOH-terminal substitutions in the serpin C1 inhibitor that cause loop overinsertion and subsequent multimerization. J Biol Chem 270 (6):2579–2587

Ellgaard L, Helenius A (2003) Quality control in the endoplasmic reticulum. Nat Rev Mol Cell Biol 4(3):181–191

Elliott PR, Bilton D, Lomas DA (1998) Lung polymers in Z alpha1-antitrypsin deficiency-related emphysema. Am J Respir Cell Mol Biol 18(5):670–674. doi:10.1165/ajrcmb.18.5.3065

Eriksson S (1965) Studies in alpha 1-antitrypsin deficiency. Acta Med Scand Suppl 432:1–85

Eriksson S, Carlson J, Velez R (1986) Risk of cirrhosis and primary liver cancer in alpha 1-antitrypsin deficiency. N Engl J Med 314(12):736–739. doi:10.1056/NEJM198603203141202

Faber JP, Poller W, Olek K, Baumann U, Carlson J, Lindmark B, Eriksson S (1993) The molecular basis of alpha 1-antichymotrypsin deficiency in a heterozygote with liver and lung disease. J Hepatol 18(3):313–321

Faber JP, Poller W, Weidinger S, Kirchgesser M, Schwaab R, Bidlingmaier F, Olek K (1994) Identification and DNA sequence analysis of 15 new alpha 1-antitrypsin variants, including two PI*Q0 alleles and one deficient PI*M allele. Am J Hum Genet 55(6):1113–1121

Fagioli C, Sitia R (2001) Glycoprotein quality control in the endoplasmic reticulum. Mannose trimming by endoplasmic reticulum mannosidase I times the proteasomal degradation of unassembled immunoglobulin subunits. J Biol Chem 276(16):12885–12892. doi:10.1074/jbc.M009603200

Fewell SW, Travers KJ, Weissman JS, Brodsky JL (2001) The action of molecular chaperones in the early secretory pathway. Annu Rev Genet 35:149–191

Flynn GC, Pohl J, Flocco MT, Rothman JE (1991) Peptide-binding specificity of the molecular chaperone BIP. Nature 353(6346):726–730. doi:10.1038/353726a0

Gardner BM, Walter P (2011) Unfolded proteins are Ire1-activating ligands that directly induce the unfolded protein response. Science 333(6051):1891–1894. doi:10.1126/science.1209126

Gething MJ, McCammon K, Sambrook J (1986) Expression of wild-type and mutant forms of influenza hemagglutinin: the role of folding in intracellular transport. Cell 46(6):939–950

Graham KS, Le A, Sifers RN (1990) Accumulation of the insoluble PiZ variant of human alpha 1-antitrypsin within the hepatic endoplasmic reticulum does not elevate the steady-state level of grp78/BiP. J Biol Chem 265(33):20463–20468

Granell S, Baldini G, Mohammad S, Nicolin V, Narducci P, Storrie B, Baldini G (2008) Sequestration of mutated alpha1-antitrypsin into inclusion bodies is a cell-protective mechanism to maintain endoplasmic reticulum function. Mol Biol Cell 19(2):572–586. doi:10.1091/mbc.E07-06-0587

Harding HP, Zhang Y, Ron D (1999) Protein translation and folding are coupled by an endoplasmic-reticulum-resident kinase. Nature 397(6716):271–274. doi:10.1038/16729

Harding HP, Novoa I, Zhang Y, Zeng H, Wek R, Schapira M, Ron D (2000) Regulated translation initiation controls stress-induced gene expression in mammalian cells. Mol Cell 6 (5):1099–1108

Haze K, Yoshida H, Yanagi H, Yura T, Mori K (1999) Mammalian transcription factor ATF6 is synthesized as a transmembrane protein and activated by proteolysis in response to endoplasmic reticulum stress. Mol Biol Cell 10(11):3787–3799

Hebert DN, Foellmer B, Helenius A (1995) Glucose trimming and reglucosylation determine glycoprotein association with calnexin in the endoplasmic reticulum. Cell 81(3):425–433

Hidvegi T, Schmidt BZ, Hale P, Perlmutter DH (2005) Accumulation of mutant alpha1-antitrypsin Z in the endoplasmic reticulum activates caspases-4 and -12, NFkappaB, and BAP31 but not the unfolded protein response. J Biol Chem 280(47):39002–39015. doi:10.1074/jbc.M508652200, M508652200 [pii]

Hidvegi T, Ewing M, Hale P, Dippold C, Beckett C, Kemp C, Maurice N, Mukherjee A, Goldbach C, Watkins S, Michalopoulos G, Perlmutter DH (2010) An autophagy-enhancing drug promotes degradation of mutant alpha1-antitrypsin Z and reduces hepatic fibrosis. Science 329(5988):229–232. doi:10.1126/science.1190354, science.1190354 [pii]

Hidvegi T, Mukherjee A, Ewing M, Kemp C, Perlmutter DH (2011) The role of autophagy in alpha-1-antitrypsin deficiency. Methods Enzymol 499:33–54. doi:10.1016/B978-0-12-386471-0.00003-1

Hiller MM, Finger A, Schweiger M, Wolf DH (1996) ER degradation of a misfolded luminal protein by the cytosolic ubiquitin-proteasome pathway. Science 273(5282):1725–1728

Hollien J, Lin JH, Li H, Stevens N, Walter P, Weissman JS (2009) Regulated Ire1-dependent decay of messenger RNAs in mammalian cells. J Cell Biol 186(3):323–331. doi:10.1083/jcb.200903014

Hosokawa N, Tremblay LO, You Z, Herscovics A, Wada I, Nagata K (2003) Enhancement of endoplasmic reticulum (ER) degradation of misfolded Null Hong Kong alpha1-antitrypsin by human ER mannosidase I. J Biol Chem 278(28):26287–26294

Hosokawa N, Wada I, Natsuka Y, Nagata K (2006) EDEM accelerates ERAD by preventing aberrant dimer formation of misfolded alpha1-antitrypsin. Genes Cells 11(5):465–476. doi:10.1111/j.1365-2443.2006.00957.x

Hosokawa N, Tremblay LO, Sleno B, Kamiya Y, Wada I, Nagata K, Kato K, Herscovics A (2010) EDEM1 accelerates the trimming of alpha1,2-linked mannose on the C branch of N-glycans. Glycobiology 20(5):567–575. doi:10.1093/glycob/cwq001

Iannotti MJ, Figard L, Sokac AM, Sifers RN (2014) A Golgi-localized mannosidase (MAN1B1) plays a non-enzymatic gatekeeper role in protein biosynthetic quality control. J Biol Chem 289 (17):11844–11858. doi:10.1074/jbc.M114.552091

Irving JA, Ekeowa UI, Belorgey D, Haq I, Gooptu B, Miranda E, Perez J, Roussel BD, Ordonez A, Dalton LE, Thomas SE, Marciniak SJ, Parfrey H, Chilvers ER, Teckman JH, Alam S, Mahadeva R, Rashid ST, Vallier L, Lomas DA (2011) The serpinopathies studying serpin polymerization in vivo. Methods Enzymol 501:421–466. doi:10.1016/B978-0-12-385950-1.00018-3

Jiang HY, Wek SA, McGrath BC, Scheuner D, Kaufman RJ, Cavener DR, Wek RC (2003) Phosphorylation of the alpha subunit of eukaryotic initiation factor 2 is required for activation of NF-kappaB in response to diverse cellular stresses. Mol Cell Biol 23(16):5651–5663

Kamimoto T, Shoji S, Hidvegi T, Mizushima N, Umebayashi K, Perlmutter DH, Yoshimori T (2006) Intracellular inclusions containing mutant alpha1-antitrypsin Z are propagated in the absence of autophagic activity. J Biol Chem 281(7):4467–4476. doi:10.1074/jbc.M509409200, M509409200 [pii]

Kegel KB, Kim M, Sapp E, McIntyre C, Castano JG, Aronin N, DiFiglia M (2000) Huntingtin expression stimulates endosomal-lysosomal activity, endosome tubulation, and autophagy. J Neurosci 20(19):7268–7278

Kimata Y, Oikawa D, Shimizu Y, Ishiwata-Kimata Y, Kohno K (2004) A role for BiP as an adjustor for the endoplasmic reticulum stress-sensing protein Ire1. J Cell Biol 167(3):445–456. doi:10.1083/jcb.200405153

Knorre A, Wagner M, Schaefer HE, Colledge WH, Pahl HL (2002) DeltaF508-CFTR causes constitutive NF-kappaB activation through an ER-overload response in cystic fibrosis lungs. Biol Chem 383(2):271–282. doi:10.1515/BC.2002.029

Kozutsumi Y, Segal M, Normington K, Gething MJ, Sambrook J (1988) The presence of malfolded proteins in the endoplasmic reticulum signals the induction of glucose-regulated proteins. Nature 332(6163):462–464. doi:10.1038/332462a0

Kroeger H, Miranda E, MacLeod I, Perez J, Crowther DC, Marciniak SJ, Lomas DA (2009) Endoplasmic reticulum-associated degradation (ERAD) and autophagy cooperate to degrade polymerogenic mutant serpins. J Biol Chem 284(34):22793–22802. doi:10.1074/jbc.M109.027102, M109.027102 [pii]

Kruse KB, Brodsky JL, McCracken AA (2006) Characterization of an ERAD gene as VPS30/ATG6 reveals two alternative and functionally distinct protein quality control pathways: one for soluble Z variant of human alpha-1 proteinase inhibitor (A1PiZ) and another for aggregates of A1PiZ. Mol Biol Cell 17(1):203–212. doi:10.1091/mbc.E04-09-0779, E04-09-0779 [pii]

Lawless MW, Greene CM, Mulgrew A, Taggart CC, O'Neill SJ, McElvaney NG (2004) Activation of endoplasmic reticulum-specific stress responses associated with the conformational disease Z alpha 1-antitrypsin deficiency. J Immunol 172(9):5722–5726

Lawrence T (2009) The nuclear factor NF-kappaB pathway in inflammation. Cold Spring Harb Perspect Biol 1(6):a001651. doi:10.1101/cshperspect.a001651

Le A, Graham KS, Sifers RN (1990) Intracellular degradation of the transport-impaired human PiZ alpha 1-antitrypsin variant. Biochemical mapping of the degradative event among compartments of the secretory pathway. J Biol Chem 265(23):14001–14007

Le A, Ferrell GA, Dishon DS, Le QQ, Sifers RN (1992) Soluble aggregates of the human PiZ alpha 1-antitrypsin variant are degraded within the endoplasmic reticulum by a mechanism sensitive to inhibitors of protein synthesis. J Biol Chem 267(2):1072–1080

Lievremont JP, Rizzuto R, Hendershot L, Meldolesi J (1997) BiP, a major chaperone protein of the endoplasmic reticulum lumen, plays a direct and important role in the storage of the rapidly exchanging pool of Ca2+. J Biol Chem 272(49):30873–30879

Liu Y, Choudhury P, Cabral CM, Sifers RN (1997) Intracellular disposal of incompletely folded human alpha1-antitrypsin involves release from calnexin and post-translational trimming of asparagine-linked oligosaccharides. J Biol Chem 272(12):7946–7951

Liu Y, Choudhury P, Cabral CM, Sifers RN (1999) Oligosaccharide modification in the early secretory pathway directs the selection of a misfolded glycoprotein for degradation by the proteasome. J Biol Chem 274(9):5861–5867

Lomas DA, Mahadeva R (2002) Alpha1-antitrypsin polymerization and the serpinopathies: pathobiology and prospects for therapy. J Clin Invest 110(11):1585–1590

Lomas DA, Evans DL, Finch JT, Carrell RW (1992) The mechanism of Z alpha 1-antitrypsin accumulation in the liver. Nature 357(6379):605–607

Lu M, Lawrence DA, Marsters S, Acosta-Alvear D, Kimmig P, Mendez AS, Paton AW, Paton JC, Walter P, Ashkenazi A (2014a) Cell death. Opposing unfolded-protein-response signals converge on death receptor 5 to control apoptosis. Science 345(6192):98–101. doi:10.1126/science.1254312

Lu Y, Liang FX, Wang X (2014b) A synthetic biology approach identifies the mammalian UPR RNA ligase RtcB. Mol Cell 55(5):758–770. doi:10.1016/j.molcel.2014.06.032

Marciniak SJ, Lomas DA (2010) Alpha1-Antitrypsin Deficiency and Autophagy. N Engl J Med 363(19):1863–1864

Marciniak SJ, Ron D (2006) Endoplasmic reticulum stress signaling in disease. Physiol Rev 86 (4):1133–1149. doi:10.1152/physrev.00015.2006

Marciniak SJ, Yun CY, Oyadomari S, Novoa I, Zhang Y, Jungreis R, Nagata K, Harding HP, Ron D (2004) CHOP induces death by promoting protein synthesis and oxidation in the stressed endoplasmic reticulum. Genes Dev 18(24):3066–3077. doi:10.1101/gad.1250704

McCracken AA, Brodsky JL (1996) Assembly of ER-associated protein degradation in vitro: dependence on cytosol, calnexin, and ATP. J Cell Biol 132(3):291–298

Meusser B, Hirsch C, Jarosch E, Sommer T (2005) ERAD: the long road to destruction. Nat Cell Biol 7(8):766–772

Miranda E, MacLeod I, Davies MJ, Perez J, Romisch K, Crowther DC, Lomas DA (2008) The intracellular accumulation of polymeric neuroserpin explains the severity of the dementia FENIB. Hum Mol Genet 17(11):1527–1539. doi:10.1093/hmg/ddn041, ddn041 [pii]

Miranda E, Perez J, Ekeowa UI, Hadzic N, Kalsheker N, Gooptu B, Portmann B, Belorgey D, Hill M, Chambers S, Teckman J, Alexander GJ, Marciniak SJ, Lomas DA (2010) A novel monoclonal antibody to characterize pathogenic polymers in liver disease associated with alpha1-antitrypsin deficiency. Hepatology 52(3):1078–1088. doi:10.1002/hep.23760

Molinari M, Calanca V, Galli C, Lucca P, Paganetti P (2003) Role of EDEM in the release of misfolded glycoproteins from the calnexin cycle. Science 299(5611):1397–1400

Muensch H, Gaidulis L, Kueppers F, So SY, Escano G, Kidd VJ, Woo SL (1986) Complete absence of serum alpha-1-antitrypsin in conjunction with an apparently normal gene structure. Am J Hum Genet 38(6):898–907

Novoa I, Zeng H, Harding HP, Ron D (2001) Feedback inhibition of the unfolded protein response by GADD34-mediated dephosphorylation of eIF2alpha. J Cell Biol 153(5):1011–1022

Novoa I, Zhang Y, Zeng H, Jungreis R, Harding HP, Ron D (2003) Stress-induced gene expression requires programmed recovery from translational repression. EMBO J 22(5):1180–1187. doi:10.1093/emboj/cdg112

Ordonez A, Snapp EL, Tan L, Miranda E, Marciniak SJ, Lomas DA (2013) Endoplasmic reticulum polymers impair luminal protein mobility and sensitize to cellular stress in alpha1-antitrypsin deficiency. Hepatology 57(5):2049–2060. doi:10.1002/hep.26173

Ou WJ, Cameron PH, Thomas DY, Bergeron JJ (1993) Association of folding intermediates of glycoproteins with calnexin during protein maturation. Nature 364(6440):771–776. doi:10.1038/364771a0

Pahl HL, Baeuerle PA (1995) A novel signal transduction pathway from the endoplasmic reticulum to the nucleus is mediated by transcription factor NF-kappa B. EMBO J 14(11):2580–2588

Pahl HL, Baeuerle PA (1996) Activation of NF-kappa B by ER stress requires both Ca2+ and reactive oxygen intermediates as messengers. FEBS Lett 392(2):129–136

Pahl HL, Sester M, Burgert HG, Baeuerle PA (1996) Activation of transcription factor NF-kappaB by the adenovirus E3/19K protein requires its ER retention. J Cell Biol 132(4):511–522

Pan S, Huang L, McPherson J, Muzny D, Rouhani F, Brantly M, Gibbs R, Sifers RN (2009) Single nucleotide polymorphism-mediated translational suppression of endoplasmic reticulum mannosidase I modifies the onset of end-stage liver disease in alpha1-antitrypsin deficiency. Hepatology 50(1):275–281. doi:10.1002/hep.22974

Pan S, Wang S, Utama B, Huang L, Blok N, Estes MK, Moremen KW, Sifers RN (2011) Golgi localization of ERManI defines spatial separation of the mammalian glycoprotein quality control system. Mol Biol Cell 22(16):2810–2822. doi:10.1091/mbc.E11-02-0118

Pan S, Cheng X, Sifers RN (2013) Golgi-situated endoplasmic reticulum alpha-1, 2-mannosidase contributes to the retrieval of ERAD substrates through a direct interaction with gamma-COP. Mol Biol Cell 24(8):1111–1121. doi:10.1091/mbc.E12-12-0886

Perlmutter DH (2006) The role of autophagy in alpha-1-antitrypsin deficiency: a specific cellular response in genetic diseases associated with aggregation-prone proteins. Autophagy 2 (4):258–263

Picard V, Dautzenberg MD, Villoutreix BO, Orliaguet G, Alhenc-Gelas M, Aiach M (2003) Antithrombin Phe229Leu: a new homozygous variant leading to spontaneous antithrombin polymerization in vivo associated with severe childhood thrombosis. Blood 102(3):919–925. doi:10.1182/blood-2002-11-3391

Pincus D, Chevalier MW, Aragon T, van Anken E, Vidal SE, El-Samad H, Walter P (2010) BiP binding to the ER-stress sensor Ire1 tunes the homeostatic behavior of the unfolded protein response. PLoS Biol 8(7), e1000415. doi:10.1371/journal.pbio.1000415

Poller W, Faber JP, Weidinger S, Tief K, Scholz S, Fischer M, Olek K, Kirchgesser M, Heidtmann HH (1993) A leucine-to-proline substitution causes a defective alpha 1-antichymotrypsin allele associated with familial obstructive lung disease. Genomics 17(3):740–743. doi:10.1006/geno.1993.1396

Qu D, Teckman JH, Omura S, Perlmutter DH (1996) Degradation of a mutant secretory protein, alpha1-antitrypsin Z, in the endoplasmic reticulum requires proteasome activity. J Biol Chem 271(37):22791–22795

Rapoport TA (2007) Protein translocation across the eukaryotic endoplasmic reticulum and bacterial plasma membranes. Nature 450(7170):663–669. doi:10.1038/nature06384

Ravikumar B, Sarkar S, Davies JE, Futter M, Garcia-Arencibia M, Green-Thompson ZW, Jimenez-Sanchez M, Korolchuk VI, Lichtenberg M, Luo S, Massey DC, Menzies FM, Moreau K, Narayanan U, Renna M, Siddiqi FH, Underwood BR, Winslow AR, Rubinsztein DC (2010) Regulation of mammalian autophagy in physiology and pathophysiology. Physiol Rev 90(4):1383–1435. doi:10.1152/physrev.00030.2009

Ron D, Walter P (2007) Signal integration in the endoplasmic reticulum unfolded protein response. Nat Rev Mol Cell Biol 8(7):519–529. doi:10.1038/nrm2199

Roussel BD, Newton TM, Malzer E, Simecek N, Haq I, Thomas SE, Burr ML, Lehner PJ, Crowther DC, Marciniak SJ, Lomas DA (2013) Sterol metabolism regulates neuroserpin polymer degradation in the absence of the unfolded protein response in the dementia FENIB. Hum Mol Genet 22(22):4616–4626. doi:10.1093/hmg/ddt310

Rudnick DA, Liao Y, An JK, Muglia LJ, Perlmutter DH, Teckman JH (2004) Analyses of hepatocellular proliferation in a mouse model of alpha-1-antitrypsin deficiency. Hepatology 39(4):1048–1055. doi:10.1002/hep.20118

Rutkowski DT, Kaufman RJ (2004) A trip to the ER: coping with stress. Trends Cell Biol 14 (1):20–28

Saleh M, Vaillancourt JP, Graham RK, Huyck M, Srinivasula SM, Alnemri ES, Steinberg MH, Nolan V, Baldwin CT, Hotchkiss RS, Buchman TG, Zehnbauer BA, Hayden MR, Farrer LA, Roy S, Nicholson DW (2004) Differential modulation of endotoxin responsiveness by human caspase-12 polymorphisms. Nature 429(6987):75–79. doi:10.1038/nature02451

Schamel WW, Kuppig S, Becker B, Gimborn K, Hauri HP, Reth M (2003) A high-molecular-weight complex of membrane proteins BAP29/BAP31 is involved in the retention of membrane-bound IgD in the endoplasmic reticulum. Proc Natl Acad Sci USA 100 (17):9861–9866. doi:10.1073/pnas.1633363100

Schipanski A, Lange S, Segref A, Gutschmidt A, Lomas DA, Miranda E, Schweizer M, Hoppe T, Glatzel M (2013) A novel interaction between aging and ER overload in a protein conformational dementia. Genetics 193(3):865–876. doi:10.1534/genetics.112.149088

Schmidt BZ, Perlmutter DH (2005) Grp78, Grp94, and Grp170 interact with alpha1-antitrypsin mutants that are retained in the endoplasmic reticulum. Am J Physiol Gastrointest Liver Physiol 289(3):G444–455. doi:10.1152/ajpgi.00237.2004

Schuck S, Prinz WA, Thorn KS, Voss C, Walter P (2009) Membrane expansion alleviates endoplasmic reticulum stress independently of the unfolded protein response. J Cell Biol 187 (4):525–536. doi:10.1083/jcb.200907074

Shen J, Chen X, Hendershot L, Prywes R (2002) ER stress regulation of ATF6 localization by dissociation of BiP/GRP78 binding and unmasking of Golgi localization signals. Dev Cell 3 (1):99–111

Sidrauski C, Walter P (1997) The transmembrane kinase Ire1p is a site-specific endonuclease that initiates mRNA splicing in the unfolded protein response. Cell 90(6):1031–1039

Sifers RN (1992) Protein transport. Z and the insoluble answer. Nature 357(6379):541–542. doi:10.1038/357541a0

Sifers RN, Brashears-Macatee S, Kidd VJ, Muensch H, Woo SL (1988) A frameshift mutation results in a truncated alpha 1-antitrypsin that is retained within the rough endoplasmic reticulum. J Biol Chem 263(15):7330–7335
So JS, Hur KY, Tarrio M, Ruda V, Frank-Kamenetsky M, Fitzgerald K, Koteliansky V, Lichtman AH, Iwawaki T, Glimcher LH, Lee AH (2012) Silencing of lipid metabolism genes through IRE1alpha-mediated mRNA decay lowers plasma lipids in mice. Cell Metab 16(4):487–499. doi:10.1016/j.cmet.2012.09.004
Sveger T (1976) Liver disease in alpha1-antitrypsin deficiency detected by screening of 200,000 infants. N Engl J Med 294(24):1316–1321. doi:10.1056/NEJM197606102942404
Tabas I, Ron D (2011) Integrating the mechanisms of apoptosis induced by endoplasmic reticulum stress. Nat Cell Biol 13(3):184–190. doi:10.1038/ncb0311-184
Tam AB, Mercado EL, Hoffmann A, Niwa M (2012) ER stress activates NF-kappaB by integrating functions of basal IKK activity, IRE1 and PERK. PLoS One 7(10), e45078. doi:10.1371/journal.pone.0045078
Tan L, Dickens JA, Demeo DL, Miranda E, Perez J, Rashid ST, Day J, Ordonez A, Marciniak SJ, Haq I, Barker AF, Campbell EJ, Eden E, McElvaney NG, Rennard SI, Sandhaus RA, Stocks JM, Stoller JK, Strange C, Turino G, Rouhani FN, Brantly M, Lomas DA (2014) Circulating polymers in alpha1-antitrypsin deficiency. Eur Respir J 43(5):1501–1504. doi:10.1183/09031936.00111213
Teckman JH, Perlmutter DH (2000) Retention of mutant alpha(1)-antitrypsin Z in endoplasmic reticulum is associated with an autophagic response. Am J Physiol Gastrointest Liver Physiol 279(5):G961–974
Teckman JH, Burrows J, Hidvegi T, Schmidt B, Hale PD, Perlmutter DH (2001) The proteasome participates in degradation of mutant alpha 1-antitrypsin Z in the endoplasmic reticulum of hepatoma-derived hepatocytes. J Biol Chem 276(48):44865–44872. doi:10.1074/jbc.M103703200, M103703200 [pii]
Teckman JH, An JK, Blomenkamp K, Schmidt B, Perlmutter D (2004) Mitochondrial autophagy and injury in the liver in alpha 1-antitrypsin deficiency. Am J Physiol Gastrointest Liver Physiol 286(5):G851–862. doi:10.1152/ajpgi.00175.2003
Travers KJ, Patil CK, Wodicka L, Lockhart DJ, Weissman JS, Walter P (2000) Functional and genomic analyses reveal an essential coordination between the unfolded protein response and ER-associated degradation. Cell 101(3):249–258
Urano F, Wang X, Bertolotti A, Zhang Y, Chung P, Harding HP, Ron D (2000) Coupling of stress in the ER to activation of JNK protein kinases by transmembrane protein kinase IRE1. Science 287(5453):664–666
van't Wout EF, Dickens JA, van Schadewijk A, Haq I, Kwok HF, Ordonez A, Murphy G, Stolk J, Lomas DA, Hiemstra PS, Marciniak SJ (2014) Increased ERK signalling promotes inflammatory signalling in primary airway epithelial cells expressing Z alpha1-antitrypsin. Hum Mol Genet 23(4):929–941. doi:10.1093/hmg/ddt487
van't Wout EF, van Schadewijk A, Lomas DA, Stolk J, Marciniak SJ, Hiemstra PS (2015) Function of monocytes and monocyte-derived macrophages in alpha1-antitrypsin deficiency. Eur Respir J 45:365–376. doi:10.1183/09031936.00046114
Vembar SS, Brodsky JL (2008) One step at a time: endoplasmic reticulum-associated degradation. Nat Rev Mol Cell Biol 9(12):944–957. doi:10.1038/nrm2546
Voeltz GK, Rolls MM, Rapoport TA (2002) Structural organization of the endoplasmic reticulum. EMBO Rep 3(10):944–950. doi:10.1093/embo-reports/kvf202
Walter P, Ron D (2011) The unfolded protein response: from stress pathway to homeostatic regulation. Science 334(6059):1081–1086. doi:10.1126/science.1209038
Wang H, Li Q, Shen Y, Sun A, Zhu X, Fang S, Shen Y (2011) The ubiquitin ligase Hrd1 promotes degradation of the Z variant alpha 1-antitrypsin and increases its solubility. Mol Cell Biochem 346(1–2):137–145. doi:10.1007/s11010-010-0600-9

Werner ED, Brodsky JL, McCracken AA (1996) Proteasome-dependent endoplasmic reticulum-associated protein degradation: an unconventional route to a familiar fate. Proc Natl Acad Sci USA 93(24):13797–13801

Wu Y, Whitman I, Molmenti E, Moore K, Hippenmeyer P, Perlmutter DH (1994) A lag in intracellular degradation of mutant alpha 1-antitrypsin correlates with the liver disease phenotype in homozygous PiZZ alpha 1-antitrypsin deficiency. Proc Natl Acad Sci USA 91 (19):9014–9018

Wu Y, Swulius MT, Moremen KW, Sifers RN (2003) Elucidation of the molecular logic by which misfolded alpha 1-antitrypsin is preferentially selected for degradation. Proc Natl Acad Sci USA 100(14):8229–8234. doi:10.1073/pnas.1430537100

Yamamoto K, Sato T, Matsui T, Sato M, Okada T, Yoshida H, Harada A, Mori K (2007) Transcriptional induction of mammalian ER quality control proteins is mediated by single or combined action of ATF6alpha and XBP1. Dev Cell 13(3):365–376. doi:10.1016/j.devcel.2007.07.018, S1534-5807(07)00300-0 [pii]

Yamasaki M, Li W, Johnson DJ, Huntington JA (2008) Crystal structure of a stable dimer reveals the molecular basis of serpin polymerization. Nature 455(7217):1255–1258. doi:10.1038/nature07394, nature07394 [pii]

Yamasaki M, Sendall TJ, Pearce MC, Whisstock JC, Huntington JA (2011) Molecular basis of alpha1-antitrypsin deficiency revealed by the structure of a domain-swapped trimer. EMBO Rep 12(10):1011–1017. doi:10.1038/embor.2011.171

Ye Y, Shibata Y, Yun C, Ron D, Rapoport TA (2004) A membrane protein complex mediates retro-translocation from the ER lumen into the cytosol. Nature 429(6994):841–847. doi:10.1038/nature02656

Ye Y, Shibata Y, Kikkert M, van Voorden S, Wiertz E, Rapoport TA (2005) Recruitment of the p97 ATPase and ubiquitin ligases to the site of retrotranslocation at the endoplasmic reticulum membrane. Proc Natl Acad Sci USA 102(40):14132–14138. doi:10.1073/pnas.0505006102

Ying Z, Wang H, Fan H, Wang G (2011) The endoplasmic reticulum (ER)-associated degradation system regulates aggregation and degradation of mutant neuroserpin. J Biol Chem 286 (23):20835–20844. doi:10.1074/jbc.M110.200808

Yoneda T, Imaizumi K, Oono K, Yui D, Gomi F, Katayama T, Tohyama M (2001) Activation of caspase-12, an endoplastic reticulum (ER) resident caspase, through tumor necrosis factor receptor-associated factor 2-dependent mechanism in response to the ER stress. J Biol Chem 276(17):13935–13940. doi:10.1074/jbc.M010677200

Yoshida H, Matsui T, Yamamoto A, Okada T, Mori K (2001) XBP1 mRNA is induced by ATF6 and spliced by IRE1 in response to ER stress to produce a highly active transcription factor. Cell 107(7):881–891. doi:10.1016/s0092-8674(01)00611-0

Zhou M, Schekman R (1999) The engagement of Sec61p in the ER dislocation process. Mol Cell 4 (6):925–934

Serpins in *Caenorhabditis elegans* 15

Cliff J. Luke, Mark T. Miedel, Linda P. O'Reilly, Allyson Wyatt, Ryan R. Knoerdel, Stephen C. Pak, and Gary A. Silverman

Contents

Abstract

C. elegans is an extremely powerful model organism to study gene function and biology. The nematodes' genetic tractability and high degree of genetic similarity with humans make it ideal to study the biologic role of serpins. The endogenous *C. elegans* serpins are most similar to the human clade B or intracellular serpin family and have a predominantly cytoplasmic distribution. Since the nematode genes often have functional orthologues in humans, the biological role of mammalian serpins can be extrapolated from *C. elegans*. Additionally,

C.J. Luke (✉) • M.T. Miedel • L.P. O'Reilly • A. Wyatt • R.R. Knoerdel • S.C. Pak • G.A.Silverman (✉)
Departments of Pediatrics and Cell Biology, Children's Hospital of Pittsburgh of UPMC and Magee-Womens Hospital Research Institute, University of Pittsburgh School of Medicine, 4401 Penn Avenue, Pittsburgh, PA 15224, USA
e-mail: cliff.luke@chp.edu; gsilverman@upmc.edu

M. Geiger et al. (eds.), *The Serpin Family*, DOI 10.1007/978-3-319-22711-5_15

mutant-mammalian serpin genes resulting in serpinopathies can be introduced into *C. elegans* to determine the biological pathways that play a role in disease progression, as well as develop potential therapeutics to ameliorate the disease.

15.1 Introduction

Since the 1970s, *Caenorhabditis elegans* has emerged as a powerful model organism to study homologous and orthologous human protein families due to its genetic amenability, relatively fast generation times, small size, large brood size, and cost effectiveness. Moreover, the sequencing of the *C. elegans* genome (Consortium 1998) has facilitated the development of whole-genome RNAi libraries (Rual et al. 2004; Kim et al. 2005; Kamath et al. 2003), large repositories of gene knockouts (Consortium 2012), global tissue and development expression data (Hunt-Newbury et al. 2007; McKay et al. 2003; Dupuy et al. 2007), as well as proteomic and drug treatment datasets (Boyd-Kimball et al. 2006; Dong et al. 2007; Madi et al. 2003a, b; Schrimpf et al. 2009; Taskov et al. 2005; van Rossum et al. 2001; Gosai et al. 2010; Leung et al. 2013; Benson et al. 2014; O'Reilly et al. 2014a, c). Thus, the resources and information available to the *C. elegans* community to determine biological function of genes have expanded exponentially. Although morphologically distinct from mammalian species, ~40 % of the *C. elegans* genome has human orthologues, with many conserved pathways (Shaye and Greenwald 2011), suggesting that the nematode has more in common with *Homo sapiens* than its anatomy dictates. In support of this, several groups have used *C. elegans* to express human genes with known disease-causing mutations and have successfully recapitulated human phenotypes in the worm. Thus *C. elegans* is an important model organism for studying biological function of genes, especially human orthologues with unknown function. Thus, *C. elegans* has proved to be a valuable model organism for studying numerous disease-related genes.

The serpin superfamily is divided into 13 clades. Most of the biological functions of the serpin superfamily have been determined from human disease-causing mutations with the serpin gene itself (e.g. α1-antitrypsin deficiency). These serpins result in what has now been termed serpinopathies. However, many serpin members have yet to have an ascribed biological function due to either no known mutations resulting in an obvious disease state or a lack of tools and resources available. Additionally, those that have mutations resulting in a serpinopathy are not always clearly linked to a known biological pathway. Thus, using *C. elegans* as a model organism can provide invaluable insight into biologic functions of serpins or determine the cellular pathways involved in an individual serpinopathy that can aid therapeutic intervention.

15.1.1 The *C. elegans* Serpins

After the complete sequence of the genome was published (Consortium 1998), *C. elegans* serpin genes were first identified in silico using genefinder algorithms (Solovyev and Salamov 1997) and homology to human serpins (Whisstock et al. 1999; Kruger et al. 2002). Phylogenetic analysis of the all known serpin primary amino acid sequences placed the *C. elegans* serpins into their own phylogenetic clade L (Silverman et al. 2001). However, *C. elegans* serpins show the strongest sequence homology to members of clade B (or intracellular) serpins, especially as they lack C-terminal extensions, an alanine at the P10 position, and a N-terminal signal peptide (Luke et al. 2006; Pak et al. 2004). The entire *C. elegans* serpin family resides on a single chromosome, V. Initially, ten *C. elegans* serpin genes were identified, all with the prefix *srp-* followed by their respective number. *srp-4* was originally identified by cDNA analysis (Reboul et al. 2001) being almost identical to *srp-3* differing by only 1 amino acid at the P1 position (Pak et al. 2004). Additionally, as all the other *C. elegans* serpin genes are positioned on chromosome V in pairs (*srp-1* and *-2*, *srp-5* and *-6*, *srp-7* and *-8*, *srp-9* and *-10*), it was tempting to speculate that *srp-3* had a genomic partner in close proximity. However, extensive RT-PCR and genomic analysis has shown the *srp-4* cDNA to be a molecular biology-generated artifact (Pak et al. 2004). Of the remaining nine *C. elegans* serpin genes, five have been determined to be true protease inhibitors (*srp-1*, *-2*, *-3*, *-6*, and *-7*), three have been classed as pseudogenes (*srp-5*, *-9*, and *-10*), and *srp-8* lacks a reactive center loop consistent with an inhibitory serpin (Luke et al. 2006, 2007; Pak et al. 2004, 2006). Interestingly, in contrast to the inhibitory *C. elegans* serpins, these non-RCL encoding serpins and pseudogenes do not have orthologous counterparts in other nematode species, suggesting that these may be unique to *C. elegans* resulting from a more recent duplication event (Luke et al. 2006).

15.2 Individual Serpins

15.2.1 Inhibitory-Type Serpins

15.2.1.1 SRP-1

Based on genefinder predictions (Solovyev and Salamov 1997), *srp-1* (C05E4.3) was predicted to be a protein-coding gene with five exons. Reverse-transcriptase PCR amplified a full-length cDNA from total wild-type RNA, which was subsequently named *srp-1* (Pak et al. 2004). The protein product of *srp-1* is a 366-amino-acid protein of predicted molecular mass of ~41.3 kDa. Primary amino acid analysis predicted the SRP-1 protein to be an intracellular serpin that inhibits the proteolytic activity of proteases (Luke et al. 2006; Pak et al. 2004; Whisstock et al. 1999; Kruger et al. 2002). The N-terminus of the protein is predicted to have several interesting protein motifs, several phosphorylation sites for glycogen synthase kinase 3 (GSK3) and casein kinase (CK) 1 and 2, an LC3 interacting domain, and an N-linked glycosylation site (Table 15.1) (Dinkel et al. 2014). While SRP-1's protease targets have yet to be identified in vitro or in vivo, it is predicted to inhibit

Table 15.1 The endogenous inhibitory serpins in *C. elegans*

C. elegans Serpin		SRP-1	SRP-2	SRP-3	SRP-6	SRP-7			
						a	b	c	d
AA	#[a]	366	359	362	375	366	366	365	373
	PI[a]	6.02	4.49	8.51	5.20	5.50	5.95	6.17	5.38
	MW[a]	41265.75	39664.67	40618.25	41695.73	40417.38	40362.42	40512.56	41219.25
Primary AA Motifs[b]		BIR; GSK3; CK1; CK2; LIR; N-Glyc; PTB	BIR; N-Glyc; PIKK	BIR;NRD; Nend; ERret	BIR; 14-3-3; CK1; CK2; NEK2	BIR	BIR	BIR	KEN; BIR
Subcellular locale		ER [1]	Cytoplasm [2]	Cytoplasm[c]	Cytoplasm [3]	Cytoplasm[c]	Cytoplasm[c]	Cytoplasm[c]	Cytoplasm[c]
Proteases inhibited		Lysosomal cysteine	Granzyme B and CatG, -L, -S and -V [2]	Chymotrypsin and cathepsin G [4]	CatL, -K and -V and calpain II [3]	Lysosomal cysteine and trypsin-like serine	Trypsin-like serine	Trypsin-like serine	Lysosomal cysteine and trypsin-like serine
P4-P4'		IFFT↓SASS	VQLE↓MMIM	AVPM↓SARM	FSLT↓SVFI	ISLK↓SAMF	FVRK↓SARP	IERC↓RKKM	ISLK↓SAMF
Tissue distribution		ArC^{Lva}; BWM^{Ad} [1, 5]	BWM [6]; PhN [2]; Hyp [2, 7]; SoC [2]; SeC [2]; Int [2] ASM [2]	Int^{Lva}; BWM$^{Lva;Ad}$ [4]	SoC; PIV; VHyp; Spe; Int [3]	Unknown			
Life stage expression		Lva; Ad [1, 5]	Emb; Lva; Ad [2, 6, 7]	Lva; Ad [4]	Lva; Ad [3]	Unknown			
Biological Function		Unknown	Development [2]	Unknown	Lysosomal dependent necrotic cell death [3]	UPR [8, 9]			

Emb, embryo; Lva, larval; Ad, adult; BIR, baculoviral IAP repeat; GSK3, GSK3 phosphorylation site; CK1, casein kinase 1 phosphorylation site; CK2, casein kinase 2 phosphorylation site; LIR, LC3-interacting region; N-Glyc, N-linked glycosylation site; PTB, phosphotyrosine-binding domain; PIKK, phosphoinositide-3-OH-kinase-related kinase phosphorylation site; NRD, N-Arg dibasic convertase (nardilysin) cleavage site; Nend, N-degron NBox domain; ERret, Endoplasmic reticulum retention signal; 14-3-3, 14-3-3 protein-binding domain; NEK2, NEK2 phosphorylation site; KEN, APCC-binding destruction motif KEN box domain; ArC, arcade cells; BWM, body wall muscle; PhN, phasmid neurons; Hyp, hypodermis; SoC, socket cells; SeC, seam cells; Cut, cuticle; Int, intestine; PIV, pharyngeal-intestinal valve; VHyp; vulval hypoderm; Spe, spermatheca Grey boxes indicated predicted data. [a]Predicted from primary amino acid sequence using MacVector Software (v13.5.0). [b]Motif prediction using the ELM protein site predication server (http://elm.eu.org; Dinkel et al. 2014). Only those outside the structural motifs are listed; [c]predicted using PSORT II database server (http://psort.hgc.jp/)

1. Meissner et al. (2011)
2. Pak et al. (2004)
3. Luke et al. (2007)
4. Pak et al. (2006)
5. McKay et al. (2003)
6. Schipanski et al. (2013)
7. Hunt-Newbury et al. (2007)
8. Shen et al. (2005)
9. Urano et al. (2002)
10. Dinkel et al. (2014)

the lysosomal cysteine proteases as it contains a phenylalanine at the P2 position, which is favored by this class of protease (Bromme et al. 1993, 1994, 1996, 1999).

Using transgenic GFP-tagged expression analysis, SRP-1 was shown to be expressed in both larval and adult stages of *C. elegans*, specifically in the arcade cells and body wall muscle (McKay et al. 2003; Meissner et al. 2011). Recently a study was undertaken to generate a subcellular "localizome" in the body wall muscle of *C. elegans*. SRP-1 was shown to have reticular expression, reminiscent of ER localization (Meissner et al. 2011). This was a surprising finding, given the lack of a secretion tag (Luke et al. 2006) and an ER retention carboxy-terminal signal (Kapulkin et al. 2005). However, the presence of N-linked glycosylation in SRP-1 may also suggest the possibility of ER transport, possibly via nontraditional mechanisms or glycosylation in the cytoplasm (Kaji et al. 2003).

While a knockout deletion allele exists for *srp-1*, this null strain shows no overt phenotypes. However, upon exposure to *Staphylococcus aureus*, *srp-1(ok262)* showed a reduced life-span when compared to wild type (Pak, Luke and Silverman, Unpublished data). However the biological significance of these findings has yet to be determined.

15.2.1.2 SRP-2

srp-2 (C05E4.1) also encodes an ovalbumin-like serpin of 359 amino acids, with a predicted molecular mass of ~39.7 kDa, and is a functional protease inhibitor. Transcriptionally, *srp-2* is unique in the *C. elegans* serpins as it is the only one of this family shown to contain the splice leader, SL1 (Pak et al. 2004). The role of splice leaders in *C. elegans* is unknown but is thought to be involved in mRNA stabilization and/or translation regulation (Maroney et al. 1995; Lall et al. 2004). The *srp-2* gene contains six exons, although the first exon has no translated codons, and is positioned on chromosome V in relatively close proximity to *srp-1*. At a primary amino acid level, *srp-2* is predicted to have an N-linked glycosylation site (Table 15.1), although, unlike the other *C. elegans* inhibitory serpins, a glycan moiety was not detected in lectin affinity pulldowns and subsequent mass spectrometry analysis (Kaji et al. 2003). *SRP-2* is also predicted to have a phosphoinositide-3-OH-kinase phosphorylation site (Table 15.1) (Dinkel et al. 2014). Phylogenetic analysis of SRP-2 shows it has the highest degree of identity with human serpins SERPINB1 and -B8 at the primary amino acid level, while human SERPINB7 showed the highest overall similarity (Silverman et al. 2015).

Recombinant SRP-2 was purified from *Escherichia coli* and assayed against a panel of proteases to identify its targets in vitro. SRP-2 inhibited granzyme B and the cathepsins G, L, S, and V, indicating that it is a cross-class inhibitor of both serine and cysteine proteases. Analysis of the interaction of recombinant SRP-2 and human granzyme B and cathepsin V indicated that it formed an SDS-stable complex with both enzymes and inhibited with a k_a value of 2.3×10^4 and 2.9×10^4 M^{-1} s^{-1}, respectively. Interestingly, analysis of the cleavage site of these two enzymes shows that granzyme B is cleaved between the P1–P1′ (Glu-Met), whereas cathepsin V is cleaved between the P1–P2′ (Met-Met) residues (Pak et al. 2004).

Transcriptional fusions expressing GFP under the control of the *srp-2* promoter suggested that this gene was expressed in multiple cell types throughout development. However, cell-type expression varies throughout development. In L1 and L2 larva, GFP expression was seen in the hypoderm, the socket cells, the alae, the seam cells, and the posterior intestinal cells. In the adult hermaphrodite, strong expression was visible in the hypodermal cells surrounding the pharynx and vulva, the phasmid neurons, and the fibrous organelles (Pak et al. 2004). Additional transcriptional expression analyses of *srp-2* have also indicated expression in the adult hypoderm (Hunt-Newbury et al. 2007; McKay et al. 2003) and seam cells (http://www.wormbase.org/species/all/expr_pattern/Expr53#021--10). Expression of full-length N-terminal GFP::SRP-2 fusions under the control of the *srp-2* promoter indicates expression in the body wall muscle (Schipanski et al. 2013) and hypodermis (Silverman et al. 2015).

While SRP-2 has a diverse inhibitory profile in vitro and expression pattern in vivo, a null deletion mutant strain, *srp-2(ok350)*, shows no overt phenotype. However, overexpression of *srp-2* resulted in larval arrest or slow-growing animals. These data suggests a potential role for *srp-2* in postembryonic development (Pak et al. 2004).

15.2.1.3 SRP-3

srp-3 (Y32G9A.4) was originally predicted to encode a 362-amino-acid protein by genefinder prediction algorithms (Solovyev and Salamov 1997). The *srp-3* gene only contains four exons and, unlike its other serpin counterparts, does not have a genetic partner in close proximity on chromosome V (Pak et al. 2004). Primary amino acid and phylogenetic analyses indicated *srp-3* to be a probable functional inhibitory serpin (Whisstock et al. 1999; Kruger et al. 2002; Pak et al. 2004; Luke et al. 2006). Using protein-prediction databases, SRP-3 is indicated to have residue in the cytoplasm; however, it is also predicted to have an ER retention signal. Phylogenetically, SRP-3 is closest to human SERPINB1 (http://www.wormbase.org/species/c_elegans/gene/WBGene00005644).

To determine whether SRP-3 was able to function as an inhibitory serpin, recombinant GST-SRP-3 inhibitory activity was tested against various cysteine and serine proteases (Pak et al. 2006). The reactive center of SRP-3 (Pro-Met·Ser-Ala) is similar to that of human α1-antitrypsin (Pro-Met·Ser-Ile). In these biochemical studies, SRP-3 completely inhibited the activity of human cathepsin G and chymotrypsin, but did not inhibit other serine proteases nor any of the cysteine proteases. rGST-SRP-3 formed SDS-stable complexes with both cathepsin G and chymotrypsin, in vitro. The stoichiometry of inhibition for the interaction of rGST-SRP-3 with both cathepsin G and chymotrypsin was approximately 1:1 with association constants (k_a) of $1.3 \times 10^5\ \mathrm{m}^{-1}\ \mathrm{s}^{-1}$ for the interaction with cathepsin G and $0.7 \times 10^5\ \mathrm{m}^{-1}\ \mathrm{s}^{-1}$ for chymotrypsin. Analysis of the cleavage site by matrix-assisted laser desorption/ionization (MALDI) mass spectrometry revealed that the P1–P1′ Met-Ser residues of SRP-3 served as the reactive center with both cathepsin G and chymotrypsin (Pak et al. 2006). Taken together, these findings describing the biochemical characterization of SRP-3 are typical for a serpin::target protease interaction (Gettins 2002).

The expression pattern of both *LacZ* and GFP promoter fusions of SRP-3 is localized in muscle cells in both larval and adult animals (Pak et al. 2006). In early developmental stages (L2), SRP-3 was observed in both the posterior intestinal and anterior muscle cells, while in late larval (L4) and adult stages, expression was restricted to the body wall muscle cells of the anterior region of the animal.

These studies provide insight into both the biochemical activity and expression pattern of SRP-3. However, little is still known regarding the biological function of

SRP-3 inhibitory function within the muscle cell of *C. elegans* and how this regulation of proteolytic activity may contribute to the maintenance of overall muscle cell homeostasis. Interestingly, human SERPINB6, which is produced in multiple cell types (keratinocytes, monocytes, mast cells) including muscle, and SRP-3 in *C. elegans* both inhibit chymotrypsin-like proteases (Coughlin et al. 1993; Strik et al. 2004). This overlap suggests that there is conservation of tissue-specific inhibitory function among these serpins and that common roles may exist regarding the function of these protease inhibitors in muscle cells.

15.2.1.4 SRP-6

SRP-6 has been the best characterized of all the *C. elegans* serpins. As with the other serpins, *srp*-6 (C036.18) was originally identified by the genefinder algorithm and subsequently confirmed by cDNA expression analysis (Reboul et al. 2001; Pak et al. 2004; Luke et al. 2007). The gene structure contains six exons, the first exon containing a small N-terminal extension of MSDSSSDEK. This extension is predicted to contain multiple casein kinase I and II and NEK2 phosphorylation sites as well as a 14-3-3 binding motif (Dinkel et al. 2014).

Similar to the inhibitory profile of human SERPINB3, biochemical analysis demonstrated that recombinant 6xHis::SRP-6 inhibits lysosomal cysteine proteases, but not serine proteases; however, unlike all cysteine protease-inhibitory serpins investigated to date, SRP-6 also inhibited the calcium-activated cytoplasmic cysteine protease, calpain-2 (Luke et al. 2007). The stoichiometry of inhibition was 1:1 for the interaction with cathepsin L and 1:1.5 for calpain-2, with k_a values of $1 \times 10^4\ \mathrm{m}^{-1}\ \mathrm{s}^{-1}$ and $2.3 \times 10^4\ \mathrm{m}^{-1}\ \mathrm{s}^{-1}$, respectively. Additionally, SDS-PAGE analysis showed that in vitro radiolabeled synthesized SRP-6 could form an SDS-stable complex with both cathepsin L and calpain-2 when recombinant SRP-6 was incubated with either protease.

A full-length SRP-6::GFP transgene showed expression within the nucleus and cytoplasm of intestinal cells in adult animals. In addition to the intestine, SRP-6 expression was also visualized in other tissues including socket cells, pharyngeal-intestinal valve, vulval hypoderm, spermatheca, and the spermathecal-uterine junction (Luke et al. 2007).

Like the other serpin-null strains, a deletion in the *srp*-6 gene (*srp-6(ok319)*) did not result in any overt phenotypes under normal conditions. However, once exposed to noxious stimuli, the intestinal cells of *srp-6* null animals underwent a necrotic cell death (NCD) routine, which resulted in death of the animals (Luke et al. 2007). This NCD was dependent on multiple factors including calcium release from the ER and full lysosomal rupture. This NCD routine was truly dependent on the absence of SRP-6 alone as it could be phenocopied by *srp-6(RNAi)* and rescued using a full-length *srp-6* gene. Additionally, intestinal expression of SRP-6 was enough to rescue the phenotype fully. Moreover, mutation of the RCL with the rescue construct at either the P14 or P2-P1 positions abolished this rescue, suggesting that it was the protease-inhibitory activity of SRP-6 responsible for the protection of intestinal cells against this stress-induced NCD pathway (Schick et al. 1998). Consistent with these findings, treatment of *srp-6* null animals with the cysteine protease inhibitor E-64d also protected against necrosis, suggesting cysteine protease activity involvement in this pathway. Using a reverse genetic approach, it was demonstrated that treatment of *srp-6* null animals with RNAi's directed against specific *C. elegans* calpains (*tra-3*; *clp-5*; *W05G11.4*) and lysosomal cysteine proteases (*cpl-1*; *cpr-6*) protected these animals from undergoing necrosis in response to various stressors. Finally, electron and fluorescence microscopy studies show that lysosomal rupture was required for the necrosis phenotype found in *srp-6* null animals (Luke et al. 2007; Zou et al. 2014). Since lysosomes did not rupture in wild-type animals when exposed to stress, this suggests that SRP-6 protects these animals against lysosomal rupture. However, since *SRP-6* was a lysosomal cysteine protease inhibitor in vitro, lysosomes were artificially damaged in both wild-type and *srp-6* null animals using the photosensizing compound acridine orange, which causes lysosomal rupture upon exposure to blue light. Full lysosomal rupture within the intestine of *srp-6* null animals resulted in nearly 100 % of animals dying. However, this death was significantly reduced in wild-type animals, suggesting that SRP-6 also functioned in the cytoprotection of intestinal cells post-lysosomal damage. Astoundingly, surviving wild-type animals post-blue light exposure reformed their lysosomal compartments, suggesting that fulminant lysosomal rupture was not the end point of this pathway (Luke et al. 2007).

Taken together, these findings demonstrate that cellular necrosis was not a passive process, but was triggered by an ordered, SRP-6-regulated, protease-driven stress response pathway involving the lysosomal compartment. Thus, these studies highlight the concept that intracellular serpins can function as pro-survival factors that play a cytoprotective role against multiple stressors that result in necrosis in *C. elegans* (Luke et al. 2007).

15.2.1.5 SRP-7

While SRP-7 is one of the least-characterized serpins in *C. elegans*, it is potentially one of the most interesting. *The srp-7* gene (F20D6.4) was predicted by the genefinder algorithm to contain five exons with all the structural hallmarks of an inhibitory serpin in its translated amino acid sequence (Solovyev and Salamov 1997). However, it is unique within this class of serpins as it encodes at least three different isoforms as a result of alternately spliced RCL-containing exons all confirmed by cDNA analysis (Pak et al. 2004). Denoted SRP-7a, b, and c, these chimeric serpins potentially have different target specificity and inhibit different proteases as they contain different amino acids from P4-P4' (Table 15.1). Recently, using a more comprehensive cDNA analysis, an additional 5′ exon has been identified upstream of the original exon 1, and thus another variant has been termed SRP-7d (http://www.wormbase.org/species/c_elegans/gene/WBGene00005648). This small new exon 1 gives rise to a N-terminal extension of MSDPKEN, which contains a KEN box domain, which in other proteins, has been shown to direct polypeptides to the ubiquitin-proteasome system for degradation (Chao et al. 2012; Pfleger and Kirschner 2000). While not confirmed by RT-PCR, the presence of this small first exon also suggests a possibility of two additional isoforms SRP-7e and f, again with potentially different target protease activity.

The presence of the basic amino acids at or within close proximity to the P1 position in the RCLs of the various isoforms (SRP-7a–c) suggests that all the isoforms have the ability to inhibit trypsin-like serine proteases. The presence of the hydrophobic leucine residue at the P2 position of SRP-7a also suggests that this isoform has the ability to inhibit the papain-like cysteine proteases as well. Interestingly, SRP-7c has a cysteine at the P1 position, similar to the human intracellular serpins SERPINB1, -6, -8, and -9, which all have this amino acid at or close to the potential cleavage site.

While many of the in vivo cell biology and in vitro biochemical assays remain uncharacterized, the *srp-7* gene has been implicated in the endoplasmic reticulum unfolded protein response (UPR) by genetic screen analyses in *C. elegans* (Urano et al. 2002; Shen et al. 2005). In these studies, SRP-7 was induced at least twofold by tunicamycin treatment in wild-type animals, but this upregulation was abolished in the absence of XBP-1 and IRE-1, both UPR-specific gene regulators. These data and the presence of a KEN box motif suggest a role of SRP-7 in the UPS pathway and regulation of turnover of misfolded substrates.

15.2.2 Other Serpin-Like Genes in *C. elegans*

Of the nine intracellular serpin genes in the *C. elegans* genome, four genes (*srp-5*, *-8*, *-9*, *and -10*) are unlikely to be inhibitory serpins, either due to their genomic structure (*srp-5*, *-9*, *and -10*) or primary amino acid sequence of the RCL (*srp-8*) (Luke et al. 2006).

Analysis of genomic *srp-5* shows a high degree of identity with *srp-6* with most of the variation found in the potential RCL (Luke et al. 2006). However, compared to *srp-6*, there is a single nucleotide deletion in exon 2, which would cause a frameshift in the translation and early termination. Initial genefinder predictions had an alternate splice site to maintain in order to circumvent this frameshift and maintain a whole translatable transcript. However, this serpin transcript would be missing vital structural motifs rendering *SRP*-5 unlikely to function as a true serpin. Thus *srp-5* has been labeled a pseudogene (http://www.wormbase.org/species/c_elegans/gene/WBGene00005646). Interestingly, a deletion in the *srp-5* gene (*srp-5 (tm1347)*) is reported as being lethal (NBRP, Japan, http://www.shigen.nig.ac.jp/c.elegans/); however, this is more likely to be due to alternate mutations within the mutagenized genome, as *srp-6(RNAi)* would also knockdown *srp-5* as a secondary target, and no overt lethality issues without stress have been observed (Luke et al. 2007).

srp-9 and *-10* are a serpin gene pair found in close genomic proximity to each other at the telomeric region of chromosome V. Both these genes contain nonsense mutations in the genomic DNA that would lead to premature stop codons in exon 1 for *srp-9* and exon 3 for *srp-10* (Luke et al. 2006). Initial genefinder predictions indicated alternative splicing to circumvent these early termination codons; however, these would also encode for serpins missing key structural motifs that would likely result in a nonfunctional serpin. Additionally, cDNAs generated for these genes did not concur with the genefinder prediction and retained the early termination codon (Luke et al. 2006). Interestingly, using transcriptional fusion transgenic *C. elegans*, the *srp-9* promoter directed GFP expression in the intestine of larval animals, suggesting that the promoter was active (Hunt-Newbury et al. 2007).

While *srp-5*, *-9*, and *-10* are all classed as pseudogenes and contain early termination codons, cDNAs could be generated for all three genes (Luke et al. 2006), and the RCL region of these genes remains a potential inhibitory serpin motif. The lack of additional mutations, especially in the RCL region, and the absence of these pseudogenes in other nematode species give strong precedent that these genes arose from a recent duplication event with *C. elegans*.

The *srp-8* gene encodes for a full-length serpin-like gene with high homology to SRP-7. The presence of *srp-8* transcripts has also been identified and they concur with the genefinder predictions (Lamesch et al. 2004). However, variations in the amino acid sequence within the RCL make this serpin unlikely to function as a protease inhibitor but still may have alternative biological roles.

15.2.3 Using *C. elegans* to Model Human Serpinopathies

15.2.3.1 *C. elegans* Versus Other Model Organisms

Since its introduction to biology in the late 1960s, the soil nematode *C. elegans* has played a critical role in elucidating many important genetic pathways involved in development (Brenner 1974), cell death (Hengartner and Horvitz 1994a, b), and aging (Kenyon 2010). The multicellular and multi-organ system complexity in the *C. elegans* model organism recapitulates more efficaciously higher complex multicellular organisms than in vitro or unicellular systems, and therefore is able to more faithfully model human diseases. As many cellular processes are conserved, *C. elegans* has been extensively used to model complex human diseases including Alzheimer's disease (Levitan et al. 1996; Levitan and Greenwald 1995; Wittenburg et al. 2000), Parkinson's disease (Braungart et al. 2004), diabetes (Ogg et al. 1997), Duchenne muscular dystrophy (Gieseler et al. 2000; Grisoni et al. 2003), and cancer (Sternberg and Han 1998).

Due to the benefits of using this model organism listed in the introduction, *C. elegans* has been successfully used for whole organism-based high-content screening (O'Reilly et al. 2014b, c) and whole-genome RNAi screens (Sin et al. 2014).

15.2.3.2 Examples of Modeled Serpinopathies Studied in *C. elegans*

SERPINA1 (α1-antitrypsin) is produced in hepatocytes and circulates to the lungs where it protects the lung tissue from neutrophil elastase. In α1-antitrypsin deficiency (ATD), the accumulation of misfolded protein aggregates within the ER of hepatocytes can be cytotoxic and may lead to liver cirrhosis and increase susceptibility to hepatocellular carcinomas. In addition, the retention of aggregated protein in the liver results in a reduced level of active AT in circulation, which may lead to lung diseases (COPD and emphysema) (Silverman et al. 2013). The gain-of-function disease state of ATD has been successfully modeled in *C. elegans*. Fluorescent protein-tagged human SERPINA1 with the most common genetic variant (ATZ) has been expressed in the *C. elegans* intestine (Gosai et al. 2010). This humanized *C. elegans* model of ATD effectively recapitulates the human disease; ATZ accumulates as polymers within the ER of the intestinal cells, is poorly secreted, and decreases longevity (Long et al. 2014). This model has also been used to successfully screen for promising inhibitors of ATZ aggregate accumulation (Gosai et al. 2010) and identify other novel genetic factors influencing AT accumulation, by whole-genome RNAi library and EMS mutagenesis screen (Long et al. 2014; O'Reilly et al. 2014b).

Recently, *srp-2* has been suggested as model of the neuroserpin serpinopathy familial encephalopathy with neuroserpin inclusion bodies (FENIB) (Schipanski et al. 2013). Neuroserpin aggregation within neurons due to missense mutations in the protein results in progressive brain dysfunction, dementia, and seizures. The same point mutation in neuroserpin that causes FENIB (H320R) also results in aggregation of *srp-2* when introduced by mutagenesis (Schipanski et al. 2013; Silverman et al. 2015). Downregulation of the unfolded protein response (UPR) was shown to increase *srp-2* aggregation in neurons. However, a more in-depth phylogenetic analysis shows that SRP-2 has closer primary amino acid identity and similarity to the clade B serpins (Silverman et al. 2015). Moreover, SRP-2 does not contain any of the neuroserpin-targeting sequences. Additionally, in intestinal cells SRP-2::GFP expression did not colocalize with the ER marker, DsRed::KDEL, suggesting that SRP-2 is entirely cytoplasmic (Silverman et al. 2015). Thus, it seems unlikely that SRP-2 could be a relevant model for neuroserpin serpinopathies in *C. elegans*; however, as with the α1-antitrypsin *C. elegans* model, it remains to be determined if human or mouse neuroserpin mutants could be expressed in *C. elegans* to more faithfully recapitulate this serpinopathy.

15.3 Conclusions

Here we present the current state of knowledge of *C. elegans* serpins. *C. elegans* expresses five serpins which are verified as true protease inhibitors, one potentially non-inhibitory serpin and three pseudogenes However, there are still much to be learned about these intracellular serpins, such as the endogenous protease targets of each of the serpins, the functionality of the non-RCL encoding serpins, and the pathways that each serpin is involved in.

While *C. elegans* possesses many of the characteristics of a good model organism to study intracellular serpin gene function, there has also been recent move to model human serpinopathies of the secreted serpins. Thus, the *C. elegans* model organism is likely to be extremely helpful for studying many of the conserved pathways involved in proteostasis and disease progression. *C. elegans*' surprising genetic similarity to mammalian systems, its ease of culture, short life-span, and genetic amenability, as well as the collaborative *C. elegans* community, make this organism an ideal starting point for drug discovery, biologic function studies, and serpin biology in general.

Acknowledgments This study was supported by grants from the National Institutes of Health (DK079806 and DK081422). Some nematode strains were provided by the Caenorhabditis Genetics Center (CGC), which is funded by NIH Office of Research Infrastructure Programs (P40 OD010440). The authors would also like to thank the entire *C. elegans* community for embracing and showing us the power of the worm.

References

Benson JA, Cummings EE, O'Reilly LP, Lee M-H (2014) A high-content assay for identifying small molecules that reprogram *C. elegans* germ cell fate. Methods 68(3):529–535

Boyd-Kimball D, Poon HF, Lynn BC, Cai J, Pierce WM Jr, Klein JB, Ferguson J, Link CD, Butterfield DA (2006) Proteomic identification of proteins specifically oxidized in *Caenorhabditis elegans* expressing human Abeta(1-42): implications for Alzheimer's disease. Neurobiol Aging 27(9):1239–1249

Braungart E, Gerlach M, Riederer P, Baumeister R, Hoener MC (2004) *Caenorhabditis elegans* MPP+ model of Parkinson's disease for high-throughput drug screenings. Neurodegener Dis 1(4-5):175–183. doi:10.1159/000080983

Brenner S (1974) The genetics of *Caenorhabditis elegans*. Genetics 77(1):71–94

Bromme D, Bonneau PR, Lachance P, Wiederanders B, Kirschke H, Peters C, Thomas DY, Storer AC, Vernet T (1993) Functional expression of human cathepsin S in *Saccharomyces cerevisiae*. J Biol Chem 268(7):4832–4838

Bromme D, Bonneau PR, Lachance P, Storer AC (1994) Engineering the S_2 subsite specificity of human cathepsin S to a cathepsin L- and cathepsin B-like specificity. J Biol Chem 269:30238–30242

Bromme D, Okamoto K, Wang BB, Biroc S (1996) Human cathepsin O2, a matrix protein-degrading cysteine protease expressed in osteoclasts. Functional expression of human cathepsin O2 in Spodoptera frugiperda and characterization of the enzyme. J Biol Chem 271 (4):2126–2132

Bromme D, Li Z, Barnes M, Mehler E (1999) Human cathepsin V functional expression, tissue distribution, electrostatic surface potential, enzymatic characterization, and chromosomal localization. Biochemistry 38(8):2377–2385

Chao WC, Kulkarni K, Zhang Z, Kong EH, Barford D (2012) Structure of the mitotic checkpoint complex. Nature 484(7393):208–213. doi:10.1038/nature10896

Consortium CDM (2012) Large-scale screening for targeted knockouts in the *Caenorhabditis elegans* genome. G3 2(11):1415–1425. doi:10.1534/g3.112.003830

Consortium TCeS (1998) Genome sequence of the nematode *C. elegans*: a platform for investigating biology. Science 282(5396):2012–2018

Coughlin P, Sun J, Cerruti L, Salem HH, Bird P (1993) Cloning and molecular characterization of a human intracellular serine proteinase inhibitor. Proc Natl Acad Sci USA 90(20):9417–9421

Dinkel H, Van Roey K, Michael S, Davey NE, Weatheritt RJ, Born D, Speck T, Kruger D, Grebnev G, Kuban M, Strumillo M, Uyar B, Budd A, Altenberg B, Seiler M, Chemes LB, Glavina J, Sanchez IE, Diella F, Gibson TJ (2014) The eukaryotic linear motif resource ELM: 10 years and counting. Nucleic Acids Res 42(Database issue):D259–D266. doi:10.1093/nar/gkt1047

Dong MQ, Venable JD, Au N, Xu T, Park SK, Cociorva D, Johnson JR, Dillin A, Yates JR III (2007) Quantitative mass spectrometry identifies insulin signaling targets in *C. elegans*. Science 317(5838):660–663

Dupuy D, Bertin N, Hidalgo CA, Venkatesan K, Tu D, Lee D, Rosenberg J, Svrzikapa N, Blanc A, Carnec A, Carvunis AR, Pulak R, Shingles J, Reece-Hoyes J, Hunt-Newbury R, Viveiros R, Mohler WA, Tasan M, Roth FP, Le Peuch C, Hope IA, Johnsen R, Moerman DG, Barabasi AL, Baillie D, Vidal M (2007) Genome-scale analysis of in vivo spatiotemporal promoter activity in *Caenorhabditis elegans*. Nat Biotechnol 25(6):663–668

Gettins PG (2002) Serpin structure, mechanism, and function. Chem Rev 102(12):4751–4803

Gieseler K, Grisoni K, Segalat L (2000) Genetic suppression of phenotypes arising from mutations in dystrophin-related genes in *Caenorhabditis elegans*. Curr Biol 10(18):1092–1097

Gosai SJ, Kwak JH, Luke CJ, Long OS, King DE, Kovatch KJ, Johnston PA, Shun TY, Lazo JS, Perlmutter DH, Silverman GA, Pak SC (2010) Automated high-content live animal drug screening using *C. elegans* expressing the aggregation prone serpin α1-antitrypsin Z. PLoS One 5(11), e15460. doi:10.1371/journal.pone.0015460

Grisoni K, Gieseler K, Mariol MC, Martin E, Carre-Pierrat M, Moulder G, Barstead R, Segalat L (2003) The stn-1 syntrophin gene of *C. elegans* is functionally related to dystrophin and dystrobrevin. J Mol Biol 332(5):1037–1046

Hengartner MO, Horvitz HR (1994a) Activation of *C. elegans* cell death protein CED-9 by an amino-acid substitution in a domain conserved in Bcl-2. Nature 369(6478):318–320

Hengartner MO, Horvitz HR (1994b) *C. elegans* cell survival gene ced-9 encodes a functional homolog of the mammalian proto-oncogene bcl-2. Cell 76(4):665–676

Hunt-Newbury R, Viveiros R, Johnsen R, Mah A, Anastas D, Fang L, Halfnight E, Lee D, Lin J, Lorch A, McKay S, Okada HM, Pan J, Schulz AK, Tu D, Wong K, Zhao Z, Alexeyenko A, Burglin T, Sonnhammer E, Schnabel R, Jones SJ, Marra MA, Baillie DL, Moerman DG (2007) High-throughput in vivo analysis of gene expression in *Caenorhabditis elegans*. PLoS Biol 5(9), e237

Kaji H, Saito H, Yamauchi Y, Shinkawa T, Taoka M, Hirabayashi J, Kasai K, Takahashi N, Isobe T (2003) Lectin affinity capture, isotope-coded tagging and mass spectrometry to identify N-linked glycoproteins. Nat Biotechnol 21(6):667–672

Kamath RS, Fraser AG, Dong Y, Poulin G, Durbin R, Gotta M, Kanapin A, Le Bot N, Moreno S, Sohrmann M, Welchman DP, Zipperlen P, Ahringer J (2003) Systematic functional analysis of the *Caenorhabditis elegans* genome using RNAi. Nature 421(6920):231–237. doi:10.1038/nature01278

Kapulkin V, Hiester BG, Link CD (2005) Compensatory regulation among ER chaperones in *C. elegans*. FEBS Lett 579(14):3063–3068

Kenyon CJ (2010) The genetics of ageing. Nature 464(7288):504–512. doi:10.1038/nature08980, nature08980 [pii]

Kim JK, Gabel HW, Kamath RS, Tewari M, Pasquinelli A, Rual JF, Kennedy S, Dybbs M, Bertin N, Kaplan JM, Vidal M, Ruvkun G (2005) Functional genomic analysis of RNA interference in *C. elegans*. Science 308(5725):1164–1167. doi:10.1126/science.1109267

Kruger O, Ladewig J, Koster K, Ragg H (2002) Widespread occurrence of serpin genes with multiple reactive centre-containing exon cassettes in insects and nematodes. Gene 293 (1-2):97–105

Lall S, Friedman CC, Jankowska-Anyszka M, Stepinski J, Darzynkiewicz E, Davis RE (2004) Contribution of trans-splicing, 5'-leader length, cap-poly(A) synergism, and initiation factors to nematode translation in an Ascaris suum embryo cell-free system. J Biol Chem 279 (44):45573–45585. doi:10.1074/jbc.M407475200

Lamesch P, Milstein S, Hao T, Rosenberg J, Li N, Sequerra R, Bosak S, Doucette-Stamm L, Vandenhaute J, Hill DE, Vidal M (2004) *C. elegans* ORFeome version 3.1: increasing the coverage of ORFeome resources with improved gene predictions. Genome Res 14 (10B):2064–2069. doi:10.1101/gr.2496804

Leung CK, Wang Y, Malany S, Deonarine A, Nguyen K, Vasile S, Choe KP (2013) An ultra high-throughput, whole-animal screen for small molecule modulators of a specific genetic pathway in *Caenorhabditis elegans*. PLoS One 8(4), e62166. doi:10.1371/journal.pone.0062166

Levitan D, Greenwald I (1995) Facilitation of lin-12-mediated signalling by sel-12, a *Caenorhabditis elegans* S182 Alzheimer's disease gene. Nature 377(6547):351–354. doi:10.1038/377351a0

Levitan D, Doyle TG, Brousseau D, Lee MK, Thinakaran G, Slunt HH, Sisodia SS, Greenwald I (1996) Assessment of normal and mutant human presenilin function in *Caenorhabditis elegans*. Proc Natl Acad Sci USA 93(25):14940–14944

Long OS, Benson JA, Kwak JH, Luke CJ, Gosai SJ, O'Reilly LP, Wang Y, Li J, Vetica AC, Miedel MT, Stolz DB, Watkins SC, Zuchner S, Perlmutter DH, Silverman GA, Pak SC (2014) A *C. elegans* model of human alpha1-antitrypsin deficiency links components of the RNAi pathway to misfolded protein turnover. Hum Mol Genet 23(19):5109–5122. doi:10.1093/hmg/ddu235

Luke CJ, Pak SC, Askew DJ, Askew YS, Smith JE, Silverman GA (2006) Selective conservation of the RSL-encoding, proteinase inhibitory-type, clade 1 serpins in *Caenorhabditis* species. Front Biosci 11:581–594

Luke CJ, Pak SC, Askew YS, Naviglia TL, Askew DJ, Nobar SM, Vetica AC, Long OS, Watkins SC, Stolz DB, Barstead RJ, Moulder GL, Bromme D, Silverman GA (2007) An intracellular serpin regulates necrosis by inhibiting the induction and sequelae of lysosomal injury. Cell 130(6):1108–1119

Madi A, Mikkat S, Ringel B, Thiesen HJ, Glocker MO (2003a) Profiling stage-dependent changes of protein expression in *Caenorhabditis elegans* by mass spectrometric proteome analysis leads to the identification of stage-specific marker proteins. Electrophoresis 24(11):1809–1817

Madi A, Mikkat S, Ringel B, Ulbrich M, Thiesen HJ, Glocker MO (2003b) Mass spectrometric proteome analysis for profiling temperature-dependent changes of protein expression in wild-type *Caenorhabditis elegans*. Proteomics 3(8):1526–1534

Maroney PA, Denker JA, Darzynkiewicz E, Laneve R, Nilsen TW (1995) Most mRNAs in the nematode *Ascaris lumbricoides* are trans-spliced: a role for spliced leader addition in translational efficiency. RNA 1(7):714–723

McKay SJ, Johnsen R, Khattra J, Asano J, Baillie DL, Chan S, Dube N, Fang L, Goszczynski B, Ha E, Halfnight E, Hollebakken R, Huang P, Hung K, Jensen V, Jones SJ, Kai H, Li D, Mah A, Marra M, McGhee J, Newbury R, Pouzyrev A, Riddle DL, Sonnhammer E, Tian H, Tu D, Tyson JR, Vatcher G, Warner A, Wong K, Zhao Z, Moerman DG (2003) Gene expression profiling of cells, tissues, and developmental stages of the nematode *C. elegans*. Cold Spring Harb Symp Quant Biol 68:159–169

Meissner B, Rogalski T, Viveiros R, Warner A, Plastino L, Lorch A, Granger L, Segalat L, Moerman DG (2011) Determining the sub-cellular localization of proteins within *Caenorhabditis elegans* body wall muscle. PLoS One 6(5), e19937. doi:10.1371/journal.pone.0019937

O'Reilly LP, Benson JA, Cummings EE, Perlmutter DH, Silverman GA, Pak SC (2014a) Worming our way to novel drug discovery with the *Caenorhabditis elegans* proteostasis network, stress response and insulin-signaling pathways. Expert Opin Drug Discov 9 (9):1021–1032. doi:10.1517/17460441.2014.930125

O'Reilly LP, Long OS, Cobanoglu MC, Benson JA, Luke CJ, Miedel MT, Hale P, Perlmutter DH, Bahar I, Silverman GA, Pak SC (2014b) A genome-wide RNAi screen identifies potential drug targets in a *C. elegans* model of alpha1-antitrypsin deficiency. Hum Mol Genet 23 (19):5123–5132. doi:10.1093/hmg/ddu236

O'Reilly LP, Luke CJ, Perlmutter DH, Silverman GA, Pak SC (2014c) *C. elegans* in high-throughput drug discovery. Adv Drug Deliv Rev 69–70:247–253. doi:10.1016/j.addr.2013.12.001

Ogg S, Paradis S, Gottlieb S, Patterson GI, Lee L, Tissenbaum HA, Ruvkun G (1997) The Fork head transcription factor DAF-16 transduces insulin-like metabolic and longevity signals in *C. elegans*. Nature 389(6654):994–999. doi:10.1038/40194

Pak SC, Kumar V, Tsu C, Luke CJ, Askew YS, Askew DJ, Mills DR, Bromme D, Silverman GA (2004) Srp-2 is a cross-class inhibitor that participates in post-embryonic development of the nematode *Caenorhabditis elegans*: initial characterization of the Clade L serpins. J Biol Chem 279:15448–15459

Pak SC, Tsu C, Luke CJ, Askew YS, Silverman GA (2006) The *Caenorhabditis elegans* muscle specific serpin, SRP-3 neutralizes chymotrypsin-like serine peptidases. Biochemistry 45 (14):4474–4480

Pfleger CM, Kirschner MW (2000) The KEN box: an APC recognition signal distinct from the D box targeted by Cdh1. Genes Dev 14(6):655–665

Reboul J, Vaglio P, Tzellas N, Thierry-Mieg N, Moore T, Jackson C, Shin-i T, Kohara Y, Thierry-Mieg D, Thierry-Mieg J, Lee H, Hitti J, Doucette-Stamm L, Hartley JL, Temple GF, Brasch MA, Vandenhaute J, Lamesch PE, Hill DE, Vidal M (2001) Open-reading-frame sequence tags (OSTs) support the existence of at least 17,300 genes in *C. elegans*. Nat Genet 27(3):332–336

Rual JF, Ceron J, Koreth J, Hao T, Nicot AS, Hirozane-Kishikawa T, Vandenhaute J, Orkin SH, Hill DE, van den Heuvel S, Vidal M (2004) Toward improving *Caenorhabditis elegans* phenome mapping with an ORFeome-based RNAi library. Genome Res 14(10B):2162–2168. doi:10.1101/gr.2505604

Schick C, Bromme D, Bartuski AJ, Uemura Y, Schechter NM, Silverman GA (1998) The reactive site loop of the serpin SCCA1 is essential for cysteine proteinase inhibition. Proc Natl Acad Sci USA 95:13465–13470

Schipanski A, Lange S, Segref A, Gutschmidt A, Lomas DA, Miranda E, Schweizer M, Hoppe T, Glatzel M (2013) A novel interaction between aging and ER overload in a protein conformational dementia. Genetics 193(3):865–876. doi:10.1534/genetics.112.149088

Schrimpf SP, Weiss M, Reiter L, Ahrens CH, Jovanovic M, Malmstrom J, Brunner E, Mohanty S, Lercher MJ, Hunziker PE, Aebersold R, von Mering C, Hengartner MO (2009) Comparative functional analysis of the *Caenorhabditis elegans* and *Drosophila melanogaster* proteomes. PLoS Biol 7(3), e48. doi:10.1371/journal.pbio.1000048, 08-PLBI-RA-2342 [pii]

Shaye DD, Greenwald I (2011) OrthoList: a compendium of *C. elegans* genes with human orthologs. PLoS One 6(5), e20085. doi:10.1371/journal.pone.0020085

Shen X, Ellis RE, Sakaki K, Kaufman RJ (2005) Genetic interactions due to constitutive and inducible gene regulation mediated by the unfolded protein response in *C. elegans*. PLoS Genet 1(3), e37. doi:10.1371/journal.pgen.0010037

Silverman GA, Bird PI, Carrell RW, Church FC, Coughlin PB, Gettins PG, Irving JA, Lomas DA, Luke CJ, Moyer RW, Pemberton PA, Remold-O'Donnell E, Salvesen GS, Travis J, Whisstock JC (2001) The serpins are an expanding superfamily of structurally similar but functionally diverse proteins. Evolution, mechanism of inhibition, novel functions, and a revised nomenclature. J Biol Chem 276(36):33293–33296. doi:10.1074/jbc.R100016200

Silverman GA, Pak SC, Perlmutter DH (2013) Disorders of protein misfolding: alpha-1-antitrypsin deficiency as prototype. J Pediatr 163(2):320–326. doi:10.1016/j.jpeds.2013.03.077

Silverman RM, Cummings EE, O'Reilly LP, Miedel MT, Silverman GA, Luke CJ, Perlmutter DH, Pak SC (2015) The aggregation-prone intracellular serpin SRP-2 fails to transit the ER in *Caenorhabditis elegans*. Genetics 200(1):207–219. doi:10.1534/genetics.115.176180

Sin O, Michels H, Nollen EA (2014) Genetic screens in *Caenorhabditis elegans* models for neurodegenerative diseases. Biochim Biophys Acta 1842(10):1951–1959. doi:10.1016/j.bbadis.2014.01.015

Solovyev V, Salamov A (1997) The Gene-Finder computer tools for analysis of human and model organisms genome sequences. Proceedings/International conference on intelligent systems for molecular biology; ISMB international conference on intelligent systems for molecular biology 5:294–302

Sternberg PW, Han M (1998) Genetics of RAS signaling in *C. elegans*. Trends Genet 14 (11):466–472

Strik MC, Wolbink A, Wouters D, Bladergroen BA, Verlaan AR, van Houdt IS, Hijlkema S, Hack CE, Kummer JA (2004) Intracellular serpin SERPINB6 (PI6) is abundantly expressed by human mast cells and forms complexes with beta-tryptase monomers. Blood 103 (7):2710–2717

Taskov K, Chapple C, Kryukov GV, Castellano S, Lobanov AV, Korotkov KV, Guigo R, Gladyshev VN (2005) Nematode selenoproteome: the use of the selenocysteine insertion system to decode one codon in an animal genome? Nucleic Acids Res 33(7):2227–2238

Urano F, Calfon M, Yoneda T, Yun C, Kiraly M, Clark SG, Ron D (2002) A survival pathway for *Caenorhabditis elegans* with a blocked unfolded protein response. J Cell Biol 158(4):639–646 van Rossum AJ, Brophy PM, Tait A, Barrett J, Jefferies JR (2001) Proteomic identification of glutathione S-transferases from the model nematode *Caenorhabditis elegans*. Proteomics 1(11):1463–1468

Whisstock JC, Irving JA, Bottomley SP, Pike RN, Lesk AM (1999) Serpins in the *Caenorhabditis elegans* genome. Proteins 36(1):31–41

Wittenburg N, Eimer S, Lakowski B, Rohrig S, Rudolph C, Baumeister R (2000) Presenilin is required for proper morphology and function of neurons in *C. elegans*. Nature 406 (6793):306–309. doi:10.1038/35018575

Zou XJ, Ma YC, Guo LE, Liu WX, Liu MJ, Zou CG, Zhou Y, Zhang JF (2014) A lysosome-targeted fluorescent chemodosimeter for monitoring endogenous and exogenous hydrogen sulfide by in vivo imaging. Chem Commun (Camb) 50(89):13833–13836. doi:10.1039/c4cc05539j

The Dual Role of Serpins and Tissue-Type Plasminogen Activator During Stroke

16

Benoit D. Roussel, Morgane Louessard, Sara Martinez de Lizarrondo, and Denis Vivien

Contents

Abstract

Serpins are well-known inhibitors of plasminogen activators, key components of fibrinolysis. Within the last decades, tissue-type plasminogen activator (tPA) became more than a fibrinolytic agent and has been demonstrated to be a neuromodulator with physiological and pathological roles within the brain. Indeed, tPA is involved in brain plasticity, learning and memory and development. But in a stroke context, tPA contributes to the increase of intraneuronal calcium, by interacting with the GluN1 subunit of the NMDA receptor, leading to an increase in excitotoxicity. It also promotes blood-brain barrier leakage and inflammation, but in certain conditions tPA can also exert beneficial effects in stroke.

B.D. Roussel (✉) • M. Louessard • S.M. de Lizarrondo • D. Vivien
INSERM, INSERM U919, University Caen Basse Normandie, Serine Proteases and Pathophysiology of the Neurovascular Unit, Public Interest Group CYCERON, Boulevard Becquerel, 14074 Caen, France
e-mail: broussel@cyceron.fr

M. Geiger et al. (eds.), *The Serpin Family*, DOI 10.1007/978-3-319-22711-5_16

These complex effects of the protease can be inhibited by several serpins, such as neuroserpin, plasminogen activator inhibitor (PAI)-1 and protease nexin (PN)-1. In this chapter, we will focus on the differential effects of the plasminogen activator tPA and make an overview of the literature about serpins' effects on the cerebral roles of tPA.

16.1 Introduction

Stroke is the third cause of death and the second cause of disability in adults in industrialized countries. Worldwide, a stroke occurs every 5 s. 25 % of these patients die within a month and only 50 % are alive after 1 year with unfortunately a permanent handicap in most of the case. Schematically, we can divide stroke into two major subgroups: haemorrhagic stroke that represents 20 % of the total stroke events and the more often represented ischaemic stroke.

Cerebral ischaemia is a complex pathology that starts with a vascular event leading to cerebral consequences. Ischaemia results from severe reductions in cerebral blood flow, due to the occlusion of cerebral and/or extracerebral vessels supplying nervous tissues, or a cardiac arrest. The drop of cerebral blood flow leads to a depletion of oxygen and energy in neurones and thus to metabolic stress, ionic perturbations and ischaemic cell death (Siesjo et al. 1989; Dirnagl et al. 1999). The molecular events triggering neuronal death are complex, involving excitotoxicity (Vivien et al. 2011; Nicole et al. 2001), inflammation (Kaindl et al. 2012), apoptosis (Broughton et al. 2009; Liot et al. 2006) and endoplasmic reticulum (ER) stress (Roussel et al. 2013).

Since 1995, medical authorities allow the use of only one drug to treat ischaemic stroke: the recombinant form of tissue-type plasminogen activator (tPA) (NINDS 1995) (Fig. 16.1). When injected intravenously, tPA promotes fibrinolysis and thus clot lysis via the conversion of the inactive fibrin-bound zymogen plasminogen into plasmin (Collen and Lijnen 1991). However, tPA is also an endogenous protein produced by various cell types, including neurones (Krystosek and Seeds 1981) that display various roles in physiological brain functions, such as learning and memory processes (Huang et al. 1996; Matys et al. 2004; Obiang et al. 2012) or synaptic plasticity (Baranes et al. 1998), and in pathological conditions like cerebral ischaemia [for review (Yepes et al. 2009)]. On top of that, it was described that exogenous tPA injected for thrombolysis can cross the blood-brain barrier (BBB) and reach the brain parenchyma to exert its deleterious effects, including BBB leakage and neurotoxicity (Benchenane et al. 2005b).

In this context, serine protease inhibitor (serpin) superfamily plays an important role in the regulation of tPA functions and dysfunctions [for review (Roussel et al. 2011a)]. Three well-known serpins can inhibit tPA: neuroserpin, plasminogen activator inhibitor (PAI)-1 and protease nexin (PN)-1.

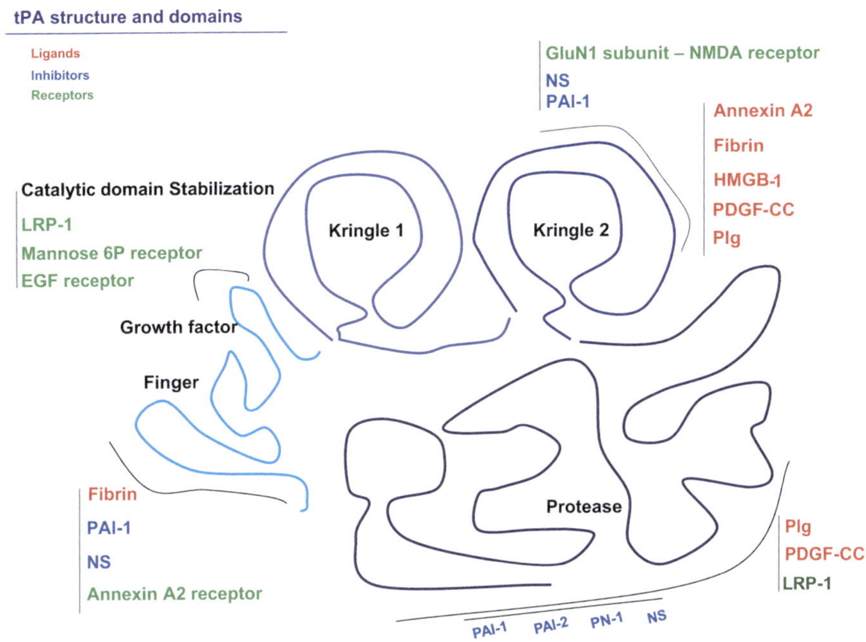

Fig. 16.1 Functions of tissue-type plasminogen activator (tPA). tPA has five domains, including a finger domain, an epidermal growth factor (EGF)-like domain, two kringle domains and a serine protease domain. Its functions are related to specific interactions of these domains to different ligands, including serpin inhibition. HMGB-1, high-mobility group box-1 protein; LRP, lipoprotein receptor-related protein; mannose-6-phosphate receptor; NMDA receptor, *N*-methyl-D-aspartate receptor; NS, neuroserpin; PAI-1, plasminogen activator inhibitor type-1; PAI-2, plasminogen activator inhibitor type-2; PDGF-CC, platelet-derived growth factor-CC, plg, plasminogen; PN-1, protease nexin-1

16.2 The tPA/Plasminogen/Plasmin System and Its Inhibitors in the Vascular Compartment

The most known function of the tPA/plasminogen/plasmin system in the vascular compartment is the dissolution of fibrin clots, the cleaved insoluble form of fibrinogen. Fibrinolysis takes place when undesirable fibrin is formed or when a haemostatic thrombus prevents efficient brain perfusion. Plasminogen activation by its activators is the initial step in fibrinolysis. tPA and urokinase-type plasminogen activator (uPA) are the two main physiological plasminogen activators. Besides fibrinolysis, this system and its regulators display other pathophysiological roles during cerebral ischaemia.

Fibrinolysis is the main mechanism (although probably not the only one as ADAMTS-13 cleaves the von Willebrand factor, and thrombectomy becomes an alternative) leading to arterial recanalization after acute ischaemic stroke (Rijken and Sakharov 2001). Intravenous administration of recombinant tPA further activates this mechanism and is, therefore, efficient to treat stroke by promoting clot dissolution (Millan et al. 2010; Adams et al. 2007). The ability of tPA to activate plasminogen is regulated by several proteins in the blood, including PAI-1, PN-1 and thrombin-activatable fibrinolysis inhibitor (TAFI) (Sashindranath et al. 2012; Gils and Declerck 2004; Bouton et al. 2012; Vercauteren et al. 2013). Circulating tPA and serpin-tPA complexes are cleared by the liver after high-affinity binding to low-density lipoprotein-related receptors (LRPs) or mannose receptors (Biessen et al. 1997; Smedsrod and Einarsson 1990).

In the vasculature, tPA is synthesized, stocked in granules and released from endothelial cells as an active form, either constitutively or in a regulated fashion following various kinds of stimuli (Urano and Suzuki 2012). PAI-1 is the main regulator of tPA activity in the vasculature and influences the efficiency of both endogenous and pharmacologically induced fibrinolysis (Rijken and Lijnen 2009). Since PAI-1 exists in plasma in molar excess over tPA, most of the tPA circulates in an inactive, complexed form with PAI-1: only a small fraction of this total tPA circulates as an active enzyme (Suzuki et al. 2011). Accordingly, PAI-1-deficient mice display reduced ischaemic damages following thrombotic stroke as compared to their wild-type (WT) littermates due to faster clot dissolution (Nagai et al. 2005). Moreover, ischaemic stroke patients with low plasmatic levels of PAI-1 present a better arterial recanalization when tPA is injected (Ribo et al. 2004); and another prospective study shows the PAI-1/tPA ratio as an independent risk factor for the development of a first stroke (Johansson et al. 2000).

Compared to PAI-1, plasmatic levels of neuroserpin are very low, suggesting that it does not significantly impact the fibrinolytic system in vivo. However, a recent study provides first clinical evidence that neuroserpin, as previously demonstrated in experimental models, could display a critical role during cerebral ischaemia (Yepes et al. 2000). They show an association between serum levels of neuroserpin and functional outcome after ischaemic stroke (Rodriguez-Gonzalez et al. 2011b). Moreover, there is a significant correlation between neuroserpin serum levels and the incidence of haemorrhagic transformation in patients treated with tPA (Rodriguez-Gonzalez et al. 2011b).

Platelet PN-1 strongly influences fibrinolysis and may play a role in stroke outcome by inhibiting tPA, plasmin and thrombin (Boulaftali et al. 2010). It has been demonstrated that PN-1 is a major determinant of the resistance of platelet-rich clots to tPA-induced fibrinolysis. Accordingly, PN-1-deficient mice present improved reperfusion in a $FeCl_3$ thrombotic model with tPA administration (Boulaftali et al. 2010). Unlike PAI-1, PN-1 expression is very weak in plasma. However, its concentration increases in the presence of platelets (Bouton et al. 2012). To date, the respective contributions of PAI-1 and PN-1 in thrombus resistance to tPA remain unknown.

Plasminogen activator inhibitor type-2 (PAI-2) is another member of the serpin family, able to inhibit both uPA and tPA (Lobov et al. 2008). PAI-2 is tenfold more efficient to inhibit uPA than tPA. PAI-2 is not usually associated with fibrinolysis because it is normally undetectable in the blood. Moreover, fibrin-bound tPA is protected from PAI-2 inhibition, as for PAI-1. Its role in ischaemic stroke pathophysiology remains so far unknown.

The inhibition of the fibrinolytic system can also occur directly at the plasmin level by alpha2-antiplasmin (α2-AP) and PN-1. The main difference between PN-1 and α2-AP is dependent on their way to inhibit plasmin: α2-AP is a specific plasmin inhibitor, able to bind free plasmin, but the rate of inactivation slows down very significantly when plasmin is bound to fibrin (Schaller and Gerber 2011). In contrast, PN-1 is able to inhibit both fibrin-bound and free plasmin (Boulaftali et al. 2011). Factor XIII plays a key role in α2-AP-mediated inhibition of clot lysis, since it cross-links α2-AP to fibrin inside the clot (Mosesson et al. 2008; Fraser et al. 2011). Interestingly, lower concentrations of α2-AP are associated with arterial recanalization after tPA administration in ischaemic stroke patients (Marti-Fabregas et al. 2005).

TAFI (thrombin-activatable fibrinolysis inhibitor) is another inhibitor of the fibrinolytic system, acting by removal of the C-terminal lysine from fibrin. Indeed, the fibrinolytic system is initiated after the formation of fibrin when both plasminogen and tPA bind at the fibrin surface. This binding is mediated by specific interactions with C-terminal lysine residues of partially degraded fibrin (Declerck 2011). Thus, TAFI-mediated removal of plasminogen and tPA lysine-binding sites present in the thrombus prevent plasminogen activation and are important regulator of fibrinolysis (Gils and Declerck 2004). Accordingly, genetic deficiency in TAFI or TAFI inhibition using matrix metalloproteinase-10 (MMP-10) improves outcome after thrombotic stroke in mice (Orbe et al. 2011). Recent studies in humans suggest that plasmatic levels of TAFI correlate with stroke severity (Montaner et al. 2003), affect the recanalization rate after thrombolysis (Brouns et al. 2009) and are associated with stroke risk (Leebeek et al. 2005).

Since fibrinolysis is thought to be the main physiological mechanism preventing excessive thrombosis, a number of studies investigated the interrelationships between genetic variants of fibrinolysis-related genes (mainly tPA (PLAT) and PAI-1) and the risk of stroke or myocardial infarction (Jood et al. 2005). Analyses regarding functional PAI-1 gene polymorphisms have shown that PAI-1 4G/5G genotype increases re-occlusion rates, poor functional outcome after stroke thrombolytic therapy (Fernandez-Cadenas et al. 2010) and the risk of future ischaemic stroke (Wiklund et al. 2005). However, another study has failed to confirm this association in young patients (de Paula Sabino et al. 2011).

16.3 The tPA/Plasminogen Axis and Its Inhibitors at the Blood-Brain Barrier

Despite its demonstrated clinical efficiency, tPA-mediated thrombolysis displays serious side effects at the vascular level, which affects patient outcomes. The most documented side effects are haemorrhagic transformation (Wardlaw 2012) and brain oedema (Strbian et al. 2013), which are thought to be caused by tPA-induced BBB permeability/leakage.

Experimental and clinical studies suggest that tPA adverse effects are mediated through activation of metalloproteinases (MMPs) that promotes BBB leakage by degrading components of the endothelial layer, tight junctions and the extracellular matrix (Lijnen 2001; Briasoulis et al. 2012; Wang et al. 2003). Laser microdissection and microarray techniques have demonstrated that several MMPs (including MMP-2, MMP-3 and MMP-9) are up-regulated in the infarcted brain tissue after stroke in humans (Cuadrado et al. 2009). Evidence argues for a crucial role of two particular MMPs, MMP-2 and MMP-9 (collagenases), in BBB disruption during ischaemic stroke (Lakhan et al. 2013). Accordingly, MMP-3 (or stromelysin-1) appears to play an important role in intracerebral haemorrhage (ICH) induced by tPA treatment of ischaemic stroke in mice (Suzuki et al. 2007).

The mechanisms by which tPA induces MMP overexpression during stroke remain unclear. Numerous studies argue for a role of LRP-1 receptor. LRP-1 is a receptor that binds various ligands and presents a very high affinity for tPA (Etique et al. 2013). In the neurovascular unit, LRP is found in endothelial cells, neurones and astrocytes (Herz and Strickland 2001; Polavarapu et al. 2007). Binding of tPA to LRP has been shown to induce the overexpression of MMP-3 and/or MMP-9 by the brain [especially endothelial cells, astrocytes and neurones (Suzuki 2010)]. Interestingly, LRP expression has been reported to be time-dependently up-regulated in endothelial cells exposed to ischaemia (Zhang et al. 2007; Suzuki et al. 2009). Therefore, the increased intracerebral haemorrhage risk associated with delayed tPA treatment may be explained by an increased LRP expression. According to this hypothesis, adjuvant treatment with RAP (receptor-associated protein) reduces the rate of haemorrhagic transformation in a photothrombotic mouse model of stroke with late tPA administration (Suzuki et al. 2007). Recent studies also suggest that tPA promotes BBB leakage via a mechanism independent of plasminogen (Yepes et al. 2003) and by LRP-mediated activation of latent platelet-derived growth factor (PDGF)-CC (Su et al. 2008). This effect requires the kringle 2 domain of tPA and can be inhibited by a PDGF receptor-alpha (PDGFRα) antagonist (Su et al. 2008). All these mechanisms are plasminogen independent but require tPA proteolytic activity and are, therefore, susceptible to serpin-mediated blockade.

It should be emphasized that if the deleterious role of tPA on the BBB is consensual, the protective effect of tPA inhibition by serpin has been recently challenged. In a mouse model of traumatic brain injury, Sashindranath and coworkers first demonstrated that tPA dose-dependently increases neurovascular damages. Surprisingly, they also showed that tPA inhibition by intracerebral administration of PAI-1 further potentiated those damages. The mechanism may involve tPA/PAI-1 complex-mediated signalling through LRP receptors leading to MMP-3 overexpression and BBB leakage (Sashindranath et al. 2012). This appears contradictory with other studies showing that PAI-1 coming from the peripheral vasculature and the brain is a regulator of the BBB, helping the barrier function of the endothelial tight junctions (Dohgu et al. 2011). Others also proved that PAI-1 overexpression increased tPA/PAI-1 complex levels in the affected brain after middle cerebral artery (MCA) ligation model, suggesting a role of extravasated PAI-1 that would counteract tPA-derived neurotoxicity (Nagai et al. 2005). Furthermore, the use of PAI-1-derived peptides that inhibited the tPA's effect on BBB function without modifying its proteolytic activity improved outcome in thromboembolic models of stroke (Abu Fanne et al. 2010). Others have also shown that neuroserpin administration reduced MMP-9 activity and this was associated with a lower BBB leakage (Zhang et al. 2002). Therefore, the mechanisms by which serpins exert their actions on the BBB are still not fully understood.

Fibrinolysis-independent role for plasminogen/plasmin during stroke has been less extensively studied. Since plasmin is required for MMP activation, it may play a crucial role in the deleterious effects of tPA on the BBB (Gur-Wahnon et al. 2013). Besides, plasmin may also act on endothelial protease-activated receptor-1 (PAR-1), which displays both pro- and anti-inflammatory roles (Coughlin 2001; Greenidge et al. 2013). Other plasmin-mediated mechanisms leading to BBB leakage have been described including cleavage of monocyte chemoattractant protein-1 (MCP-1) and Rho-kinase pathway-dependent astrocytic retraction (Yao and Tsirka 2011). Whether inhibitors of plasmin-dependent deleterious effects may be used to improve thrombolysis outcome remains to be demonstrated.

During stroke, the tPA/plasminogen/plasmin axis plays also a crucial role on cerebrovascular inflammation. Indeed, tPA administration leads to adhesion molecule overexpression and to a pro-thrombogenic phenotype of endothelial cells (Zhang et al. 2005, 2009b). The mechanism of tPA-induced inflammation remains elusive and may involve both intravascular and parenchymal effects, especially since tPA is able to cross both compromised and intact BBB after stroke (Benchenane et al. 2005a). In physiological conditions, this transcytosis is mediated by LRP receptors and is independent of tPA proteolytic activity. In hypoxic conditions, this crossing is exacerbated and becomes LRP independent (Benchenane et al. 2005b). Therefore, endogenous and exogenous plasmatic tPAs are able to reach the brain parenchyma, thereby directly acting on neurones and glial cells (Jin et al. 2010).

Altogether, these results demonstrate that tPA displays both beneficial and deleterious effects at the vascular level during cerebral ischaemia. On one hand, tPA facilitates arterial recanalization but on the other hand, it worsens BBB damages (Benchenane et al. 2004). Therefore, the net benefit of tPA-mediated thrombolysis on stroke outcome is dependent on the fine equilibrium between these opposite roles, with blood-derived serpins as mediators.

16.4 The tPA/Plasminogen/Plasmin System and Its Inhibitors in the Brain Parenchyma

These fibrinolytic actors are also present in the brain parenchyma with various roles in physiological and pathological conditions. tPA has mainly been described in neurones (Rogove et al. 1999; Docagne et al. 1999) but is also present in astrocytes, microglia and oligodendrocytes. tPA can be found in the hippocampus, hypothalamus, cerebellum, corpus callosum and amygdala (Docagne et al. 1999; Salles and Strickland 2002; Correa et al. 2011). The presence of plasminogen in the brain parenchyma has been debated for years, but it has been described at the surface of hippocampal neurones (Ledesma et al. 2000), and there are evidences of an implication of the tPA/plasmin ax in neuronal growth cone (Krystosek and Seeds 1981), migration of cerebellar granule neurones during development (Friedman and Seeds 1995) and synaptic remodelling (Baranes et al. 1998). Inhibitors of this system are also present in the brain parenchyma. PAI-1 co-localizes with tPA in the hippocampus (Salles and Strickland 2002) but seems to be mainly expressed by astrocytes (Hultman et al. 2010; Gravanis and Tsirka 2005). Neuroserpin is mostly synthesized and secreted from neurones of the central and peripheral nervous systems (Hastings et al. 1997; Krueger et al. 1997). PN-1 is also present in astrocytes and neurones (Mansuy et al. 1993) and its expression is increased during cerebral ischaemia (Hoffmann et al. 1992).

The whole system and its regulators have been implicated in cerebral ischaemia-mediated excitotoxicity, apoptosis, ER stress and inflammation and will be discuss in each following section.

16.4.1 Excitotoxicity

Excitotoxicity designs a pathological process by which nerve cells are killed by excessive glutamate stimulation. In 1995, Tsirka and collaborators were the first to establish a link between the glutamatergic transmission and tPA (Tsirka et al. 1995): tPA-deficient mice are more resistant to intra-hippocampal kainate injections than their WT littermates. Later, it was described that tPA can bind to the GluN1 subunit of NMDA receptors and then increase its permeability to Ca^{2+} (Nicole et al. 2001; Fernandez-Monreal et al. 2004). The direct effect of tPA on GluN1 happens by a two-step docking system (Lopez-Atalaya et al. 2008), but other observations suggest an effect on GluN1 via the transformation of plasminogen into active plasmin (Matys and Strickland 2003) and an indirect cleavage of the NMDA receptor by tPA through the LRP receptor (Samson et al. 2008).

The use of serpins to counteract tPA-mediated excitotoxicity has been widely used in the literature but has not reached the clinical state. For example, it was demonstrated that the astrocytic overexpression of PAI-1 by transforming growth factor-alpha (TGF-α) rescued neurones from excitotoxicity (Gabriel et al. 2003). As explained before, in a MCA ligation model of cerebral ischaemia, the use of transgenic mice overexpressing PAI-1 shows a decrease in infarct volume, but in a thrombosis model, the inhibition of vascular tPA provokes a bigger infarct volume (Nagai et al. 2005). The use of these two models demonstrates that the inhibitory effect of PAI-1 in reperfusion leads to greater damages in the brain, while inhibition of neuronal tPA protects them from excitotoxicity. This is supported by the use of PAI-1 knock-out mice that display exacerbated brain damages (Nagai et al. 1999). It is interesting to note that a meta-analysis shows that the PAI-1 4G high expression allele in humans seems to have a protective effect in cerebral ischaemia, whereas the same allele may be associated with an increased risk of myocardial infarction (Bentley et al. 2010).

Neuroserpin has also been widely studied in cerebral ischaemia. In a rat stroke model, the intracerebral administration of neuroserpin decreases the infarct volume (Yepes et al. 2000). The direct inhibition of tPA by neuroserpin in the brain parenchyma leads to a lesser level of excitotoxicity in vitro (Lebeurrier et al. 2005; Rodriguez-Gonzalez et al. 2011b) and in mice models of ischaemia (Cinelli et al. 2001; Lebeurrier et al. 2005) but also reduces microglial activation, which in turn results in a lower production of tPA.

In contrast to the overexpression of PAI-1 and neuroserpin during cerebral ischaemia, there is no induction of PN-1 (Hultman et al. 2010). Nevertheless, PN-1 shows neuroprotective effects in cerebral ischaemia, but it seems mediated via its action on thrombin (Mirante et al. 2013).

16.4.2 Apoptosis

Serpins have also been implicated in anti-apoptotic processes. Yepes and collaborators discovered that neuroserpin decreases the number of apoptotic cells after cerebral ischaemia (Yepes et al. 2000). Once again, this effect is mediated by the direct interaction between neuroserpin and tPA (Yepes et al. 2000). This first observation was followed by other works that show either anti-apoptotic effects (Lee et al. 2007; Liot et al. 2006) or pro-apoptotic effects (Liu et al. 2004; Flavin et al. 2000) of tPA.

In the hippocampus, the delayed neuronal cell death observed after cerebral ischaemia is dependent of tPA's proteolytic activity and is mediated by plasmin via NMDA receptor-Akt phosphorylation (Echeverry et al. 2010). Another hypothesis would be the conversion of pro-brain-derived neurotrophic factor (BDNF) into its active form (Pang et al. 2004). But tPA can also act independently of its proteolytic activity, by inhibiting apoptotic cell death in a model of serum deprivation on neurones, via the recruitment of the PI3-K pathway (Liot et al. 2006). In oligodendrocytes, tPA is clearly anti-apoptotic independently of its proteolytic properties, and this effect requires the activation of the EGF receptor (Correa et al. 2011). In contrast to these results, tPA has been described as a pro-apoptotic molecule in cerebral ischaemia by activating caspase-3 via the caspase-8. This effect is partially blocked by activated protein C (Liu et al. 2004). These discrepancies on apoptosis can partially be explicated by the variety of models used for these studies: in the cerebral ischaemia field, results depend on the stroke model used (Young et al. 2007).

Regarding PAI-1, the same contradiction exists and also depends on the models used. For example, PAI-1 shows anti-apoptotic effects in various cell lines like neutrophils (Zmijewski et al. 2011) or fibroblasts (Zhang et al. 2013) and has pro-apoptotic effects in some other cell lines [for review (Balsara and Ploplis 2008)]. In the central nervous system, the deficiency of PAI-1 in astrocytes leads to apoptosis in neurones, by stimulating the release of cytochrome C and in the meantime decreasing the transcription of B-cell lymphoma 2 (Bcl-2) and B-cell lymphoma-extra large (Bcl-$X_{L)}$ (anti-apoptotic genes) while increasing pro-apoptotic Bcl-X_S and Bax mRNA (Soeda et al. 2001).

Serpins also inhibit the transformation of plasminogen into active plasmin and thus decrease anoïkis-mediated apoptosis. PN-1 protects cells from apoptosis mediated by plasminogen activation (Rossignol et al. 2004). PN-1 also protects from OGD-induced apoptosis, but this effect is mediated by the inhibition of thrombin (Mirante et al. 2013).

16.4.3 ER Stress

The endoplasmic reticulum (ER) is an important organelle where membrane and secretory proteins are folded and processed. To control the accumulation of native/aberrant proteins within the ER, a signal exists and is mediated by three sensors at the reticulum membrane: protein kinase RNA-like endoplasmic reticulum kinase (PERK), inositol-requiring protein 1 (IRE-1) and activating transcription factor 6 (ATF6) [for review (Marciniak and Ron 2006)]. Under physiological conditions, the chaperone protein Bip binds to the intra-reticulum domain of the three sensors. Under stress conditions, Bip disengages from the sensors, leading to their activation. PERK oligomerizes and then phosphorylates the α-subunit of the eukaryotic initiation factor 2 (eIF2α), leading to the shutdown of translation (Marciniak and Ron 2006). Despite this shutdown, a small proportion of proteins are actively expressed, like activating transcription factor (ATF4), C/EBP homologous protein (CHOP), growth arrest and DNA damage-inducible protein (GADD34) and other chaperones. GADD34 will be responsible for the dephosphorylation of eIF2α and the recovery of protein synthesis. In the meantime, the other sensors are also activated by ER stress. When unbound to Bip, IRE-1, an endoribonuclease, is able to cleave the mRNA X-box binding protein 1 (XBP-1). The spliced variant of XBP-1 is an active transcription factor that promotes the expression of ER stress genes (Marciniak and Ron 2006). Under stress conditions, ATF6 is released from the reticulum membrane and migrates to the Golgi where the protein is cleaved. The cleaved fragment acts as a transcription factor and activates ER stress genes (Marciniak and Ron 2006). If unfolded/aberrant proteins remain within the ER, they will be targeted by the ERAD (ER-associated degradation) for degradation by the proteasome (Needham and Brodsky 2013; Brodsky 2012; Smith et al. 2011).

ER stress has been associated with many neurodegenerative diseases, including stroke (Roussel et al. 2013). In that part, we will focus on ER stress in cerebral ischaemia and the influence of tPA and its regulators.

Many studies have clearly demonstrated the activation of ER stress in in vitro and in vivo models of ischaemia. In culture of cortical neurones under oxygen deprivation (Halterman et al. 2010), in rat astrocytes cultures under oxygen and glucose deprivation (OGD) (Benavides et al. 2005) and in primary cultures of mixed rat brain cortical cells under OGD (Badiola et al. 2011), ER stress markers such as PERK, IRE-1, GADD34, ATF4, CHOP and Bip are clearly up-regulated. ER stress-related genes and proteins are also increase in murine [mice (and rats) models of middle cerebral artery occlusion (Morimoto et al. 2007; McCaig et al. 2005)] and in a transient model of bilateral common carotid arteries occlusion (Owen et al. 2005; Tajiri et al. 2004; Doutheil et al. 1999). However, some discrepancies remain about the role of ER stress in cerebral ischaemia.

Many studies consider CHOP as a pro-apoptotic factor. One of the first studies was performed by Tajiri and colleagues in 2004 where they show that primary hippocampal neurones from CHOP(−/−) mice are more resistant to a hypoxia-reoxygenation insult than those from WT animals (Tajiri et al. 2004). CHOP is also increased in astrocytes under OGD, leading to cell death (Benavides et al. 2005). CHOP is also implicated in neuronal cell death during cerebral ischaemia in in vivo studies (Li et al. 2012; Yuan et al. 2011; Nakka et al. 2010). Despite these effects, a study shows a protective role of CHOP in hypoxia-induced neuronal death (Halterman et al. 2010): the deletion of CHOP increases the neuronal susceptibility to both hypoxic and thapsigargin-mediated injury and the overexpression of CHOP protects culture against hypoxia (Halterman et al. 2010). Moreover, CHOP leads to the expression of GADD34, and it is well described that GADD34 displays pro-survival effects after cerebral ischaemia (McCaig et al. 2005; White et al. 2004; Paschen et al. 2004).

Ischaemia induces an energetic depletion, impairing the ER calcium homeostasis. Within the cell, the ER is the main calcium pool, and disturbance of its homeostasis is known to play a role in ER stress (Nakagawa et al. 2000). tPA is also responsible for the increase in intracytosolic calcium in ischaemic conditions, by cleaving the GluN1 subunit of the NMDA receptor (Nicole et al. 2001). The combination of ER stress and ischaemic conditions in the presence of tPA might deliver a double blow of calcium within the cytoplasm that could increase cell death.

16.4.4 Inflammation

Inflammation plays a critical role in many diseases related to neurodegeneration, including stroke [for review (Nilupul Perera et al. 2006)] and Alzheimer (Eikelenboom et al. 2002; Gonzalez-Scarano and Baltuch 1999) and Parkinson diseases (Perry et al. 2010; Tansey and Goldberg 2010). It has been shown that tPA is able to activate microglia in excitotoxic conditions independently of its

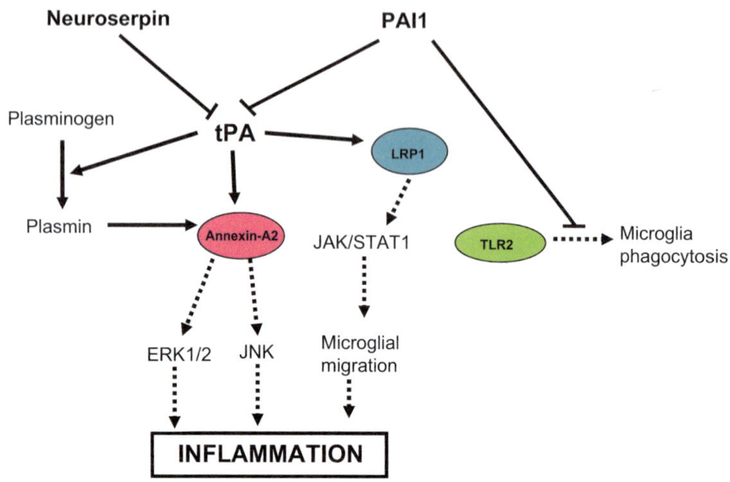

Fig. 16.2 tPA mediates microglia activation. In the injured brain, tissue-type plasminogen activator (tPA) can activate microglia by a direct interaction with the Annexin A2 receptor or by transforming plasminogen into plasmin. It results in the activation of signal-regulated kinase (ERK1/2) and c-Jun N-terminal kinase (JNK/Akt) pathways. tPA can also bind to lipoprotein receptor-related protein-1 (LRP-1) receptor, leading to the microglial migration via the Janus kinase/signal transducer and activator of transcription-1 (JAK/STAT1) pathway

proteolytic activity (Gravanis and Tsirka 2005; Gelderblom et al. 2013; Rogove et al. 1999; Pineda et al. 2012). Indeed, in a culture model of mixed cortical neurones and purified microglial cell, treatment with lipopolysaccharide (LPS) leads to a transient increase in the expression of tumor necrosis factor-alpha (TNFα) and to morphological changes of microglia. When realized in tPA−/− cultures, no microglial activation is observed after LPS treatment (Rogove et al. 1999). The microglial activation is restored in tPA−/− cultures by a pretreatment with tPA or inactive tPA. On its own, tPA is not capable to activate microglia. Similarly, in an in vivo model of kainate injection, activation of microglia is restored in tPA−/− mice when inactive tPA is co-injected (Rogove et al. 1999) (Fig. 16.2).

To further characterize the mechanisms involved in tPA-mediated microglial activation, different mutants of tPA were tested in vitro. With that method, it has been shown that the interaction of the finger domain of tPA with Annexin A2 receptor causes a molecular cascade that leads to microglia activation (Siao and Tsirka 2002). This molecular cascade involves a rapid activation of extracellular signal-regulated kinase (ERK1/2) and c-Jun N-terminal kinase (JNK) pathways and a delayed activation of protein kinase B (PKB also called Akt) pathway. It is interesting to note that inactive tPA only activates ERK1/2 and JNK pathways.

Although most studies show a non-proteolytic effect of tPA in microglial activation, others found the opposite (Godier and Hunt 2013; Rogove and Tsirka 1998; Gelderblom et al. 2013). tPA is not the only ligand for the Annexin A2 receptor; plasminogen (Kim and Hajjar 2002) or plasmin [for review (Godier and

Hunt 2013)] can also bind to it. Thereby, a subunit of Annexin A2 receptor can be cleaved by plasmin, leading to a pro-inflammatory response as a ligand for transmembrane receptors (Godier and Hunt 2013). This action of plasmin reveals a proteolytic, but indirect, effect of tPA in Annexin A2 activation-mediated inflammation.

Annexin A2 receptor is not the only mediator of tPA-induced inflammation. By using an in vivo model of MCA occlusion (MCAO), it has been demonstrated that the microglia activation is mediated by the binding of tPA to LRP-1 receptor (Zhang et al. 2009a). Moreover, PAI-1 promotes microglial migration via the LRP-1/Janus kinase (JAK)/signal transducer and activator of transcription-1 (STAT1) pathway and modulates microglial activation by suppression of the Toll-like receptor-2 (TLR-2) signalling (Jeon et al. 2012). Despite the pro-inflammatory properties of PAI-1, another serpin seems to be protective in cerebral ischaemia, by decreasing the inflammatory response. In an in vivo stroke model of temporary MCAO, neuroserpin inhibits the pro-inflammatory effects of tPA (Gelderblom et al. 2013). In this study, they subjected neuroserpin-deficient (Ns−/−) and WT mice to temporary MCAO. They show an increase of the infarct size and neurological disability in Ns−/− mice compared to WT mice 72 h after ischaemia. Ns−/− mice present a decrease of fibrin/fibrinogen level in the ischaemic hemisphere. Furthermore, in Ns−/− mice there is an increase in TNFα expression which is produced by activated microglia. Thus, they confirm the neuroprotective role of neuroserpin in ischaemia (Gelderblom et al. 2013).

The intracellular signalling cascades triggered by tPA could be blocked using JNK or Akt pathway inhibitors. Treatment in mixed glial cells with tPA and JNK inhibitor-1 (JNKI-1) decreases the induction of TNFα induced by tPA treatment alone, while treatment with tPA and an inhibitor of Akt pathway increases the expression of TNFα (Pineda et al. 2012).

16.5 New Therapeutic Strategies for Stroke

tPA is a two-side molecule: it is clearly beneficial within the vascular compartment, by promoting the dissolution of fibrin clots and restoring the cerebral blood flow, but tPA is also pro-neurotoxic by its influence on excitotoxicity (Nicole et al. 2001) (Fig. 16.3).

The identification of new strategies to enhance tPA-related fibrinolytic efficacy and reduce at the same time its associated complications (haemorrhage and neurotoxicity) is still a priority. Annexin A2, a cofactor which binds to tPA and plasminogen resulting in enhanced plasmin generation, was identified in endothelial cells (Hajjar et al. 1994). Administration of recombinant Annexin A2 can make low-dose tPA more effective in a murine stroke model, minimizing haemorrhagic transformation and infarction, indicating that this cofactor is a promising approach for ischaemic stroke treatment (Zhu et al. 2010).

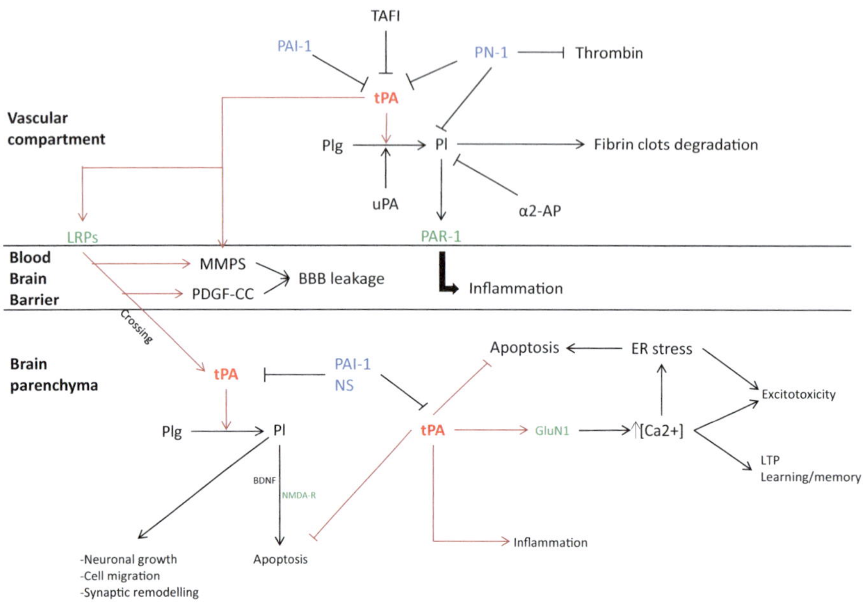

Fig. 16.3 tPA actions on both sides of the blood-brain barrier (BBB). tPA promotes clot lysis in the vascular compartment by activating plasminogen into active plasmin. tPA can also cross the BBB through LRP receptors and promotes excitotoxicity in the brain compartment during cerebral ischaemia. tPA, tissue-type plasminogen activator; PAI-1, plasminogen activator inhibitor 1; TAFI, thrombin-activatable fibrinolysis inhibitor; PN-1, protease nexin 1; plg, plasminogen; Pl, plasmin; uPA, urokinase plasminogen activator; α2-AP, alpha-2 antiplasmin; PAR-1, protease-activated receptor 1; LRP, lipoprotein receptor-related protein; BBB, blood-brain barrier; MMP, matrix metalloproteinase; PDGF-CC, platelet-derived growth factor-CC; NS, neuroserpin; BDNF, brain-derived neurotrophic factor; NMDA, *N*-methyl-D-aspartate; LTP, long-term potentiation

Another way to improve thrombolysis is to make a "safer" tPA by decreasing its pro-neurotoxicity without changing its ability to convert plasminogen into active plasmin. By generating mutants of tPA with these properties, Parcq and collaborators have proved that a non-cleavable single-chain (sc) form with restored zymogenicity (sc*-tPA) and a tPA modified in the kringle 2 (K2) lysine-binding site (K2*-tPA) are fibrinolyticly active and without NMDA receptor mediated (Parcq et al. 2013). Another possibility is a direct inhibition of the interaction between the NMDA receptor and tPA. By generating an antibody targeting the cleaved part of the GluN1 subunit (Macrez et al. 2010), it was demonstrated that the toxicity triggered by tPA during cerebral ischaemia is significantly reduced (Macrez et al. 2011). A third way to improve tPA is the co-injection with another molecule. For example, the co-injection of tPA and high-mobility group box-1 protein (HMGB-1) promotes fibrinolysis and reduces tPA-mediated neurotoxicity (Roussel et al. 2011b). It could give the opportunity to inject less tPA for the same fibrinolysis efficiency, but also decrease a toxicity that would be lower due to the lesser dose injected. However, many drugs have been tested in co-injection protocols, but few have reached the clinical level and none are used nowadays in the clinic: in 1026 neuroprotective drugs in 2006, all of them have failed in clinical trials (O'Collins et al. 2006).

Serpin's inhibition is a major topic in the field of thrombolysis since elevated levels of antifibrinolytic proteins, such as TAFI and PAI-1, are known to provoke and increase the risk for thrombosis. For the moment, different therapies have been developed. Currently, early studies on PAI-1 inhibition have been focused on monoclonal antibodies that convert PAI-1 to the latent or cleaved form and the inhibitory peptides that correspond to the reactive site of PAI-1. However, the implications of plasminogen activation go beyond of just fibrinolytic control and extend other processes. For instance, serpins are also used as molecules that block tPA-mediated toxicity. For example, a stable mutant of PAI-1 blocks tPA toxicity when injected intracerebroventricularly in a neonatal model of ischaemia (Yang et al. 2013). In vitro, neuroserpin also blocks tPA-mediated death by attenuating inflammation and BBB leakage (Rodriguez-Gonzalez et al. 2011a), by inhibiting the interaction with NMDA receptor (Lebeurrier et al. 2005; Zhang et al. 2002) and by decreasing apoptosis in cerebral ischaemia (Yepes et al. 2000). However, the use of serpins seems limited in clinic because of its inhibition of fibrinolysis (Nagai et al. 2005). Moreover, in most of the neuroprotective studies using serpins, authors apply the molecule directly in the brain to show their effects. Only few studies inject serpins in the vascular compartment or do use a thromboembolic stroke model.

This review highlights the two roles of tPA in cerebral ischaemia with a positive fibrinolytic function in the vascular compartment that restores a normal cerebral blood flow and a pro-neurotoxic effect in the brain parenchyma. Serpins have already demonstrated a good therapeutic potential to counteract the parenchymal effects of tPA; however, this strategy needs to be improved to allow a peripheral administration that will not decrease thrombolysis.

References

Abu Fanne R, Nassar T, Yarovoi S, Rayan A, Lamensdorf I, Karakoveski M, Vadim P, Jammal M, Cines DB, Higazi AA (2010) Blood-brain barrier permeability and tPA-mediated neurotoxicity. Neuropharmacology 58(7):972–980. doi:10.1016/j.neuropharm.2009.12.017

Adams HP Jr, del Zoppo G, Alberts MJ, Bhatt DL, Brass L, Furlan A, Grubb RL, Higashida RT, Jauch EC, Kidwell C, Lyden PD, Morgenstern LB, Qureshi AI, Rosenwasser RH, Scott PA, Wijdicks EF (2007) Guidelines for the early management of adults with ischemic stroke: a guideline from the American Heart Association/American Stroke Association Stroke Council, Clinical Cardiology Council, Cardiovascular Radiology and Intervention Council, and the Atherosclerotic Peripheral Vascular Disease and Quality of Care Outcomes in Research Interdisciplinary Working Groups: The American Academy of Neurology affirms the value of this guideline as an educational tool for neurologists. Circulation 115(20):e478–e534. doi:10.1161/CIRCULATIONAHA.107.181486

Badiola N, Penas C, Minano-Molina A, Barneda-Zahonero B, Fado R, Sanchez-Opazo G, Comella JX, Sabria J, Zhu C, Blomgren K, Casas C, Rodriguez-Alvarez J (2011) Induction of ER stress in response to oxygen-glucose deprivation of cortical cultures involves the activation of the PERK and IRE-1 pathways and of caspase-12. Cell Death Dis 2, e149. doi:10.1038/cddis.2011.31

Balsara RD, Ploplis VA (2008) Plasminogen activator inhibitor-1: the double-edged sword in apoptosis. Thromb Haemost 100(6):1029–1036

Baranes D, Lederfein D, Huang YY, Chen M, Bailey CH, Kandel ER (1998) Tissue plasminogen activator contributes to the late phase of LTP and to synaptic growth in the hippocampal mossy fiber pathway. Neuron 21(4):813–825

Benavides A, Pastor D, Santos P, Tranque P, Calvo S (2005) CHOP plays a pivotal role in the astrocyte death induced by oxygen and glucose deprivation. Glia 52(4):261–275. doi:10.1002/glia.20242

Benchenane K, Lopez-Atalaya JP, Fernandez-Monreal M, Touzani O, Vivien D (2004) Equivocal roles of tissue-type plasminogen activator in stroke-induced injury. Trends Neurosci 27 (3):155–160. doi:10.1016/j.tins.2003.12.011

Benchenane K, Berezowski V, Ali C, Fernandez-Monreal M, Lopez-Atalaya JP, Brillault J, Chuquet J, Nouvelot A, MacKenzie ET, Bu G, Cecchelli R, Touzani O, Vivien D (2005a) Tissue-type plasminogen activator crosses the intact blood-brain barrier by low-density lipoprotein receptor-related protein-mediated transcytosis. Circulation 111(17):2241–2249. doi:10.1161/01.CIR.0000163542.48611.A2

Benchenane K, Berezowski V, Fernandez-Monreal M, Brillault J, Valable S, Dehouck MP, Cecchelli R, Vivien D, Touzani O, Ali C (2005b) Oxygen glucose deprivation switches the transport of tPA across the blood-brain barrier from an LRP-dependent to an increased LRP-independent process. Stroke 36(5):1065–1070. doi:10.1161/01.STR.0000163050.39122.4f

Bentley P, Peck G, Smeeth L, Whittaker J, Sharma P (2010) Causal relationship of susceptibility genes to ischemic stroke: comparison to ischemic heart disease and biochemical determinants. PLoS One 5(2), e9136. doi:10.1371/journal.pone.0009136

Biessen EA, van Teijlingen M, Vietsch H, Barrett-Bergshoeff MM, Bijsterbosch MK, Rijken DC, van Berkel TJ, Kuiper J (1997) Antagonists of the mannose receptor and the LDL receptor-related protein dramatically delay the clearance of tissue plasminogen activator. Circulation 95 (1):46–52

Boulaftali Y, Adam F, Venisse L, Ollivier V, Richard B, Taieb S, Monard D, Favier R, Alessi MC, Bryckaert M, Arocas V, Jandrot-Perrus M, Bouton MC (2010) Anticoagulant and antithrombotic properties of platelet protease nexin-1. Blood 115(1):97–106. doi:10.1182/blood-2009-04-217240

Boulaftali Y, Ho-Tin-Noe B, Pena A, Loyau S, Venisse L, Francois D, Richard B, Arocas V, Collet JP, Jandrot-Perrus M, Bouton MC (2011) Platelet protease nexin-1, a serpin that strongly influences fibrinolysis and thrombolysis. Circulation 123(12):1326–1334. doi:10.1161/CIRCULATIONAHA.110.000885

Bouton MC, Boulaftali Y, Richard B, Arocas V, Michel JB, Jandrot-Perrus M (2012) Emerging role of serpinE2/protease nexin-1 in hemostasis and vascular biology. Blood 119 (11):2452–2457. doi:10.1182/blood-2011-10-387464

Briasoulis A, Tousoulis D, Papageorgiou N, Kampoli AM, Androulakis E, Antoniades C, Tsiamis E, Latsios G, Stefanadis C (2012) Novel therapeutic approaches targeting matrix metalloproteinases in cardiovascular disease. Curr Top Med Chem 12(10):1214–1221

Brodsky JL (2012) Cleaning up: ER-associated degradation to the rescue. Cell 151(6):1163–1167. doi:10.1016/j.cell.2012.11.012

Broughton BR, Reutens DC, Sobey CG (2009) Apoptotic mechanisms after cerebral ischemia. Stroke 40(5):e331–e339. doi:10.1161/STROKEAHA.108.531632

Brouns R, Heylen E, Sheorajpanday R, Willemse JL, Kunnen J, De Surgeloose D, Hendriks DF, De Deyn PP (2009) Carboxypeptidase U (TAFIa) decreases the efficacy of thrombolytic therapy in ischemic stroke patients. Clin Neurol Neurosurg 111(2):165–170. doi:10.1016/j.clineuro.2008.09.002

Cinelli P, Madani R, Tsuzuki N, Vallet P, Arras M, Zhao CN, Osterwalder T, Rulicke T, Sonderegger P (2001) Neuroserpin, a neuroprotective factor in focal ischemic stroke. Mol Cell Neurosci 18(5):443–457. doi:10.1006/mcne.2001.1028

Collen D, Lijnen HR (1991) Basic and clinical aspects of fibrinolysis and thrombolysis. Blood 78 (12):3114–3124

Correa F, Gauberti M, Parcq J, Macrez R, Hommet Y, Obiang P, Hernangomez M, Montagne A, Liot G, Guaza C, Maubert E, Ali C, Vivien D, Docagne F (2011) Tissue plasminogen activator prevents white matter damage following stroke. J Exp Med 208(6):1229–1242. doi:10.1084/jem.20101880

Coughlin SR (2001) Protease-activated receptors in vascular biology. Thromb Haemost 86 (1):298–307

Cuadrado E, Rosell A, Penalba A, Slevin M, Alvarez-Sabin J, Ortega-Aznar A, Montaner J (2009) Vascular MMP-9/TIMP-2 and neuronal MMP-10 up-regulation in human brain after stroke: a combined laser microdissection and protein array study. J Proteome Res 8(6):3191–3197. doi:10.1021/pr801012x

de Paula Sabino A, Ribeiro DD, Domingueti CP, Dos Santos MS, Gadelha T, Dusse LM, das Gracas Carvalho M, Fernandes AP (2011) Plasminogen activator inhibitor-1 4G/5G promoter polymorphism and PAI-1 plasma levels in young patients with ischemic stroke. Mol Biol Rep 38(8):5355–5360. doi:10.1007/s11033-011-0687-4

Declerck PJ (2011) Thrombin activatable fibrinolysis inhibitor. Hamostaseologie 31(3):165–166. doi:10.5482/ha-1155, 168–173

Dirnagl U, Iadecola C, Moskowitz MA (1999) Pathobiology of ischaemic stroke: an integrated view. Trends Neurosci 22(9):391–397

Docagne F, Nicole O, Marti HH, MacKenzie ET, Buisson A, Vivien D (1999) Transforming growth factor-beta1 as a regulator of the serpins/t-PA axis in cerebral ischemia. FASEB J 13 (11):1315–1324

Dohgu S, Takata F, Matsumoto J, Oda M, Harada E, Watanabe T, Nishioku T, Shuto H, Yamauchi A, Kataoka Y (2011) Autocrine and paracrine up-regulation of blood-brain barrier function by plasminogen activator inhibitor-1. Microvasc Res 81(1):103–107. doi:10.1016/j.mvr.2010.10.004

Doutheil J, Althausen S, Gissel C, Paschen W (1999) Activation of MYD116 (gadd34) expression following transient forebrain ischemia of rat: implications for a role of disturbances of endoplasmic reticulum calcium homeostasis. Brain Res Mol Brain Res 63(2):225–232, S0169328X98002769 [pii]

Echeverry R, Wu J, Haile WB, Guzman J, Yepes M (2010) Tissue-type plasminogen activator is a neuroprotectant in the mouse hippocampus. J Clin Invest 120(6):2194–2205. doi:10.1172/JCI41722

Eikelenboom P, Bate C, Van Gool WA, Hoozemans JJ, Rozemuller JM, Veerhuis R, Williams A (2002) Neuroinflammation in Alzheimer's disease and prion disease. Glia 40(2):232–239. doi:10.1002/glia.10146

Etique N, Verzeaux L, Dedieu S, Emonard H (2013) LRP-1: a checkpoint for the extracellular matrix proteolysis. BioMed Res Int 2013:152163. doi:10.1155/2013/152163

Fernandez-Cadenas I, Del Rio-Espinola A, Rubiera M, Mendioroz M, Domingues-Montanari S, Cuadrado E, Hernandez-Guillamon M, Rosell A, Ribo M, Alvarez-Sabin J, Molina CA, Montaner J (2010) PAI-1 4G/5G polymorphism is associated with brain vessel reocclusion after successful fibrinolytic therapy in ischemic stroke patients. Int J Neurosci 120(4):245–251. doi:10.3109/00207451003597169

Fernandez-Monreal M, Lopez-Atalaya JP, Benchenane K, Cacquevel M, Dulin F, Le Caer JP, Rossier J, Jarrige AC, Mackenzie ET, Colloc'h N, Ali C, Vivien D (2004) Arginine 260 of the amino-terminal domain of NR1 subunit is critical for tissue-type plasminogen activator-mediated enhancement of N-methyl-D-aspartate receptor signaling. J Biol Chem 279 (49):50850–50856. doi:10.1074/jbc.M407069200

Flavin MP, Zhao G, Ho LT (2000) Microglial tissue plasminogen activator (tPA) triggers neuronal apoptosis in vitro. Glia 29(4):347–354

Fraser SR, Booth NA, Mutch NJ (2011) The antifibrinolytic function of factor XIII is exclusively expressed through alpha(2)-antiplasmin cross-linking. Blood 117(23):6371–6374. doi:10.1182/blood-2011-02-333203

Friedman GC, Seeds NW (1995) Tissue plasminogen activator mRNA expression in granule neurons coincides with their migration in the developing cerebellum. J Comp Neurol 360(4):658–670. doi:10.1002/cne.903600410

Gabriel C, Ali C, Lesne S, Fernandez-Monreal M, Docagne F, Plawinski L, MacKenzie ET, Buisson A, Vivien D (2003) Transforming growth factor alpha-induced expression of type 1 plasminogen activator inhibitor in astrocytes rescues neurons from excitotoxicity. FASEB J 17(2):277–279. doi:10.1096/fj.02-0403fje

Gelderblom M, Neumann M, Ludewig P, Bernreuther C, Krasemann S, Arunachalam P, Gerloff C, Glatzel M, Magnus T (2013) Deficiency in serine protease inhibitor neuroserpin exacerbates ischemic brain injury by increased postischemic inflammation. PLoS One 8(5), e63118. doi:10.1371/journal.pone.0063118

Gils A, Declerck PJ (2004) Plasminogen activator inhibitor-1. Curr Med Chem 11(17):2323–2334

Godier A, Hunt BJ (2013) Plasminogen receptors and their role in the pathogenesis of inflammatory, autoimmune and malignant disease. J Thromb Haemost 11(1):26–34. doi:10.1111/jth.12064

Gonzalez-Scarano F, Baltuch G (1999) Microglia as mediators of inflammatory and degenerative diseases. Annu Rev Neurosci 22:219–240. doi:10.1146/annurev.neuro.22.1.219

Gravanis I, Tsirka SE (2005) Tissue plasminogen activator and glial function. Glia 49(2):177–183. doi:10.1002/glia.20115

Greenidge AR, Hall KR, Hambleton IR, Thomas R, Monroe DM, Landis RC (2013) Plasmin activation of glial cells through protease-activated receptor 1. Pathol Res Int 2013:314709. doi:10.1155/2013/314709

Gur-Wahnon D, Mizrachi T, Maaravi-Pinto FY, Lourbopoulos A, Grigoriadis N, Higazi AA, Brenner T (2013) The plasminogen activator system: involvement in central nervous system inflammation and a potential site for therapeutic intervention. J Neuroinflammation 10(1):124. doi:10.1186/1742-2094-10-124

Hajjar KA, Jacovina AT, Chacko J (1994) An endothelial cell receptor for plasminogen/tissue plasminogen activator. I. Identity with annexin II. J Biol Chem 269(33):21191–21197

Halterman MW, Gill M, DeJesus C, Ogihara M, Schor NF, Federoff HJ (2010) The endoplasmic reticulum stress response factor CHOP-10 protects against hypoxia-induced neuronal death. J Biol Chem 285(28):21329–21340. doi:10.1074/jbc.M109.095299

Hastings GA, Coleman TA, Haudenschild CC, Stefansson S, Smith EP, Barthlow R, Cherry S, Sandkvist M, Lawrence DA (1997) Neuroserpin, a brain-associated inhibitor of tissue plasminogen activator is localized primarily in neurons. Implications for the regulation of motor learning and neuronal survival. J Biol Chem 272(52):33062–33067

Herz J, Strickland DK (2001) LRP: a multifunctional scavenger and signaling receptor. J Clin Invest 108(6):779–784. doi:10.1172/JCI13992

Hoffmann MC, Nitsch C, Scotti AL, Reinhard E, Monard D (1992) The prolonged presence of glia-derived nexin, an endogenous protease inhibitor, in the hippocampus after ischemia-induced delayed neuronal death. Neuroscience 49(2):397–408

Huang YY, Bach ME, Lipp HP, Zhuo M, Wolfer DP, Hawkins RD, Schoonjans L, Kandel ER, Godfraind JM, Mulligan R, Collen D, Carmeliet P (1996) Mice lacking the gene encoding tissue-type plasminogen activator show a selective interference with late-phase long-term potentiation in both Schaffer collateral and mossy fiber pathways. Proc Natl Acad Sci USA 93(16):8699–8704

Hultman K, Blomstrand F, Nilsson M, Wilhelmsson U, Malmgren K, Pekny M, Kousted T, Jern C, Tjarnlund-Wolf A (2010) Expression of plasminogen activator inhibitor-1 and protease nexin-

1 in human astrocytes: response to injury-related factors. J Neurosci Res 88(11):2441–2449. doi:10.1002/jnr.22412
Jeon H, Kim JH, Lee WH, Lee MS, Suk K (2012) Plasminogen activator inhibitor type 1 regulates microglial motility and phagocytic activity. J Neuroinflammation 9:149. doi:10.1186/1742-2094-9-149
Jin R, Yang G, Li G (2010) Molecular insights and therapeutic targets for blood-brain barrier disruption in ischemic stroke: critical role of matrix metalloproteinases and tissue-type plasminogen activator. Neurobiol Dis 38(3):376–385. doi:10.1016/j.nbd.2010.03.008
Johansson L, Jansson JH, Boman K, Nilsson TK, Stegmayr B, Hallmans G (2000) Tissue plasminogen activator, plasminogen activator inhibitor-1, and tissue plasminogen activator/plasminogen activator inhibitor-1 complex as risk factors for the development of a first stroke. Stroke 31(1):26–32
Jood K, Ladenvall P, Tjarnlund-Wolf A, Ladenvall C, Andersson M, Nilsson S, Blomstrand C, Jern C (2005) Fibrinolytic gene polymorphism and ischemic stroke. Stroke 36(10):2077–2081. doi:10.1161/01.STR.0000183617.54752.69
Kaindl AM, Degos V, Peineau S, Gouadon E, Chhor V, Loron G, Le Charpentier T, Josserand J, Ali C, Vivien D, Collingridge GL, Lombet A, Issa L, Rene F, Loeffler JP, Kavelaars A, Verney C, Mantz J, Gressens P (2012) Activation of microglial N-methyl-D-aspartate receptors triggers inflammation and neuronal cell death in the developing and mature brain. Ann Neurol 72(4):536–549. doi:10.1002/ana.23626
Kim J, Hajjar KA (2002) Annexin II: a plasminogen-plasminogen activator co-receptor. Front Biosci 7:d341–d348
Krueger SR, Ghisu GP, Cinelli P, Gschwend TP, Osterwalder T, Wolfer DP, Sonderegger P (1997) Expression of neuroserpin, an inhibitor of tissue plasminogen activator, in the developing and adult nervous system of the mouse. J Neurosci 17(23):8984–8996
Krystosek A, Seeds NW (1981) Plasminogen activator release at the neuronal growth cone. Science 213(4515):1532–1534
Lakhan SE, Kirchgessner A, Tepper D, Leonard A (2013) Matrix metalloproteinases and blood-brain barrier disruption in acute ischemic stroke. Front Neurol 4:32. doi:10.3389/fneur.2013.00032
Lebeurrier N, Liot G, Lopez-Atalaya JP, Orset C, Fernandez-Monreal M, Sonderegger P, Ali C, Vivien D (2005) The brain-specific tissue-type plasminogen activator inhibitor, neuroserpin, protects neurons against excitotoxicity both in vitro and in vivo. Mol Cell Neurosci 30(4):552–558. doi:10.1016/j.mcn.2005.09.005
Ledesma MD, Da Silva JS, Crassaerts K, Delacourte A, De Strooper B, Dotti CG (2000) Brain plasmin enhances APP alpha-cleavage and Abeta degradation and is reduced in Alzheimer's disease brains. EMBO Rep 1(6):530–535. doi:10.1093/embo-reports/kvd107
Lee CZ, Xue Z, Zhu Y, Yang GY, Young WL (2007) Matrix metalloproteinase-9 inhibition attenuates vascular endothelial growth factor-induced intracerebral hemorrhage. Stroke 38(9):2563–2568. doi:10.1161/STROKEAHA.106.481515
Leebeek FW, Goor MP, Guimaraes AH, Brouwers GJ, Maat MP, Dippel DW, Rijken DC (2005) High functional levels of thrombin-activatable fibrinolysis inhibitor are associated with an increased risk of first ischemic stroke. J Thromb Haemost 3(10):2211–2218. doi:10.1111/j.1538-7836.2005.01484.x
Li Y, Zhu W, Tao J, Xin P, Liu M, Li J, Wei M (2012) Fasudil protects the heart against ischemia-reperfusion injury by attenuating endoplasmic reticulum stress and modulating SERCA activity: the differential role for PI3K/Akt and JAK2/STAT3 signaling pathways. PLoS One 7(10), e48115. doi:10.1371/journal.pone.0048115
Lijnen HR (2001) Plasmin and matrix metalloproteinases in vascular remodeling. Thromb Haemost 86(1):324–333
Liot G, Roussel BD, Lebeurrier N, Benchenane K, Lopez-Atalaya JP, Vivien D, Ali C (2006) Tissue-type plasminogen activator rescues neurones from serum deprivation-induced apoptosis

through a mechanism independent of its proteolytic activity. J Neurochem 98(5):1458–1464. doi:10.1111/j.1471-4159.2006.03982.x

Liu D, Cheng T, Guo H, Fernandez JA, Griffin JH, Song X, Zlokovic BV (2004) Tissue plasminogen activator neurovascular toxicity is controlled by activated protein C. Nat Med 10(12):1379–1383. doi:10.1038/nm1122

Lobov S, Croucher DR, Saunders DN, Ranson M (2008) Plasminogen activator inhibitor type 2 inhibits cell surface associated tissue plasminogen activator in vitro: potential receptor interactions. Thromb Haemost 100(2):319–329

Lopez-Atalaya JP, Roussel BD, Levrat D, Parcq J, Nicole O, Hommet Y, Benchenane K, Castel H, Leprince J, Van To D, Bureau R, Rault S, Vaudry H, Petersen KU, Santos JS, Ali C, Vivien D (2008) Toward safer thrombolytic agents in stroke: molecular requirements for NMDA receptor-mediated neurotoxicity. J Cereb Blood Flow Metab 28(6):1212–1221. doi:10.1038/jcbfm.2008.14

Macrez R, Bezin L, Le Mauff B, Ali C, Vivien D (2010) Functional occurrence of the interaction of tissue plasminogen activator with the NR1 Subunit of N-methyl-D-aspartate receptors during stroke. Stroke 41(12):2950–2955. doi:10.1161/STROKEAHA.110.592360

Macrez R, Obiang P, Gauberti M, Roussel B, Baron A, Parcq J, Casse F, Hommet Y, Orset C, Agin V, Bezin L, Berrocoso TG, Petersen KU, Montaner J, Maubert E, Vivien D, Ali C (2011) Antibodies preventing the interaction of tissue-type plasminogen activator with N-methyl-D-aspartate receptors reduce stroke damages and extend the therapeutic window of thrombolysis. Stroke 42(8):2315–2322. doi:10.1161/STROKEAHA.110.606293

Mansuy IM, van der Putten H, Schmid P, Meins M, Botteri FM, Monard D (1993) Variable and multiple expression of Protease Nexin-1 during mouse organogenesis and nervous system development. Development 119(4):1119–1134

Marciniak SJ, Ron D (2006) Endoplasmic reticulum stress signaling in disease. Physiol Rev 86 (4):1133–1149. doi:10.1152/physrev.00015.2006, 86/4/1133 [pii]

Marti-Fabregas J, Borrell M, Cocho D, Belvis R, Castellanos M, Montaner J, Pagonabarraga J, Aleu A, Molina-Porcel L, Diaz-Manera J, Bravo Y, Alvarez-Sabin J, Davalos A, Fontcuberta J, Marti-Vilalta JL (2005) Hemostatic markers of recanalization in patients with ischemic stroke treated with rt-PA. Neurology 65(3):366–370. doi:10.1212/01.wnl.0000171704.50395.ba

Matys T, Strickland S (2003) Tissue plasminogen activator and NMDA receptor cleavage. Nat Med 9(4):371–372. doi:10.1038/nm0403-371, author reply 372–373

Matys T, Pawlak R, Matys E, Pavlides C, McEwen BS, Strickland S (2004) Tissue plasminogen activator promotes the effects of corticotropin-releasing factor on the amygdala and anxiety-like behavior. Proc Natl Acad Sci USA 101(46):16345–16350. doi:10.1073/pnas.0407355101

McCaig D, Imai H, Gallagher L, Graham DI, Harland J, Moira Brown S, Mhairi Macrae I (2005) Evolution of GADD34 expression after focal cerebral ischaemia. Brain Res 1034(1–2):51–61. doi:10.1016/j.brainres.2004.11.058, S0006-8993(04)01887-6 [pii]

Millan M, Dorado L, Davalos A (2010) Fibrinolytic therapy in acute stroke. Curr Cardiol Rev 6 (3):218–226. doi:10.2174/157340310791658758

Mirante O, Price M, Puentes W, Castillo X, Benakis C, Thevenet J, Monard D, Hirt L (2013) Endogenous protease nexin-1 protects against cerebral ischemia. Int J Mol Sci 14 (8):16719–16731. doi:10.3390/ijms140816719

Montaner J, Ribo M, Monasterio J, Molina CA, Alvarez-Sabin J (2003) Thrombin-activatable fibrinolysis inhibitor levels in the acute phase of ischemic stroke. Stroke 34(4):1038–1040. doi:10.1161/01.STR.0000063139.06585.45

Morimoto N, Oida Y, Shimazawa M, Miura M, Kudo T, Imaizumi K, Hara H (2007) Involvement of endoplasmic reticulum stress after middle cerebral artery occlusion in mice. Neuroscience 147(4):957–967. doi:10.1016/j.neuroscience.2007.04.017, S0306-4522(07)00470-8 [pii]

Mosesson MW, Siebenlist KR, Hernandez I, Lee KN, Christiansen VJ, McKee PA (2008) Evidence that alpha2-antiplasmin becomes covalently ligated to plasma fibrinogen in the circulation: a new role for plasma factor XIII in fibrinolysis regulation. J Thromb Haemost 6 (9):1565–1570. doi:10.1111/j.1538-7836.2008.03056.x

Nagai N, De Mol M, Lijnen HR, Carmeliet P, Collen D (1999) Role of plasminogen system components in focal cerebral ischemic infarction: a gene targeting and gene transfer study in mice. Circulation 99(18):2440–2444

Nagai N, Suzuki Y, Van Hoef B, Lijnen HR, Collen D (2005) Effects of plasminogen activator inhibitor-1 on ischemic brain injury in permanent and thrombotic middle cerebral artery occlusion models in mice. J Thromb Haemost 3(7):1379–1384. doi:10.1111/j.1538-7836.2005.01466.x

Nakagawa T, Zhu H, Morishima N, Li E, Xu J, Yankner BA, Yuan J (2000) Caspase-12 mediates endoplasmic-reticulum-specific apoptosis and cytotoxicity by amyloid-beta. Nature 403 (6765):98–103. doi:10.1038/47513

Nakka VP, Gusain A, Raghubir R (2010) Endoplasmic reticulum stress plays critical role in brain damage after cerebral ischemia/reperfusion in rats. Neurotox Res 17(2):189–202. doi:10.1007/s12640-009-9110-5

Needham PG, Brodsky JL (2013) How early studies on secreted and membrane protein quality control gave rise to the ER associated degradation (ERAD) pathway: the early history of ERAD. Biochim Biophys Acta 1833(11):2447–2457. doi:10.1016/j.bbamcr.2013.03.018

Nicole O, Docagne F, Ali C, Margaill I, Carmeliet P, MacKenzie ET, Vivien D, Buisson A (2001) The proteolytic activity of tissue-plasminogen activator enhances NMDA receptor-mediated signaling. Nat Med 7(1):59–64. doi:10.1038/83358

Nilupul Perera M, Ma HK, Arakawa S, Howells DW, Markus R, Rowe CC, Donnan GA (2006) Inflammation following stroke. J Clin Neurosci 13(1):1–8. doi:10.1016/j.jocn.2005.07.005

O'Collins VE, Macleod MR, Donnan GA, Horky LL, van der Worp BH, Howells DW (2006) 1,026 experimental treatments in acute stroke. Ann Neurol 59(3):467–477. doi:10.1002/ana.20741

Obiang P, Macrez R, Jullienne A, Bertrand T, Lesept F, Ali C, Maubert E, Vivien D, Agin V (2012) GluN2D subunit-containing NMDA receptors control tissue plasminogen activator-mediated spatial memory. J Neurosci 32(37):12726–12734. doi:10.1523/JNEUROSCI.6202-11.2012

Orbe J, Barrenetxe J, Rodriguez JA, Vivien D, Orset C, Parks WC, Birkland TP, Serrano R, Purroy A, Martinez de Lizarrondo S, Angles-Cano E, Paramo JA (2011) Matrix metalloproteinase-10 effectively reduces infarct size in experimental stroke by enhancing fibrinolysis via a thrombin-activatable fibrinolysis inhibitor-mediated mechanism. Circulation 124(25):2909–2919. doi:10.1161/CIRCULATIONAHA.111.047100

Owen CR, Kumar R, Zhang P, McGrath BC, Cavener DR, Krause GS (2005) PERK is responsible for the increased phosphorylation of eIF2alpha and the severe inhibition of protein synthesis after transient global brain ischemia. J Neurochem 94(5):1235–1242. doi:10.1111/j.1471-4159.2005.03276.x, JNC3276 [pii]

Pang PT, Teng HK, Zaitsev E, Woo NT, Sakata K, Zhen S, Teng KK, Yung WH, Hempstead BL, Lu B (2004) Cleavage of proBDNF by tPA/plasmin is essential for long-term hippocampal plasticity. Science 306(5695):487–491. doi:10.1126/science.1100135

Parcq J, Bertrand T, Baron AF, Hommet Y, Angles-Cano E, Vivien D (2013) Molecular requirements for safer generation of thrombolytics by bioengineering the tissue-type plasminogen activator A chain. J Thromb Haemost 11(3):539–546. doi:10.1111/jth.12128

Paschen W, Hayashi T, Saito A, Chan PH (2004) GADD34 protein levels increase after transient ischemia in the cortex but not in the CA1 subfield: implications for post-ischemic recovery of protein synthesis in ischemia-resistant cells. J Neurochem 90(3):694–701. doi:10.1111/j.1471-4159.2004.02555.x, JNC2555 [pii]

Perry VH, Nicoll JA, Holmes C (2010) Microglia in neurodegenerative disease. Nat Rev Neurol 6 (4):193–201. doi:10.1038/nrneurol.2010.17

Pineda D, Ampurdanes C, Medina MG, Serratosa J, Tusell JM, Saura J, Planas AM, Navarro P (2012) Tissue plasminogen activator induces microglial inflammation via a noncatalytic molecular mechanism involving activation of mitogen-activated protein kinases and Akt

signaling pathways and AnnexinA2 and Galectin-1 receptors. Glia 60(4):526–540. doi:10.1002/glia.22284
Polavarapu R, Gongora MC, Yi H, Ranganthan S, Lawrence DA, Strickland D, Yepes M (2007) Tissue-type plasminogen activator-mediated shedding of astrocytic low-density lipoprotein receptor-related protein increases the permeability of the neurovascular unit. Blood 109 (8):3270–3278. doi:10.1182/blood-2006-08-043125
Ribo M, Montaner J, Molina CA, Arenillas JF, Santamarina E, Alvarez-Sabin J (2004) Admission fibrinolytic profile predicts clot lysis resistance in stroke patients treated with tissue plasminogen activator. Thromb Haemost 91(6):1146–1151. doi:10.1267/THRO04061146
Rijken DC, Lijnen HR (2009) New insights into the molecular mechanisms of the fibrinolytic system. J Thromb Haemost 7(1):4–13. doi:10.1111/j.1538-7836.2008.03220.x
Rijken DC, Sakharov DV (2001) Basic principles in thrombolysis: regulatory role of plasminogen. Thromb Res 103(Suppl 1):S41–S49
Rodriguez-Gonzalez R, Agulla J, Perez-Mato M, Sobrino T, Castillo J (2011a) Neuroprotective effect of neuroserpin in rat primary cortical cultures after oxygen and glucose deprivation and tPA. Neurochem Int 58(3):337–343. doi:10.1016/j.neuint.2010.12.006
Rodriguez-Gonzalez R, Millan M, Sobrino T, Miranda E, Brea D, de la Ossa NP, Blanco M, Perez J, Dorado L, Castellanos M, Lomas DA, Moro MA, Davalos A, Castillo J (2011b) The natural tissue plasminogen activator inhibitor neuroserpin and acute ischaemic stroke outcome. Thromb Haemost 105(3):421–429. doi:10.1160/TH10-09-0621
Rogove AD, Tsirka SE (1998) Neurotoxic responses by microglia elicited by excitotoxic injury in the mouse hippocampus. Curr Biol 8(1):19–25
Rogove AD, Siao C, Keyt B, Strickland S, Tsirka SE (1999) Activation of microglia reveals a non-proteolytic cytokine function for tissue plasminogen activator in the central nervous system. J Cell Sci 112(Pt 22):4007–4016
Rossignol P, Ho-Tin-Noe B, Vranckx R, Bouton MC, Meilhac O, Lijnen HR, Guillin MC, Michel JB, Angles-Cano E (2004) Protease nexin-1 inhibits plasminogen activation-induced apoptosis of adherent cells. J Biol Chem 279(11):10346–10356. doi:10.1074/jbc.M310964200
Roussel BD, Irving JA, Ekeowa UI, Belorgey D, Haq I, Ordonez A, Kruppa AJ, Duvoix A, Rashid ST, Crowther DC, Marciniak SJ, Lomas DA (2011a) Unravelling the twists and turns of the serpinopathies. FEBS J 278(20):3859–3867. doi:10.1111/j.1742-4658.2011.08201.x
Roussel BD, Mysiorek C, Rouhiainen A, Jullienne A, Parcq J, Hommet Y, Culot M, Berezowski V, Cecchelli R, Rauvala H, Vivien D, Ali C (2011b) HMGB-1 promotes fibrinolysis and reduces neurotoxicity mediated by tissue plasminogen activator. J Cell Sci 124 (Pt 12):2070–2076. doi:10.1242/jcs.084392
Roussel BD, Kruppa AJ, Miranda E, Crowther DC, Lomas DA, Marciniak SJ (2013) Endoplasmic reticulum dysfunction in neurological disease. Lancet Neurol 12(1):105–118. doi:10.1016/S1474-4422(12)70238-7
Salles FJ, Strickland S (2002) Localization and regulation of the tissue plasminogen activator-plasmin system in the hippocampus. J Neurosci 22(6):2125–2134
Samson AL, Nevin ST, Croucher D, Niego B, Daniel PB, Weiss TW, Moreno E, Monard D, Lawrence DA, Medcalf RL (2008) Tissue-type plasminogen activator requires a co-receptor to enhance NMDA receptor function. J Neurochem 107(4):1091–1101. doi:10.1111/j.1471-4159.2008.05687.x
Sashindranath M, Sales E, Daglas M, Freeman R, Samson AL, Cops EJ, Beckham S, Galle A, McLean C, Morganti-Kossmann C, Rosenfeld JV, Madani R, Vassalli JD, Su EJ, Lawrence DA, Medcalf RL (2012) The tissue-type plasminogen activator-plasminogen activator inhibitor 1 complex promotes neurovascular injury in brain trauma: evidence from mice and humans. Brain 135(Pt 11):3251–3264. doi:10.1093/brain/aws178
Schaller J, Gerber SS (2011) The plasmin-antiplasmin system: structural and functional aspects. Cell Mol Life Sci 68(5):785–801. doi:10.1007/s00018-010-0566-5
Siao CJ, Tsirka SE (2002) Tissue plasminogen activator mediates microglial activation via its finger domain through annexin II. J Neurosci 22(9):3352–3358

Siesjo BK, Agardh CD, Bengtsson F (1989) Free radicals and brain damage. Cerebrovasc Brain Metab Rev 1(3):165–211

Smedsrod B, Einarsson M (1990) Clearance of tissue plasminogen activator by mannose and galactose receptors in the liver. Thromb Haemost 63(1):60–66

Smith MH, Ploegh HL, Weissman JS (2011) Road to ruin: targeting proteins for degradation in the endoplasmic reticulum. Science 334(6059):1086–1090. doi:10.1126/science.1209235

Soeda S, Oda M, Ochiai T, Shimeno H (2001) Deficient release of plasminogen activator inhibitor-1 from astrocytes triggers apoptosis in neuronal cells. Brain Res Mol Brain Res 91(1-2):96–103

Strbian D, Meretoja A, Putaala J, Kaste M, Tatlisumak T (2013) Cerebral edema in acute ischemic stroke patients treated with intravenous thrombolysis. Int J Stroke 8(7):529–534. doi:10.1111/j.1747-4949.2012.00781.x

Su EJ, Fredriksson L, Geyer M, Folestad E, Cale J, Andrae J, Gao Y, Pietras K, Mann K, Yepes M, Strickland DK, Betsholtz C, Eriksson U, Lawrence DA (2008) Activation of PDGF-CC by tissue plasminogen activator impairs blood-brain barrier integrity during ischemic stroke. Nat Med 14(7):731–737. doi:10.1038/nm1787

Suzuki Y (2010) Role of tissue-type plasminogen activator in ischemic stroke. J Pharmacol Sci 113(3):203–207

Suzuki Y, Nagai N, Umemura K, Collen D, Lijnen HR (2007) Stromelysin-1 (MMP-3) is critical for intracranial bleeding after t-PA treatment of stroke in mice. J Thromb Haemost 5 (8):1732–1739. doi:10.1111/j.1538-7836.2007.02628.x

Suzuki Y, Nagai N, Yamakawa K, Kawakami J, Lijnen HR, Umemura K (2009) Tissue-type plasminogen activator (t-PA) induces stromelysin-1 (MMP-3) in endothelial cells through activation of lipoprotein receptor-related protein. Blood 114(15):3352–3358. doi:10.1182/blood-2009-02-203919

Suzuki Y, Yasui H, Brzoska T, Mogami H, Urano T (2011) Surface-retained tPA is essential for effective fibrinolysis on vascular endothelial cells. Blood 118(11):3182–3185. doi:10.1182/blood-2011-05-353912

Tajiri S, Oyadomari S, Yano S, Morioka M, Gotoh T, Hamada JI, Ushio Y, Mori M (2004) Ischemia-induced neuronal cell death is mediated by the endoplasmic reticulum stress pathway involving CHOP. Cell Death Differ 11(4):403–415. doi:10.1038/sj.cdd.4401365, 4401365 [pii]

Tansey MG, Goldberg MS (2010) Neuroinflammation in Parkinson's disease: its role in neuronal death and implications for therapeutic intervention. Neurobiol Dis 37(3):510–518. doi:10.1016/j.nbd.2009.11.004

Tsirka SE, Gualandris A, Amaral DG, Strickland S (1995) Excitotoxin-induced neuronal degeneration and seizure are mediated by tissue plasminogen activator. Nature 377(6547):340–344. doi:10.1038/377340a0

Urano T, Suzuki Y (2012) Accelerated fibrinolysis and its propagation on vascular endothelial cells by secreted and retained tPA. J Biomed Biotechnol 2012:208108. doi:10.1155/2012/208108

Vercauteren E, Gils A, Declerck PJ (2013) Thrombin activatable fibrinolysis inhibitor: a putative target to enhance fibrinolysis. Semin Thromb Hemost 39(4):365–372. doi:10.1055/s-0033-1334488

Vivien D, Gauberti M, Montagne A, Defer G, Touze E (2011) Impact of tissue plasminogen activator on the neurovascular unit: from clinical data to experimental evidence. J Cereb Blood Flow Metab 31(11):2119–2134. doi:10.1038/jcbfm.2011.127

Wang X, Lee SR, Arai K, Tsuji K, Rebeck GW, Lo EH (2003) Lipoprotein receptor-mediated induction of matrix metalloproteinase by tissue plasminogen activator. Nat Med 9 (10):1313–1317. doi:10.1038/nm926

Wardlaw JM (2012) Prediction of haematoma expansion with the CTA spot sign: a useful biomarker? Lancet Neurol 11(4):294–295. doi:10.1016/S1474-4422(12)70045-5

White F, McCaig D, Brown SM, Graham DI, Harland J, Macrae IM (2004) Up-regulation of a growth arrest and DNA damage protein (GADD34) in the ischaemic human brain: implications

for protein synthesis regulation and DNA repair. Neuropathol Appl Neurobiol 30(6):683–691. doi:10.1111/j.1365-2990.2004.00584.x
Wiklund PG, Nilsson L, Ardnor SN, Eriksson P, Johansson L, Stegmayr B, Hamsten A, Holmberg D, Asplund K (2005) Plasminogen activator inhibitor-1 4G/5G polymorphism and risk of stroke: replicated findings in two nested case-control studies based on independent cohorts. Stroke 36(8):1661–1665. doi:10.1161/01.STR.0000174485.10277.24
Yang D, Sun YY, Nemkul N, Baumann JM, Shereen A, Dunn RS, Wills-Karp M, Lawrence DA, Lindquist DM, Kuan CY (2013) Plasminogen activator inhibitor-1 mitigates brain injury in a rat model of infection-sensitized neonatal hypoxia-ischemia. Cereb Cortex 23(5):1218–1229. doi:10.1093/cercor/bhs115
Yao Y, Tsirka SE (2011) Truncation of monocyte chemoattractant protein 1 by plasmin promotes blood-brain barrier disruption. J Cell Sci 124(Pt 9):1486–1495. doi:10.1242/jcs.082834
Yepes M, Sandkvist M, Wong MK, Coleman TA, Smith E, Cohan SL, Lawrence DA (2000) Neuroserpin reduces cerebral infarct volume and protects neurons from ischemia-induced apoptosis. Blood 96(2):569–576
Yepes M, Sandkvist M, Moore EG, Bugge TH, Strickland DK, Lawrence DA (2003) Tissue-type plasminogen activator induces opening of the blood-brain barrier via the LDL receptor-related protein. J Clin Invest 112(10):1533–1540. doi:10.1172/JCI19212
Yepes M, Roussel BD, Ali C, Vivien D (2009) Tissue-type plasminogen activator in the ischemic brain: more than a thrombolytic. Trends Neurosci 32(1):48–55. doi:10.1016/j.tins.2008.09.006
Young AR, Ali C, Duretete A, Vivien D (2007) Neuroprotection and stroke: time for a compromise. J Neurochem 103(4):1302–1309. doi:10.1111/j.1471-4159.2007.04866.x
Yuan Y, Guo Q, Ye Z, Pingping X, Wang N, Song Z (2011) Ischemic postconditioning protects brain from ischemia/reperfusion injury by attenuating endoplasmic reticulum stress-induced apoptosis through PI3K-Akt pathway. Brain Res 1367:85–93. doi:10.1016/j.brainres.2010.10.017
Zhang Z, Zhang L, Yepes M, Jiang Q, Li Q, Arniego P, Coleman TA, Lawrence DA, Chopp M (2002) Adjuvant treatment with neuroserpin increases the therapeutic window for tissue-type plasminogen activator administration in a rat model of embolic stroke. Circulation 106 (6):740–745
Zhang L, Zhang ZG, Ding GL, Jiang Q, Liu X, Meng H, Hozeska A, Zhang C, Li L, Morris D, Zhang RL, Lu M, Chopp M (2005) Multitargeted effects of statin-enhanced thrombolytic therapy for stroke with recombinant human tissue-type plasminogen activator in the rat. Circulation 112(22):3486–3494. doi:10.1161/CIRCULATIONAHA.104.516757
Zhang X, Polavarapu R, She H, Mao Z, Yepes M (2007) Tissue-type plasminogen activator and the low-density lipoprotein receptor-related protein mediate cerebral ischemia-induced nuclear factor-kappaB pathway activation. Am J Pathol 171(4):1281–1290. doi:10.2353/ajpath.2007.070472
Zhang C, An J, Haile WB, Echeverry R, Strickland DK, Yepes M (2009a) Microglial low-density lipoprotein receptor-related protein 1 mediates the effect of tissue-type plasminogen activator on matrix metalloproteinase-9 activity in the ischemic brain. J Cereb Blood Flow Metab 29 (12):1946–1954. doi:10.1038/jcbfm.2009.174
Zhang L, Chopp M, Jia L, Cui Y, Lu M, Zhang ZG (2009b) Atorvastatin extends the therapeutic window for tPA to 6 h after the onset of embolic stroke in rats. J Cereb Blood Flow Metab 29 (11):1816–1824. doi:10.1038/jcbfm.2009.105
Zhang YP, Wang WL, Liu J, Li WB, Bai LL, Yuan YD, Song SX (2013) Plasminogen activator inhibitor-1 promotes the proliferation and inhibits the apoptosis of pulmonary fibroblasts by Ca (2+) signaling. Thromb Res 131(1):64–71. doi:10.1016/j.thromres.2012.09.003
Zhu H, Fan X, Yu Z, Liu J, Murata Y, Lu J, Zhao S, Hajjar KA, Lo EH, Wang X (2010) Annexin A2 combined with low-dose tPA improves thrombolytic therapy in a rat model of focal embolic stroke. J Cereb Blood Flow Metab 30(6):1137–1146. doi:10.1038/jcbfm.2009.279
Zmijewski JW, Bae HB, Deshane JS, Peterson CB, Chaplin DD, Abraham E (2011) Inhibition of neutrophil apoptosis by PAI-1. Am J Physiol Lung Cell Mol Physiol 301(2):L247–254. doi:10.1152/ajplung.00075.2011

Serpins, Viruses, and the Virome: New Directions in Therapy

17

Sriram Ambadapadi, Hao Chen, Donghang Zheng, Liying Liu, Erbin Dai, Ganesh Munuswamy-Ramanujam, Sufi Morshed, Jorge Fuentes, Adisson Fortunel, Mohammad Al-Ani, Lakshmyya Kesavalu, Grant McFadden, and Alexandra Lucas

S. Ambadapadi, Ph.D. • D. Zheng, M.D., Ph.D. • S. Morshed, Ph.D.
Division of Cardiovascular Medicine, College of Medicine, University of Florida, Gainesville, FL, USA

H. Chen, M.D., Ph.D.
Division of Cardiovascular Medicine, College of Medicine, University of Florida, Gainesville, FL, USA

Department of Tumor Surgery, College of Medicine, Lanzhou University, Lanzhou, China

L. Liu, M.D. • E. Dai, M.D.
Brigham & Women's Hospital, Harvard University, Boston, MA, USA

G. Munuswamy-Ramanujam, Ph.D.
Interdisciplinary School of Indian System of Medicine, SRM University, Kattankulathur, Tamil Nadu, India

J. Fuentes, M.D. • M. Al-Ani, M.D.
Department of Medicine, University of Florida, Gainesville, FL, USA

A. Fortunel, B.Sc.
Department of Biochemistry, University of Florida, Gainesville, FL, USA

L. Kesavalu, B.V.Sc., M.Sc.
Department of Periodontology/Oral Biology, College Dentistry, University of Florida, Gainesville, FL, USA

G. McFadden, Ph.D.
Department of Molecular Genetics and Microbiology, University of Florida, Gainesville, FL, USA

A. Lucas, M.D., F.R.C.P.(C.) (✉)
Division of Cardiovascular Medicine, College of Medicine, University of Florida, Gainesville, FL, USA

Department of Molecular Genetics and Microbiology, University of Florida, Gainesville, FL, USA

Ethel Smith Chair Vasculitis Research, Section Head Vascular Research, Divisions Cardiovascular Medicine and Rheumatology, University of Florida, 1600 SW Archer Road, 100277, Gainesville, FL 32610-0277, USA
e-mail: alexandra.lucas@medicine.ufl.edu

M. Geiger et al. (eds.), *The Serpin Family*, DOI 10.1007/978-3-319-22711-5_17

Contents

Abstract

Serine protease inhibitors, termed serpins, regulate myriad physiological processes in the mammalian body from thrombotic and thrombolytic pathways to inflammation, angiogenesis, hormone transport, and hypertension. The large percentage of serpins among the plasma proteins in the circulating blood as in the case of plasminogen activators, the functional redundancy of serpins, and also the debilitating serpinopathies of antithrombin III (SERPINC1), neuroserpin (SERPINI1), and alpha-1 antitrypsin (SERPINA1) provide evidence of the importance of serpins and their widespread impact in normal physiological homeostasis. Inflammation, also termed innate immunity, interacts closely with and both regulates and is regulated by thrombotic and thrombolytic serine proteases. Activation of the coagulation proteases is, in turn, controlled by serpins. Apoptosis is also modulated by serpins with cross-class inhibitory activity for cysteine and serine proteases. Excessive inflammation and cell death processes are now recognized as interacting with the thrombotic and thrombolytic proteases. Viruses have evolved to communicate and control these processes by encoding their own serpins, which confer on them the ability to evade host immune defenses. This chapter provides an introduction to the viral serpins derived from poxvirus origins that have been shown not only to be essential for successful viral infection but, in some cases, as for Serp-1 and Serp-2, to have the potential to mitigate inflammatory disease in animal models. Serp-1 has in fact been successfully tested in a small phase 2A clinical trial in unstable angina patients with coronary stent implants. A tandem discussion of mammalian serpins with actions similar to those of the viral serpins is also presented to emphasize potential evolutionary relationships between viral and mammalian serpins. The anti-inflammatory serpins hold the potential to be effective in disease states such as atheroma, sepsis, cancer, and wound healing given that these conditions are all associated with aberrant inflammatory responses and with dysregulation of thrombotic, thrombolytic, and apoptotic protease cascades. The capacity of viral serpins to provide antiviral protection by modulating the virome as well as possible therapeutic effects of serpin metabolites in inflammation will also be discussed. In summary, viral serpins have evolved over many millions of years and provide a unique and highly potent reservoir for both the study of serpin modulation of inflammatory responses and for new therapeutic approaches to inflammatory and even infectious diseases.

Serpins
Serpent molecule
Serpiginous trap
Strikes
Secures
Stabilizes
The ultimate protease regulatory weapon
Alexandra Lucas, MD

Preface

Current research indicates that wound healing, atherosclerosis, sepsis, and cancer share pathogenic as well as curative processes. In disease states, many of these pathways are driven by an excess or an imbalance in the normal innate immune response. Activation of innate immune, or inflammatory, responses is also associated with activation of proteases that drive the coagulation, the inflammasome, and the apoptotic pathways. These serine proteases are regulated by inhibitors, termed serpins.

Activation of serine proteases in clot-forming and clot-dissolving cascades can increase inflammatory responses. These interactions between the thrombotic (clotting) and thrombolytic (clot lysis) proteases are now recognized as central mediators in pathogenic states where there is excess inflammation. The thrombotic and thrombolytic cascades are composed of sequentially activated serine protease cascades that are, as noted regulated by serpins. Similar interactions are also seen between the apoptotic (cell death), inflammasome, and innate immune pathways. The serine protease granzyme B and select cysteine proteases (caspases) drive cell death and also activation of the inflammasome. Some serpins can function as cross-class inhibitors, inhibiting not only serine proteases, but also cysteine proteases in the apoptotic and inflammasome pathways.

Large DNA viruses have evolved highly effective immune-modulating mechanisms over millions of years that serve as a protective shield against host immune responses to viral infection. These viruses have identified serpins as pivotal immune-modulating agents. Virus-derived serpins demonstrate the impact of serpin-mediated inhibition of protease pathways on host immune responses, displaying powerful blockades of the host responses to viral infection. Viral serpins may also provide a mechanism for blockade of invasion by other infectious agents, potentially modifying the host microbiome. With this chapter, we will discuss the viral serpins, their immunomodulatory mechanisms, and their potential as a new class of anti-inflammatory therapeutic and for identification of new therapeutic directions. We begin with the discovery of innate immunity followed by discussions of coagulation, apoptosis, and inflammasome pathways and the interaction of these protease pathways with inflammation. We will then provide an overview of serpin biology and individual viral serpins, both in viral infections and as translational therapeutics. The potential roles for viral serpins in the virome will also be discussed. In translation research on viral proteins as therapeutics, the known viral serpins have demonstrated the capacity to alter the course of disease in a broad array of disease models. One such myxomaviral serpin, Serp-1, has demonstrated efficacy in animal models of atherosclerosis, inflammatory vasculitis, transplant rejection, arthritis, cancer, and viral sepsis illustrating a very wide range of disease-modifying actions. This same serpin has been safely and effectively tested in one small phase 2a clinical trial. *In sum, we would suggest that analysis of serpin immunomodulation has extraordinary potential for improving understanding of inflammatory mechanisms in disease and providing new directions for treatment of inflammation-based disorders. Current work on viral serpins has only begun to tap this natural biological resource.*

17.1 Central Roles of Innate Immunity in Healing and Disease: The History

The innate immune system plays a pivotal role in inflammation and is increasingly reported as fundamental for both healing processes in man and pathogenic processes in disease. This early cellular response occurs within hours, long before antibodies are upregulated in acquired immunity. Innate immune responses are responsible for 80 % of healing and antimicrobial responses (Zetterstrom 2009; Ross et al. 1977, 1984; Lowe 2003; Bennett et al. 2009). Inflammatory cell activation at sites of local injury has been reported by researchers from Mechnikov and Virchow to Russell Ross. Ilya Ilyich Mechnikov (Institut Pasteur, France; 1845–1916) was the first to discover the innate immune cellular response. He reported a very early massing of mononuclear cells in the transparent starfish larvae *bipinnaria* that occurred after insertion of tangerine thorns. Mechnikov later confirmed that there was also an early inflammatory cell response in *Daphnia* water fleas infected with microbes. Rudolf Virchow (Germany, 1821–1902) was a multitalented pathologist and public health physician, biologist, and scientist. Virchow was among the first to note a connection between inflammatory cell invasion in atherosclerotic plaques in the arterial wall and also to observe the connection between endothelial cell injury, hypercoagulability, and thrombosis in venous thrombosis and pulmonary emboli (although Virchow's triad was only named as such many years later) (Lowe 2003; Bennett et al. 2009). Finally, Russell Ross (University of Washington, Seattle, USA; 1929–1999) formulated the "response to injury hypothesis" in which he stated that atherosclerotic plaque develops in response to multiple different forms of endothelial injury in the artery from hypertension, diabetes, and hyperlipidemia to surgical injury during percutaneous angioplasty. These three seminal observations provide the basis for what is now known as innate immunity, beginning with endothelial cell injury and culminating in platelet activation and adherence and inflammatory cell activation, adhesion, and invasion (Ross et al. 1977, 1984). Excess inflammation is now understood to be associated with disease progression in wound healing, atheroma, sepsis, and also cancer.

With endothelial damage and attendant inflammatory cell responses, there is upregulation of coagulation protease activity on the surface of activated platelets, on damaged endothelium, on mononuclear cells, and in the surrounding matrix. The inflammatory and coagulation systems modulate one another stimulating further inflammation and also clotting and/or bleeding (Carmeliet et al. 1997; Hu et al. 2015; Fu et al. 2014; Chen and Dorling 2009; Chen et al. 2008). With cellular injury, there is also cell death and cell suicide (apoptosis). Apoptotic cells can also induce activation of proteases in the thrombotic cascade as well as release of cytokines that induce further activation of inflammatory cells (Hiebert and Granville 2012; Wen et al. 2012; Masters 2013; Wensink et al. 2015). This same coactivation of thrombotic, thrombolytic, and inflammatory cascades is seen in atherosclerotic and inflammatory vascular disease, wound healing, uncontrolled septic states with disseminated intravascular coagulation (DIC), and also cancer. In recent studies, inflammatory macrophage and neutrophils, designated tumor-associated macrophage and neutrophils (TAM and TAN), and myeloid-derived suppressor cells (MDSC) have been reported to initiate and/or promote local cancer invasion and metastases. Additionally, work by Maden and Scott on wound healing in the newt axolotl (*Ambystoma mexicanum*) and the African spiny mouse (*Acomys*) has demonstrated associations with some of these same thrombotic, thrombolytic, inflammatory, and connective tissue pathways in tissue regeneration during wound healing (Fig. 17.1). The spiny mice along with the axolotl have very active regeneration systems when compared to wild-type mice, and their work has detected associated changes in macrophage, collagen XII, matrix metalloproteinases (MMPs), and several growth factors (Seifert et al. 2012; Lopez et al. 2014; Maden 2009). MMPs are activated from the pro-form, and selected growth factors are released from connective tissues by plasminogen activators and plasmin.

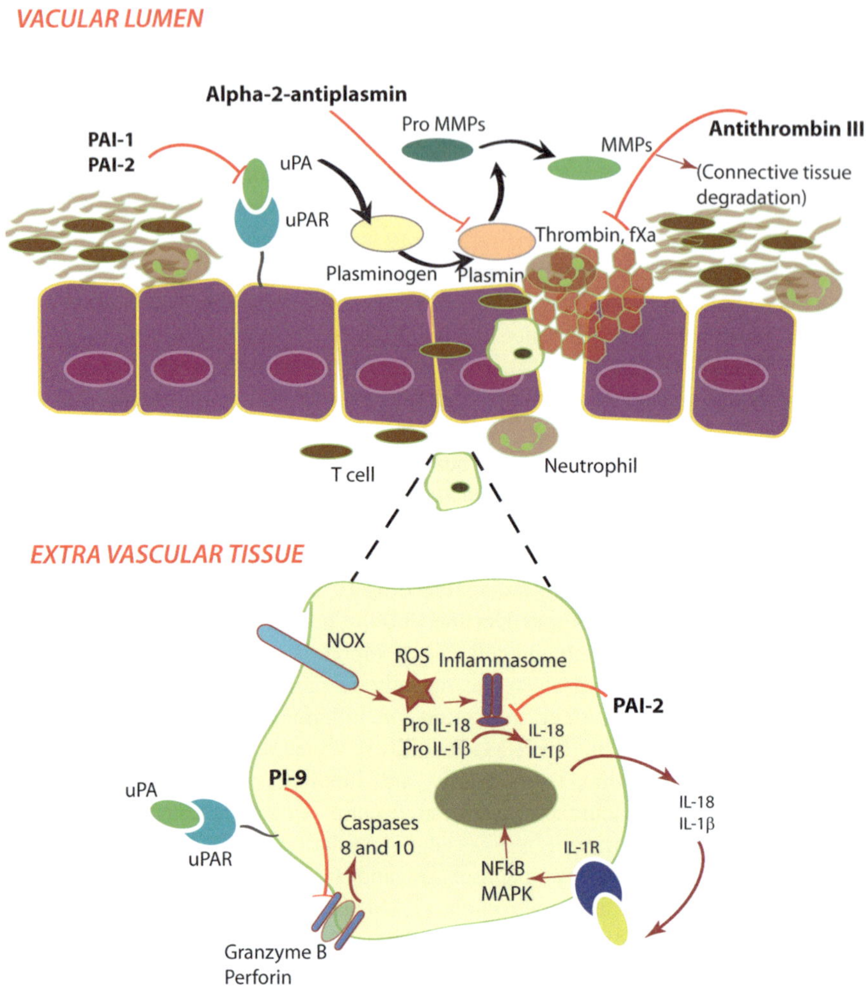

Fig. 17.1 Interaction of platelets, monocytes, and T cells with areas of endothelial damage. Many of the cytokine, chemokine, protease, and connective tissue responses are shared in processes from wound healing to atherosclerosis, sepsis, and cancer

17.2 Inflammation and Mammalian Serpins

In the following section, protease pathways that initiate inflammatory cell responses will be discussed followed by an overview of the serpins that regulate these pathways, their biology, the known viral serpins, and their protease targets. We will then discuss the immune-modulating functions of known viral serpins and their capacity to modify pathogenesis in disease. The potential for serpin-mediated immune modulation in therapeutic applications as well as the concept of serpin-directed changes in the virome. In each case, we will discuss what is known about the role of serpins in specific pathologies, and where possible, the regulatory actions of mammalian serpins will be contrasted with viral serpins that target similar pathways.

17.2.1 Coagulation Pathways and Inflammatory Responses

The thrombolytic serine proteases, specifically urokinase- and tissue-type plasminogen activators (uPA and tPA), activate plasmin. Plasmin not only functions to dissolve fibrin clots but also activates matrix metalloproteinase (MMP) enzymes, as do tPA and uPA. MMPs together with plasmin break down connective tissue, specifically collagen, elastin, and gelatin, surrounding the endothelial layer in the vascular tree (Huang et al. 2014; Gong et al. 2008). This breakdown of connective tissue layers allows cells to move freely into the arterial wall or in fact any vascular wall, in vein, arterial, lymphatic, and capillaries. uPA binds to the uPA receptor (uPAR) on the leading edge of invading monocytes, activating MMPs to allow cell ingress into tissues and also activating and releasing vascular endothelial growth factor (VEGF), transforming growth factor beta (TGFβ), and basic fibroblast growth factor (bFGF) from connective tissue stores. Interaction of uPA with its receptor, uPAR, also modifies cell activation. The uPAR is glycosylphosphatidylinositol (GPI) linked and also sits in a large lipid raft of membrane proteins that initiate signaling pathway activation and thus modifies gene expression potentially through altered signaling of adjacent lipid raft integrins and seven transmembrane G protein-coupled chemokine receptors (GPCRs). Once inhibited by serpin binding to the uPA/uPAR as for plasminogen activator inhibitor-1 (PAI-1, SERPINE1) this complex becomes internalized and can alter cell signaling through select JAK/STAT pathways (Fleetwood et al. 2014; Viswanathan et al. 2009; Kanse et al. 2004; Chen et al. 2013a).

Factor X (FXa) and thrombin bind selected protease-activated receptors (PARs) 1–4 with variable efficiency on the mononuclear cell surface. This interaction also leads to altered cytokine expression in the activated mononuclear cell (Chen and Dorling 2009; Chen et al. 2008).

17.2.2 Inflammasome Pathways in Inflammatory Responses

Caspase 1 is a cysteine protease that activates interleukin-1 beta (IL-1β) as well as IL-18, two pro-inflammatory cytokines. Nod-like receptor proteins (NLRPs) form multiprotein complexes together with the adaptor protein apoptosis-associated speck-like protein containing a CARD (ASC). Caspase 1 engages the adaptor protein ASC and dimerizes to form an active protease that cleaves pro-IL-1β producing the active, secreted cytokine (Wen et al. 2012; Masters 2013; Viswanathan et al. 2012). NLRP3 has been the most intensely investigated NLR and detects pathogen-associated molecular patterns (PAMPs) and damage-associated molecular patterns (DAMPs) and also environmental irritants, hazardous chemicals, and drugs. Genetic deletion studies of NLRP3 implicate NLRP3 in protective responses to viruses, bacteria, and fungi. NLRP3 is associated with diseases such as diabetes, atherosclerosis, and infections. The inflammasome not only initiates activation of inflammatory cytokines but also is associated with a form of cell death termed pyroptosis.

17.2.3 Apoptotic Pathways in Inflammatory Responses

Apoptosis pathways are activated both by the serine protease granzyme B (GrB) as well as through other activating pathways which then signal through cytoplasmic cysteine proteases (caspases) via sequential protease reactions. Apoptotic cells release pro-inflammatory mediators or cytokines, and some apoptotic macrophages release large packets of cytokines into the circulation (Chen et al. 2008; Wensink et al. 2015). Apoptotic cells are also reported to activate platelets, and the thrombotic cascade again engenders more inflammatory as well as clot-forming responses.

GrB is a serine protease mainly produced by cytotoxic lymphocytes and is traditionally considered part of the molecular machinery in cells that induce T cell-driven apoptosis (cell death or cell suicide) in virus-infected cells or tumor cells. GrB is released together with perforin into the immunological synapse between the cytotoxic T cell and the target cell. The granule exocytosis pathway releases cytotoxic proteins stored in lymphocyte granules which includes perforin, a pore-forming protein, and serine protease granzymes. GrB together with perforin enters the target cell and activates apoptotic pathways by cleavage of intracellular substrates. More recent work has found that GrB can also enter a cell and induce apoptosis independently of the perforin-mediated pathway.

17.3 Serpin Biology

For each of these protease-activated responses, there are virus-derived regulatory serpins that modify their activation. These viral serpins are both classical inhibitors of serine proteases, such as the coagulation enzymes and GrB, and also cross-class inhibitors that block both serine and cysteine proteases through a serpin inhibitory mechanism (Turner et al. 1999; Messud-Petit et al. 1998; Turner and Moyer 2001; Komiyama et al. 1994).

Serpins are found throughout evolution from viruses to bacteria, from horseshoe crabs, dinosaurs, and shrimp to mammals and man (Janciauskiene 2001; Lomas et al. 2005; Carrell and Travis 1985; Irving et al. 2000; Iwanaga and Kawabata 1998; Law et al. 2006; Silverman et al. 2001; Marszal and Shrake 2006; Huntington et al. 2000; Patston et al. 2004; Ye et al. 2001; Lawrence et al. 2000; Whisstock and Bottomley 2006; Liu et al. 2014; Lucas et al. 2009). *Serpins* have central regulatory roles throughout the mammalian body, serving as pivotal points in normal physiological functions, regulating protease activation from thrombotic pathways to inflammation, complement, angiogenesis, hormone transport, or even hypertension, among many others (Law et al. 2006; Silverman et al. 2001; Whisstock and Bottomley 2006). The role of serpins as central regulators of normal physiologic functions is well illustrated by the profound effects of genetic mutations in some serpins, disease states with serpin deficiency, and the newly discovered marked immunomodulating capacities for virus-derived serpins. Genetic mutations of serpins cause severe disorders, named serpinopathies. Alpha-1 antitrypsin (SERPINA1) mutations lead to severe emphysema with lung damage as well as liver and pancreatic disease (Janciauskiene et al. 2011), and antithrombin (AT, SERPINC1) deficiency causes excess thrombosis (Ishiguro et al. 2000; Conard and Samama 1986; Coller et al. 1987). Unstable or unbalanced serpin levels can be lethal in severe sepsis with DIC. Additionally, as further evidence of the profound effects of serpins on the mammalian inflammatory response, presence or absence of viral serpins can be the difference between a benign poxviral infection and a lethal infection (Upton et al. 1990), and when used in animal models, viral serpins markedly reduce inflammation-mediated disease progression (Viswanathan et al. 2006, 2009, 2012; Nash et al. 1997; Miller et al. 2000; Lucas et al. 1996, 2000, 2014a; Chen et al. 2013b, 2015).

Serpin regulation of thrombotic and thrombolytic cascades is one of the primary examples of the central roles for serpins in physiologic regulation of a very major pathway in the bloodstream. Without regulation of clot formation, there is risk of excess thrombosis and organ ischemia in heart attack and stroke, as well as hemorrhage, as seen in sepsis and DIC. Regulation of apoptosis and the inflammasome represents a second major pathway that is controlled by serpins.

The serine proteases initiate a series of sequential reactions, illustrated best in the coagulation cascade which causes clot formation and thrombosis and which in turn is regulated by AT (SERPINC1). The thrombolytic or clot lysis pathway is also a serine protease cascade regulated by serpins, specifically plasminogen activator inhibitors-1, 2, and 3 as well as antiplasmin and protease nexin. The impact of serpin regulation of protease functions in normal pathophysiologic function is clearly illustrated in disease states where there is an imbalance in either the proteases or the serpins. Serpins with cross-class inhibitory activity are also seen wherein the serpin inhibits both *serine* and *cysteine* proteases. This serpin function has the potential to suppress activation of apoptotic pathways through blockade of GrB as well as inhibition of both the apoptotic and inflammasome pathways through cysteine protease (caspase) inhibition (Turner et al. 1999; Messud-Petit et al. 1998; Turner and Moyer 2001; Komiyama et al. 1994).

Serpin inhibitory activity is defined as a "suicide inhibition" (Fig. 17.2). With serpin inhibition, the reactive site loop (RSL) presents a P1–P1′ scissile bond to target proteases as a form of bait or trap. The targeted proteases then cleave the P1–P1′ bond, and in this process, the protease springs the serpin trap, becomes bound to the RSL, and the protease is dragged across the serpin face forming a new link in the A beta-sheet (Janciauskiene 2001; Lomas et al. 2005; Carrell and Travis 1985; Silverman et al. 2001; Marszal and Shrake 2006; Huntington et al. 2000; Patston et al. 2004; Ye et al. 2001; Lawrence et al. 2000; Whisstock and Bottomley 2006; Francis et al. 2012; Rashid et al. 2015; Stocks et al. 2012; Khan et al. 2011; Olson

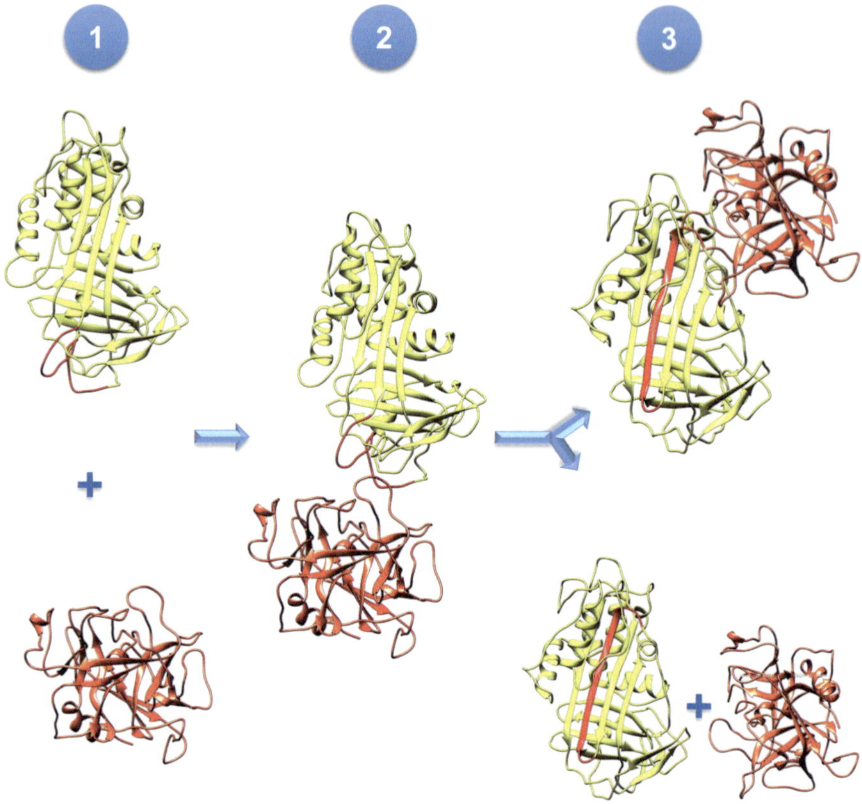

Fig. 17.2 Interaction of Serp-1 with tPA. The crystal structures with PDB IDs 3Q02, 1DVN, and 1K9O are used for homology modeling of relaxed Serp-1 (1), stressed Serp-1 (2), and Serp-1: tPA interaction (3), respectively. (1) Relaxed form of Serp-1 and free tPA (2) Proteolytic interaction between Serp-1 and tPA (3) Formation of covalently bound Serp-1: tPA complex or relaxed Serp-1 and active tPA as a result of the interaction

et al. 2010; Yamasaki et al. 2010). The targeted protease may first interact with an exosite, providing a structure that allows RSL cleavage to occur and loop insertion to proceed (Rashid et al. 2015; Stocks et al. 2012; Khan et al. 2011; Olson et al. 2010; Yamasaki et al. 2010; Shi et al. 2013; Khanaree et al. 2013). Regulation of the large numbers of serine proteases in a cascade such as the thrombotic and thrombolytic pathways thus requires an equally large number of regulatory serpins specific to each protease pathway. As serpins are suicide, one-to-one, inhibitors, there would be a need for relatively high proportional quantities of serpins to modify protease activation. Serpins are in fact reported to represent from 2% to 10 % of circulating plasma proteins.

In certain genetic abnormalities, the RSL of one serpin is inserted into the beta-sheet of an adjacent serpin causing formation of an aggregate of inactive and even pro-inflammatory serpins (Janciauskiene 2001; Lomas et al. 2005; Carrell and Travis 1985; Irving et al. 2000; Iwanaga and Kawabata 1998; Law et al. 2006;

Silverman et al. 2001; Marszal and Shrake 2006; Huntington et al. 2000; Patston et al. 2004; Ye et al. 2001; Lawrence et al. 2000; Whisstock and Bottomley 2006; Liu et al. 2014; Lucas et al. 2009; Francis et al. 2012; Rashid et al. 2015; Stocks et al. 2012; Khan et al. 2011; Olson et al. 2010; Yamasaki et al. 2010). The AAT genetic serpinopathy leads to severe lung disease as well as hepatitis and pancreatitis (Janciauskiene et al. 2011); neuroserpin genetic mutations cause epilepsy and dementia (Onda et al. 2005; Davis et al. 1999a, b; Bradshaw et al. 2001; Miranda and Lomas 2006), while mutation of C1 esterase causes angioedema, and AT mutation causes excess clot formation (Ishiguro et al. 2000; Conard and Samama 1986; Coller et al. 1987). Not all serpin/protease interactions necessarily lead to inhibition. In some cases, the serpin RSL is cleaved by the protease causing a loss of function of the serpin, while leaving the protease active. There are also serpins that lack serpin functions, the best known of which are ovalbumin, thyroxine-binding globulin, corticosteroid-binding globulin, and angiotensinogen, the precursor to the major hypertension regulating protein angiotensin II.

As noted above in addition to regulating the coagulation cascades, there are serpins involved in regulation of apoptosis, complement activation, neuron signaling, angiogenesis, and many other central response pathways. Serpins that block apoptosis are now believed to aid in development of T cells as well as hematopoietic lines in the bone marrow. Several mammalian serpins have been associated with effective inhibition of cancer growth as is reported for maspin (SERPINB5) and pigment epithelial-derived factor (PEDF; SERPINF1), although maspin's role in cancer biology has been contested (Shi et al. 2013; Khanaree et al. 2013).

17.4 Viral Serpins and Immunomodulation

The extraordinary roles of serpin regulation of key physiological pathways in innate immune responses are also illustrated by the markedly effective serpin suppression of host inflammatory response in myxoma virus infections (Viswanathan et al. 2006, 2009; Turner et al. 1999; Messud-Petit et al. 1998; Nash et al. 1997, 1998; Miller et al. 2000; Lucas et al. 1996, 2000; Chen et al. 2013b; Bedard et al. 2006; Bot et al. 2003; Christov et al. 2005; Dai et al. 2003, 2006; Davids et al. 2014; Liu et al. 2013; Lomas et al. 1993; Lucas and McFadden 2004; MacNeill et al. 2006; McFadden et al. 1995; Munuswamy-Ramanujam et al. 2010; Richardson et al. 2006; Seet et al. 2003; Tardif et al. 2010; Zalai et al. 2002). We have furthermore demonstrated very potent anti-inflammatory activity for two viral serpins, as well as one mammalian serpin (neuroserpin, SERPINI1) in a wide range of animal models of inflammatory vascular disease, transplant, cancer, and more recently mouse models of severe and lethal viral infection (Turner et al. 1999; Messud-Petit et al. 1998; Nash et al. 1997, 1998; Lomas et al. 1993; MacNeill et al. 2006; McFadden et al. 1995). In the following sections, we will discuss current knowledge on viral serpins (Table 17.1). Each viral serpin will be discussed from the perspective of

Table 17.1 Viral serpin and their mammalian counterparts regulate key physiological processes

Viral serpins	Protease targets	Virus origin	Mammalian serpins with related functions	Animal trial outcomes	Clinical trial	References
Serp-1	tPA, uPA, plasmin, FXa, thrombin	Myxoma virus	PAI-1 (SERPINE1), PAI-2 (SERPINB2), neuroserpin (SERPINI1), alpha-2 antiplasmin (SERPINF2), antithrombin (SERPINC1)	1. Reduced plaque angioplasty, stent, transplant, temporal arteritis, carotid compression 2. Reduced pancreatic cancer cell growth 3. Reduced collagen-induced arthritis 4. Improved survival MHV68 infections	1. Phase 1—safe, no adverse effects 2. Phase 2A—safe, reduced TN, CK-MB (makers myocardial damage)	Viswanathan et al. (2006, 2009, 2012), Miller et al. (2000), Lucas et al. (1996, 2000), Chen et al. (2013b, 2015), Bedard et al. (2006), Bot et al. (2003), Christov et al. (2005), Dai et al. (2003, 2006), Davids et al. (2014), Liu et al. (2013), Lucas and McFadden (2004), Munuswamy-Ramanujam et al. (2010), Richardson et al. (2006), Seet et al. (2003), Tardif et al. (2010), Zalai et al. (2002), Wang et al. (2003), Li et al. (2008)
SPI-3	tPA, uPA, plasmin, FXa, thrombin	*Vaccinia virus*, cowpox, monkeypox,	PAI-1 (SERPINE1), PAI-2, neuroserpin (SERPINI1), alpha-2 antiplasmin (SERPINF2)	1. Reduced plaque angioplasty	N/A	Turner et al. (2000)
Serpin ORF3	Plasminogen	Hepatitis E virus	Alpha-2 antiplasmin (SERPINF2)	N/A	N/A	Zhou et al. (2014)

Serp-2	GrB, caspase 1,8	Myxoma virus	PAI-2 (SERPINB2)	1. Reduced plaque aortic transplant, carotid compression 2. Reduced IR injury liver	N/A	Viswanathan et al. (2012), Turner et al. (1999), Messud-Petit et al. (1998), Turner and Moyer (2001), MacNeill et al. (2006), Thompson et al. (1993)
SPI-2, CrmA	GrB, caspase 1,8	Cowpox virus	PAI-2 (SERPINB2), PI-9 (SERPINB9)	N/A	N/A	Thompson et al. (1993), Kettle et al. (1995), Melo-Silva et al. (2011), Nathaniel et al. (2004)
SPI-1	Cathepsin G	Orthopox virus, *Vaccinia virus*, rabbitpox	N/A	N/A	N/A	Kettle et al. (1995), Brooks et al. (1995), Moon et al. (1999), Shisler et al. (1999)
M3	Chemokine not serine proteases	Mouse herpesvirus (MHV)	N/A	1. Reduced plaque aortic transplant	N/A	Alexander et al. (2002), van Berkel et al. (2000), Dai et al. (2010), Liu et al. (2004)
HESP018	Unknown	Baculovirus	N/A	N/A	N/A	Rohrmann et al. (2013)

N/A, not assessed to date; GrB, granzyme B; PAI, plasminogen activator inhibitor; CK, creatinine kinase; TN, troponin virus-mediated evasion of host inflammatory responses as well as potential efficacy in inflammatory diseases. These serpins will also be discussed with reference to mammalian serpins that regulate similar protease pathways (Table 17.1). There are in fact suggestions that these virus-derived serpins were originally acquired from the mammalian genome. Others have suggested, given the markedly greater numbers of microbes that colonize mammals, that some mammalian serpins may have originated from the viral genome. Large DNA viruses have developed powerful strategies to modify immune responses. The high level of function of many of these immune-modulating serpins is due to the fact that these serpins have evolved over many millions of years to provide effective protective strategies for virus survival. Among the poxvirus immune targets, as might be predicted, are the serine proteases in the coagulation, inflammasome, and apoptotic pathways. The poxviruses and specifically smallpox (variola) were one of the most deadly scourges before vaccination was developed. Poxviruses are highly effective viral invaders. There is still fear that variola will be developed as a bioweapon, and stocks are carefully guarded by the World Health Organization. Thus, viruses have evolved extremely effective immune-modulating strategies, and one such stratagem is the expression of viral serpins that regulate thrombotic and thrombolytic protease cascades (Fig. 17.3). A second viral survival stratagem involves the expression of cross-class serpins that block both cysteine proteases and caspases in the inflammasome pathways, as well as serine and cysteine proteases, GrB, and caspase-8, respectively, in the apoptotic cascade (Fig. 17.3).

Viral serpins significantly alter viral pathogenesis through blockade of the host immune response, the host defenses mounted against the invading virus. The impact of one myxoma viral serpin is illustrated by its role in myxoma viral infections. Myxoma virus infects European rabbits, *Oryctolagus cuniculus*, with 80–100 % mortality. In these infected animals, the virus effectively blocks the host immune response allowing for subsequent severe bacterial sepsis and early death. In a natural variant of myxomavirus or in viruses genetically engineered to lack this serpin, the infection becomes a benign local infection that is cleared within a few weeks (Nash et al. 1997, 1998; Lomas et al. 1993; McFadden et al. 1995). Thus, the myxomavirus has developed very effective immune-modulating serpins that target and block central immune response pathways demonstrating the outstanding capacity of these virus-engineered immunomodulating molecules. In some cases, these immune-modulating viral proteins indicate new or previously underappreciated protease pathways critical to normal innate and acquired immune responses.

Large DNA viruses such as the pox and herpesviruses induce many factors that are instrumental in modifying activation of innate and acquired immune responses. Thus, these larger DNA viruses can block host immune responses designed by the host to reduce viral infection and invasion. In some cases, the viruses literally hijack the host immune cell responses riding along in infected cells in order to disseminate the infectious particles throughout the host body.

The virus-derived serpins have profound effects on the outcomes of viral infections (Turner et al. 1999; Messud-Petit et al. 1998; Nash et al. 1997, 1998;

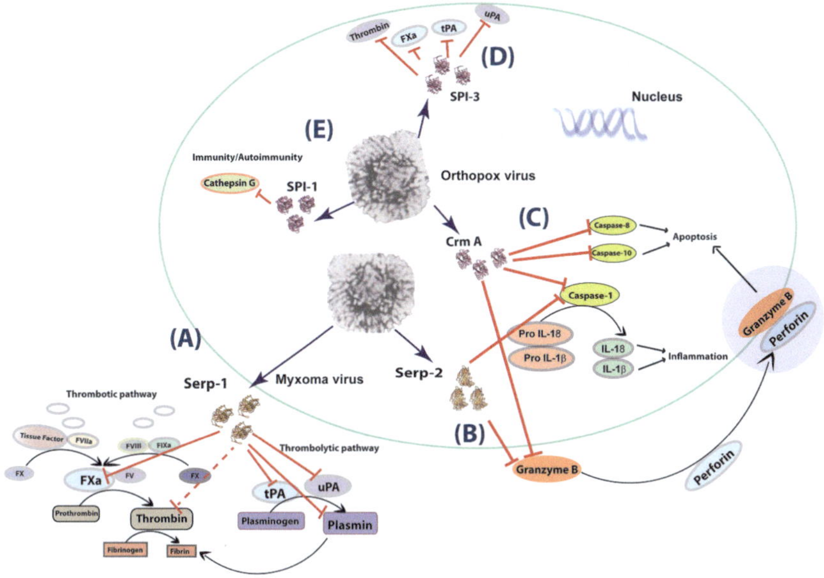

Fig. 17.3 Figure depicting poxviruses (EM micrograph of poxvirus), virus-expressed serpins, and the serpin targets. (A) Serp-1, a myxoma virus-secreted serpin, inhibits tPA, uPA, and plasmin of the thrombolytic pathway and FXa and thrombin of the thrombotic pathway. (B) Serp-2, cytosolic myxoma viral serpin, inhibits caspase-1 and its downstream inflammatory effects along with inhibiting granzyme B which along with perforin enters the cell and contributes to apoptotic cell death. (C) CrmA or SPI-2 is a cytosolic orthopox viral protein capable of inhibiting caspase-1 and granzyme B like Serp-2 along with inhibiting apoptotic caspases 8 and 10. (D) SPI-3 is a cytosolic orthopox viral protein capable of inhibiting proteases in thrombotic and thrombolytic pathways much like Serp-1. (E) SPI-1, another cytosolic orthopox viral protein, can affect the immune and autoimmune pathways of the cell by its inhibition of cathepsin G. *Arrow* indicates virus expression of serpins and targeted pathways. *Solid line with bar* indicates inhibition. *Dotted line with bar* indicates postulated inhibition

Lomas et al. 1993; MacNeill et al. 2006; McFadden et al. 1995). A few virus-derived serpins have also been tested in vivo in animal models and have been highly effective at reducing inflammation and disease progression for disorders from inflammatory vascular disease and atherosclerosis, transplant rejection and vasculopathy, cancer growth, and arthritis (Viswanathan et al. 2006, 2009, 2012; Lucas et al. 1996, 2000; Chen et al. 2013b; Bedard et al. 2006; Bot et al. 2003; Christov et al. 2005; Dai et al. 2003, 2006; Davids et al. 2014; Liu et al. 2013; Lucas and McFadden 2004; Munuswamy-Ramanujam et al. 2010; Richardson et al. 2006; Seet et al. 2003; Zalai et al. 2002). One small phase 2A clinical trial has been successfully completed, demonstrating safety and efficacy in patients with coronary stent implants in the setting of small heart attacks and unstable angina (Tardif et al. 2010).

There are three serpins encoded by myxoma virus, two of which have been studied in depth in animal models Serp-1 and Serp-2. Serp-1 is a secreted serpin that binds and inhibits thrombotic and thrombolytic serine proteases (Viswanathan et al. 2006, 2009; Nash et al. 1997, 1998; Miller et al. 2000; Lucas et al. 1996, 2000; Chen et al. 2013b; Bedard et al. 2006; Bot et al. 2003; Christov et al. 2005; Dai et al. 2003, 2006; Davids et al. 2014; Liu et al. 2013; Lomas et al. 1993; Lucas and McFadden 2004; McFadden et al. 1995; Munuswamy-Ramanujam et al. 2010; Richardson et al. 2006; Seet et al. 2003; Tardif et al. 2010; Zalai et al. 2002), and Serp-2 is an intracellular serpin that inhibits serine and cysteine proteases in the apoptotic and inflammasome pathways (Viswanathan et al. 2012; Turner et al. 1999; Messud-Petit et al. 1998; Davids et al. 2014; MacNeill et al. 2006). Both serpins markedly increase the pathogenesis of myxoma infections in rabbits.

17.4.1 Viral Serpins That Target the Thrombotic and Thrombolytic Proteases

17.4.1.1 Serp-1

The impact of Serp-1 expression on myxoma viral infection in European rabbits is clearly illustrated by the effect of the presence and absence of a functional Serp-1 gene in the virus. Wild-type myxoma with normal Serp-1 gene expression causes a lethal infection with 80–100 % mortality in European rabbits, while in the Serp-1 defective variant of myxoma virus, Shope fibroma virus, or in viruses genetically engineered to lack Serp-1(S-1 knockout myxoma virus), the infection is benign (Nash et al. 1997, 1998; Lomas et al. 1993; McFadden et al. 1995).

Serp-1 is a 55 kDa secreted serpin, the only known secreted viral serpin, that binds and inhibits tissue- and urokinase-type plasminogen activators (tPA and uPA, respectively), plasmin, and factor Xa. Serp-1 also inhibits thrombin at select concentrations of heparin in vitro. The K_{ass} for Serp-1 is low at 10^3 to 10^5 for each protease (Lomas et al. 1993). Treatment with purified Serp-1 protein has now been tested in a very wide range of models of inflammatory disorders (Viswanathan et al. 2006, 2009; Miller et al. 2000; Lucas et al. 1996, 2000; Chen et al. 2013b; Bedard et al. 2006; Bot et al. 2003; Christov et al. 2005; Dai et al. 2003, 2006; Davids et al. 2014; Liu et al. 2013; Lucas and McFadden 2004; Munuswamy-Ramanujam et al. 2010; Richardson et al. 2006; Seet et al. 2003; Zalai et al. 2002) and has been successfully tested in a small phase 2A randomized, dose escalating, double-blind clinical trial (as mentioned above) in patients with unstable angina, and non-ST elevation myocardial infarction (non-STEMI) at 7 sites in Canada and the USA (Zalai et al. 2002). Each patient had a documented coronary plaque lesion with significant stenosis considered appropriate for stent implant. After the decision was made to proceed with stent implant, patients were randomized to treatment with either sterile saline vehicle or Serp-1 (VT-111; Viron Therapeutics, Inc., London, Canada). In this clinical trial, there was a significant reduction in markers of myocardial damage (troponin and CK-MB) early after stent implant with Serp-1 treatment but with no detected decrease in plaque size on intravascular ultrasound (IVUS) interrogation. The major adverse clinical event (MACE) for Serp-1 treatment was zero, these studies thus demonstrating some efficacy and safety.

Serp-1 treatment has been demonstrated to reduce inflammation and disease progression in aortic allograft models of chronic transplant vasculopathy, a leading cause of transplant failure after the first year posttransplant (Miller et al. 2000; Lucas et al. 2000; Dai et al. 2006). Serp-1 treatment was also efficacious in acute and chronic renal (Bedard et al. 2006) and cardiac allografts (Jiang et al. 2007) and even a xenograft (Wang et al. 2003) transplant model, significantly reducing inflammation, vasculitis, arterial plaque, and scarring. In the aortic allograft model, Serp-1 reduced engrafted aortic inflammatory cell invasion and intimal plaque in engrafted mouse aortas with knockout of mammalian serpin, plasminogen activator inhibitor-1 (PAI-1), but not in engrafted uPAR-deficient mouse aortas (Dai et al. 2003). This indicated that Serp-1 inhibits vasculitis at least in part through the uPAR pathway in the aortic transplant model. More recently, Serp-1 was found to reduce growth of temporal artery (TA) xenograft implants in the abdominal aorta of severe combined immunodeficient (SCID) mice lacking T and B lymphocytes. The TA specimens were derived from human temporal artery biopsies isolated from biopsies taken from patients with suspected temporal arteritis and giant cell arteritis (Chen et al. 2015). In prior work spanning 20–25 years, Serp-1 treatment has thus proven effective in mouse, rat, rabbit, and microswine models of aortic and coronary angioplasty and stent implant, inflammatory vasculitis, and also transplant vascular disease (Viswanathan et al. 2006, 2009, 2012; Miller et al. 2000; Lucas et al. 1996, 2000; Chen et al. 2013b, 2015; Bedard et al. 2006; Bot et al. 2003; Christov et al. 2005; Dai et al. 2003, 2006; Davids et al. 2014; Liu et al. 2013; Lucas and McFadden 2004; Munuswamy-Ramanujam et al. 2010; Richardson et al. 2006; Seet et al. 2003; Tardif et al. 2010; Zalai et al. 2002; Wang et al. 2003; Li et al. 2008) with proven safety in a small clinical trial in unstable coronary plaque (Zalai et al. 2002).

Serp-1 and Collagen-Induced Arthritis Models

The rodent collagen-induced arthritis (CIA) model illustrates how fibrinolytic enzymes such as plasminogen and uPA are important components of inflammatory responses, contributing to the pathogenesis in arthritis. Increased expression of uPA and the uPA receptor (uPAR) is seen on joint macrophage and synoviocytes in RA. Thus, the thrombolytic protease pathways have a role in activating early innate immune responses in arthritis as well as in vasculitis and cancer. uPA and tPA also activate plasminogen to form plasmin, and these thrombolytic proteases then activate matrix metalloproteinases (MMPs). The MMPs are active in connective tissue degradation, which allows inflammatory cells to invade joint tissues and causes further damage. Control of these pathways by serpins represents a new regulatory approach to reducing joint inflammation and damage and potentially reducing lytic lesions. Serp-1 has been used in models of both rat and rabbit CIA and ovalbumin-induced arthritis with effective decreases in joint swelling as well as lytic lesions and DTH (Brahn et al. 2014; Maksymowych et al. 1996).

Serp-1 and Tumor-Associated Macrophages (TAM)

In a small study using human pancreatic cell lines, Serp-1 treatment also reduced tumor weight for two differing pancreatic cell lines implanted in SCID mice (Zheng et al. 2013). In this model, Serp-1 treatment reduced tumor weight and associated invasive tumor macrophages at 4 weeks. Additionally, serpin treatment reduced the pancreatic cancer-driven increase in spleen cell myeloid-derived suppressor cells that the tumor uses to block immune responses to the foreign cancer cells. Other tumor cell lines tested in this model, e.g., breast cancer and hepatoma, were not affected by treatment with this viral serpin.

Serp-1 Modulation of Unrelated Viral Infections

In one of the more recent studies, the effects of Serp-1 on vasculitis and survival in a lethal mouse herpesvirus 68 (MHV68) infection were examined, and the outcome is particularly interesting. This study demonstrates that Serp-1 treatment improves survival in lethal MHV68 infections in interferon-gamma receptor knockout mice (Chen et al. 2013b). In the MHV68 infection model, there is 80–100 % mortality over 30–40 days. With Serp-1 treatment for 10 up to 30 days, there was a significant improvement in survival ($P < 0.04$). Serp-1 also improved survival in a prior pilot study with *mouse-adapted Ebola* infection (Chen et al. 2013b), but not in Ebola-infected hamsters (personal communication, H Feldmann, NIH Rocky Mountain Labs). The improved survival is associated with a significant reduction in pulmonary hemorrhage and inflammatory consolidation, as well as reduced macrophage invasion in the lung and aorta. There is a trend toward reduced T-cell invasion, but this is not significant. Of even greater interest, there is a reduction in tissue-associated MHV68 antigen and a similar reduction in detected Ebola genomes in the two mouse models of lethal viral infections. In the MHV68 infection, the related mammalian serpin, NSP, was ineffective. NSP inhibits only thrombolytic pathways but not FXa or thrombin. These findings suggest that Serp-1 improves lung consolidation and hemorrhage as well as vasculitis through blockade of either FXa or thrombin rather than uPA/uPAR or tPA in this disease model.

Reduction in the bleeding time, a measure of hemorrhagic tendency, was also detected. The decrease in bleeding time with Serp-1 treatment was half that seen with NSP, and an associated reduction in FXa tissue levels was also detected. We also demonstrated a significant increase in interleukin-10 (IL-10) and several mammalian serpins.

Serp-1 and the Virome

This marked efficacy of a myxoma virus-derived serpin to potentially modify the disease course of two severe viral infections, one by a DNA virus and one by an RNA virus, would suggest that this serpin function may represent a form of viral defense against competing viral invaders that vie for host resources. Recent literature has demonstrated that both bacterial and viral infections can modify the immune response, whether innate or acquired, on a more broad scale. Commensal microorganisms that live on and within a host animal are now known to alter infections by other pathogens. The native mammalian host is normally colonized by a large number of bacteria and one presumes also by viruses (Lucas et al. 2014b; Norman et al. 2015; Virgin 2014). This microbiome of colonizing bacteria and the virome of colonizing viruses living inside the gastrointestinal and respiratory tracts and also on the dermis can theoretically modify host immune responses to foreign infectious pathogens. We now understand that there is approximately tenfold more DNA present in the microbiome than there is human DNA. The viral serpins may thus not only serve as a response to the host attack on the invading myxoma virus but may also allow the virus to combat other viruses invading their space. The microbiome has become an area of intense interest and investigation. The microbiome usually refers more to the bacterial colonization of the human oral, GI, or respiratory tracts, capable of modifying host inflammatory responses. The virome is also now of interest, and certainly, this newly observed function of Serp-1 treatment in modifying severe viral infections would imply potential functions for virus-derived immune-modulating agents that may work to alter the capacity of other viruses to infect and invade a mammalian host (Lucas et al. 2014b; Norman et al. 2015). We have not as yet examined whether infection with this rabbitpox virus does modify the host virome. However, we have demonstrated that treatment with a viral anti-inflammatory protein such as Serp-1 can reduce plaque growth in ApoE$^{-/-}$ mice with active periodontal bacterial infections (Lucas et al. 2014a). A recent study also demonstrated reduced plaque growth in microswine after infection with swine pox (Shimamura et al. 2012). Both these studies support, in part, a role for virus-derived serpin modulation of unrelated pathogenic microorganisms.

Serpin-Derived Reactive Site Loop (RSL) Peptide Functions

Serp-1 has also demonstrated the ability to extend immune-modulating functions via binding to host receptors such as the uPA receptor (uPAR) and interaction with associated uPAR lipid raft molecules (Viswanathan et al. 2009). In one prior study, Serp-1 was found to pull down together with the uPAR and also with filamin B. Altered filamin B expression was detected in cultured mononuclear cells by microarray and reverse transcriptase-polymerase chain reaction (RT-PCR) analysis. Filamin B is an actin-binding protein associated with the GPI-linked uPAR with a role in macrophage invasion. Small inhibitory ribonucleic acid (siRNA) knockdown of filamin decreased Serp-1 efficacy in reducing macrophage migration in vitro in Boyden chambers. The interaction of Serp-1 with uPAR and associated membrane proteins in the uPAR lipid raft can induce changes in expression of selected genes in targeted inflammatory cells, specifically human monocytes and T and endothelial cell lines.

We have also recently detected an expanded anti-inflammatory repertoire for Serp-1 in a series of naturally processed RSL-derived peptides (unpublished data), suggesting that serpins extend their anti-inflammatory functions through natural peptide metabolites. Other peptides derived from mammalian serpins have been reported to modify sepsis and infections. Angiotensinogen is a protein with serpin structure but lacking serpin inhibitory activity. Angiotensinogen is cleaved by renin and subsequently the angiotensin-converting enzyme (ACE) to form angiotensin I and angiotensin II, respectively. Angiotensin II is a potent vasoconstrictor with a central role in regulation of vascular tone and blood pressure. C36 is a C-terminal peptide-derived AAT (SERPIN A1) that induces a pro-inflammatory response in human monocytes, together with receptor interaction and the capacity to alter gene expression. One AAT C-terminal peptide is also reported to have antiviral functions, inhibiting HIV-1. The N-terminus of a second mammalian serpin, heparin cofactor II (SERPIN D1), is reported to block endotoxin activity. However, anti-inflammatory functions of RSL peptides have not been previously reported.

17.4.1.2 SPI-3

Serine protease inhibitor-like protein 3, or SPI-3, is an orthopox serpin belonging to clade O of the serpin superfamily with arginine as its P1 residue. SPI-3 is expressed by vaccinia, cowpox (Turner et al. 2000; Turner and Moyer 2008), monkeypox (Farlow et al. 2010), and also rabbitpox viruses and has a potential role in cell fusion after infection. SPI-3 may act to prevent superinfection by orthopoxviruses (Turner and Moyer 2008). Similar to the protease inhibitory spectrum of myxoma viral Serp-1, purified SPI-3 inhibits plasmin, urokinase-type, and tissue-type plasminogen activators with inhibition constants of 0.64, 0.51, 1.9 nM, an inhibitory effect that is lost with P1–P1′ mutations (Turner et al. 2000). SPI-3 also inhibits FXa and thrombin, albeit with weaker complex formation (Turner et al. 2000). *Vaccinia virus* with disrupted SPI-3 open reading frame has equivalent replication and pathogenesis to its wild-type counterpart in mice, suggesting a limited role for SPI-3 in pathogenesis (Law and Smith 1992). The SPI-3 knockout in vitro however causes fusion of the infected cells in certain cell lines (Law and Smith 1992; Zhou et al. 1992). Subsequently, similar disruptions in SPI-3 in the cowpox virus (BR strain) and rabbitpox virus (HA+ strain) were shown to result in a fusion phenotype (Turner and Moyer 1992). The fusion phenotype did not seem to depend on the serine protease inhibitory activity of SPI-3 as site-directed mutagenesis in the crucial RSL at P1 to P1′ positions, P5 to P5′ positions, or changes in P17 to P10 residues did not result in loss of activity or fusion inhibition (Turner and Moyer 1995). Of interest, substitution of SPI-3 for Serp-1 in myxoma virus did not replace Serp-1-mediated anti-inflammatory activity in rabbit infections (Wang et al. 2000). Despite the fact that SPI-3 lacks the capacity to replace Serp-1 in myxoma virus-infected rabbits, in one pilot study (unpublished work of Dr. R Moyer), SPI-3 expression from AAV vectors was able to reduce arterial plaque in a rodent atherosclerosis model.

17.4.1.3 HEV/Hepatitis E Virus Serpin

Protease and serpin interactions for open reading frame 3 (ORF3) in hepatitis E virus (HEV) have recently been identified through co-immunoprecipitation (co-IP). Binding to plasminogen and alpha-2 antiplasmin (AP; Serpin F2) was detected (Zhou et al. 2014). As noted above, conversion of plasminogen to plasmin can enhance macrophage activation and increase inflammatory cell responses to viral infections. In herpesvirus infections, enhanced ORF3 is believed to generate increased kinase and growth factor signaling and reduced mitochondrial cell death pathways with improved cell survival, with the potential to increase cell infection. ORF3 is also reported to modulate host innate immune responses. Herpesviruses increase plasmin generation which in turn enhances herpesviral infections. Whether HEV ORF3 interaction with plasminogen or AP modifies inflammatory cell responses, as for herpes, it remains to be determined.

17.4.2 Mammalian Serpins with Related Targeting of Thrombotic and Thrombolytic Cascades

A few selected mammalian serpins with functions potentially related to the viral serpins will be discussed here (Table 17.1).

17.4.2.1 Plasminogen Activator Inhibitor-1 (PAI-1; SERPIN E1)

PAI-1 is considered the main regulator for thrombolytic serine proteases in the circulation, preventing excess clot lysis and bleeding. PAI-1 binds to and inhibits the uPA/uPAR complex causing internalization and reduced uPA/uPAR pro-inflammatory activity. PAI-1 also binds tPA-reducing thrombolysis and potentially reducing tPA-mediated inflammation through binding to protease-activated receptors (PAR). In vascular injury models, Carmeliet and Collen demonstrated that PAI-1-deficient mice had increased inflammatory arterial plaque after arterial injury in some rodent models that was reduced by increased expression of PAI-1 (de Waard et al. 2002; Devy et al. 2002; Nagai et al. 1999). Other rodent models demonstrated an opposite effect for PAI-1 with increased plaque and smooth muscle cell proliferation when PAI-1 was expressed. Dichek et al. have studied other aortic injury models demonstrating that uPA and PAI-1 expression alter plaque growth and aneurysm formation in part through modification of MMP activity (de Waard et al. 2002; Devy et al. 2002). MMPs are activated from the pro-form by thrombolytic proteases, tPA, uPA, and plasmin. Other models have conversely demonstrated increased inflammatory responses with PAI-1 expression. Altered expressions of uPA, tPA, uPAR, and PAI-1 have been reported as markers for inflammation and aggressive atherosclerotic vascular disease and also cancer growth and metastasis in some tumors. PAI-1 has furthermore been found to alter angiogenesis in a dose-dependent manner. Thus, Serp-1 efficacy in moderating plaque growth and vascular inflammation does parallel the effects of PAI-1.

17.4.2.2 Plasminogen Activator Inhibitor-2 (PAI-2/SERPIN B2)

PAI-2 is a major product of activated monocyte/macrophage cells and is induced during infection and inflammation together with close associations with cancer cells and metastases (Chuang et al. 2013; Lomas and Carrell 2002; Schroder et al. 2014). PAI-2 as for PAI-1 is an inhibitor of the thrombolytic proteases but is considered a less potent inhibitor of the clot lytic cascade than PAI-1. Analyses of published microarray studies have detected higher levels of PAI-2 expression (mRNA) in monocytes from HIV-1-infected patients. PAI-2 is also purported to modulate inflammasome/IL-1β activation in macrophage and Th1/Th2 responses, but the physiological role in mammals for PAI-2 remains elusive as absence of PAI-2 does not markedly alter thrombolytic activity in vivo (Chuang et al. 2013).

17.4.2.3 Alpha-1 Antiplasmin (SERPINF2)

Plasmin is activated by tPA and uPA and can in turn activate MMPs and release select growth factors. Thus, plasmin is predicted to induce inflammatory cell invasion and activation, and AP might be expected to reduce inflammation. Deficiency of plasminogen has been reported to reduce excitotoxin-induced nerve damage in mouse models (Nagai et al. 1999). AP deficiency conversely induced increased stroke volumes in mice after middle cerebral arterial ligation. In contrast, local injection of either plasmin or AP reduced nerve damage in response to injected excitotoxin such as a glutamic acid analog.

17.4.2.4 Neuroserpin (NSP/SERPINI1)

NSP binds and inhibits tPA and uPA, but not thrombotic proteases. Genetic mutations of NSP cause a serpinopathy with development of seizures (epilepsy) and dementia (Lomas and Carrell 2002). Mice with genetic deficiency of NSP have increased stroke area and worse neurological outcome after middle cerebral artery occlusion in murine stroke models (Gelderblom et al. 2013), whereas mice-expressing NSP have been documented to have smaller size strokes. We have demonstrated that treatment of mice after aortic allograft transplant with purified NSP protein, but not with an inactive P1–P1′ mutant NSP protein, significantly reduces inflammatory arterial plaque (Munuswamy-Ramanujam et al. 2010). This reduction in plaque area is associated with a decrease in Th1/Th2 ratios. This suggests a decrease in the pro-inflammatory CD4+ T-cell population. In this model, the mammalian serpin, NSP, reduced inflammation and plaque in aortic allografts. NSP binds and inhibits tPA and uPA, but not thrombotic proteases FX and thrombin, suggesting that Serp-1 and NSP block inflammatory vasculopathy in mouse aortic transplant through blockade of uPA and/or potentially tPA activity. Work with a series of predicted naturally processed NSP RSL peptides has also displayed some inflammation-modulating ability (unpublished observations).

17.4.2.5 Antithrombin III (AT, SERPIN C1)

Thrombin is now known to induce pro-inflammatory responses via the PARs (Chen and Dorling 2009; Chu 2010; Esmon 2005). AT (SERPINC1) is the main inhibitor of thrombin and factor X (FX) in the mammalian circulation. In fact, AT is the target for the anticoagulant drug heparin, a glycosaminoglycan (GAG) that increases AT activation 100- to 1000-fold when given into the circulating blood. Mutations of AT also produce a serpinopathy, which is associated with excess thrombosis (Lomas and Carrell 2002). Patients with DIC in severe sepsis have been treated with AT alone or AT with heparin (Hoffmann et al. 2006). In two studies, AT treatment did improve outcomes in sepsis with associated hemorrhagic or thrombotic (DIC) complications, but not when heparin was given concomitantly.

17.4.3 Viral Cross-Class Serpins That Target the Apoptotic and Inflammasome Pathways

17.4.3.1 Serp-2

Serp-2 is a 34 kDa cross-class myxoma virus-derived serpin that binds and inhibits the serine protease, GrB, and also the cysteine proteases, caspases 1 and 8 (K_{as}s—0.96×10^5 for caspase 1) (Turner et al. 1999; Messud-Petit et al. 1998; Turner and Moyer 2001; MacNeill et al. 2006; Thompson et al. 1993). Thus, Serp-2 is capable of inhibiting apoptotic responses in cells through binding of GrB and caspase 8, as well as inhibiting the inflammasome through binding and inhibition of caspase 1. The apoptotic response in an infected cell represents a mechanism by which the infected cell commits suicide and destroys the infecting virus. Serp-2 when present in myxomavirus increases viral pathogenesis.

Serp-2 is an intracellular serpin that is not predicted to function as an extracellular protease inhibitor. Although it has been reported to be adherent to the extracellular surface of cells it is not generally reported to be found free in the circulation. However, viral proteins often are multifunctional and to have extended actions. We thus theorized that Serp-1 might work outside the infected cell to block inflammatory responses and additionally to block apoptosis when released by an infected cell after virus-induced lysis.

We detected a significant reduction in inflammatory plaque in allograft transplants in wild-type (C57Bl/6 mice, WT) allografts, but not in GrB-deficient aortic transplants engrafted into BALB/c mice (Viswanathan et al. 2012).

In a collaborating lab (E Biessen and I Bot, U Maastricht and Leiden U, the Netherlands, respectively) (Viswanathan et al. 2012), Serp-2, but not CrmA, inhibited aortic root plaque in ApoE$^{-/-}$ mice. Furthermore, in a liver ischemia-reperfusion (IR) injury model in normal mice, Serp-2 significantly improved survival and reduced histological changes in an ischemia reperfusion (IR) liver model in WT mice. Caspase-1-deficient mice were also protected against IR liver injury in mice, but caspase 1 deficiency was not as effective as Serp-2 treatment in wild-type (WT) livers after IR injury (unpublished observations).

17.4.3.2 SPI-2/CrmA

CrmA is an intracellular anti-apoptotic cross-class cowpox virus serpin with targets very similar proteases to Serp 2 (Messud Petit et al. 1998; Turner and Moyer 2001; MacNeill et al. 2006; Thompson et al. 1993; Ali et al. 1994; Kettle et al. 1995; Melo-Silva et al. 2011; Nathaniel et al. 2004). CrmA, however, has greater in vitro biochemical inhibitory function than Serp-2 for caspases 1 and 4, 5, 9, and 10 and also GrB (K_{as}s 170×10^5 for caspase 1). CrmA increases virulence in infected BALB/c mice and inhibits inflammation and apoptosis and in chicken embryonic chorioallantoic membrane studies (MacNeill et al. 2006; Nathaniel et al. 2004). CrmA inhibits apoptosis in swine cells. CrmA also blocks cytotoxic T cell (CTL)-induced apoptosis. However, in vivo in viral infection, Serp-2 deletion leads to a benign myxoma virus infection with 10 % or less mortality, while CrmA replacement of Serp-2 still allows 70 % mortality rather than the 90–100 % seen with Serp-

2 expression (Thompson et al. 1993). Interestingly, CrmA does not modify inflammatory changes in engrafted aortic allografts nor plaque growth in carotid crush injury, unlike Serp-2, nor did it favorably modify the host T-cell response as determined by flow cytometry analysis of spleen cells (Viswanathan et al. 2012). In vitro control of ectromelia infection, a mouse poxvirus is mediated by CTL death, and GrB is a CTL death effector for infected cells (Melo-Silva et al. 2011). Of further interest is the fact that the ectromelia mouse poxvirus which is also a member of the *Poxviridae* family encodes a CrmA serpin, a cross-class serpin that inhibits GrB and several caspases as noted above for CrmA from cowpox. Thus, here a native mouse, mammalian protease regulatory serpin PI-9, and a virus-expressed serpin CrmA target similar serine protease pathways.

17.4.4 Mammalian Serpins That Target the Inflammasome and Apoptosis: PI-9

CrmA and Serp-2 are cross-class inhibitors similar to the human serpin proteinase inhibitor 9 (PI-9, SERPINB9). GrB is inhibited by PI-9, which is reported to control GrB. Activated cytotoxic lymphocyte (CL) cell death is regulated in part by PI-9 which is an intracellular protein, and PI-9 is also found associated with GrB in circulating complexes (Bird et al. 2014). Visceral adipose tissue-derived serpin (vaspin) is another mammalian serpin that inhibits inflammation and apoptosis and is associated with sepsis (Motal et al. 2015). Among the other granzymes, GrA is inhibited by the serpin Kazal, and extracellular GrA is inhibited by AT and a2-macroglobulin. GrH is inhibited by SERPINB1. GrM is inhibited by SERPIN B4 as well as AAT and AP (de Waard et al. 2002). PAI-2 has been identified as a regulator of NLRP3 inflammasome dependent caspase-1 and its downstream effector cytokine IL-1β and is found to modulate the cellular apoptotic response to pathogen or damage related signalling (Chuang et al. 2013).

17.4.5 Viral Serpin That Targets Cathepsin G

17.4.5.1 SPI-1

SPI-1 is a 40 kD early orthopoxvirus serpin protein belonging to clade N of the serpin superfamily understood to be nonessential for the virulence of cowpox, rabbitpox, and vaccinia viruses (Brooks et al. 1995; Moon et al. 1999; Shisler et al. 1999). It was subsequently shown that a rabbitpox virus mutant with a knockout of SPI-1 reduced the host range for rabbitpox in pig kidney cells (PK-15) and human lung carcinoma cell lines (A549) (Kettle et al. 1995). SPI-1 is also reported to reduce apoptosis and to prevent CTL-induced cell killing (Moon et al. 1999). The reduction in virulence was reported to be caused by the death of the infected cells through a nonclassical or caspase-independent apoptotic pathway that did not include the terminal activation of caspase-3 (Brooks et al. 1995; Moon et al. 1999). SPI-1 forms a complex with cathepsin G, in vitro, and this activity is abolished by mutations in the RSL in the P1 and P14 residues. The serpin function of SPI-1 appeared to be important in regulating host range as these rabbitpox mutant viruses with altered P1 and P14 residues did not grow in A549 cells (Shisler et al. 1999). A vaccinia virus mutant with deleted SPI-1 also has restricted growth in human keratinocytes and A549 cells, further demonstrating the importance of SPI-1 in viral infections.

17.4.6 Other Viral Serpins

17.4.6.1 M3

MHV68 expresses a viral serpin that functions as a chemokine-binding protein with no known serpin inhibitory activity (Alexander et al. 2002; van Berkel et al. 2000). M3 protein is a known chemokine-modulating protein expressed by MHV68. M3 is a highly active chemokine modulator that binds and inhibits both the chemokine to receptor and the chemokine to glycosaminoglycan-binding functions. This M3 protein effectively inhibits aortic transplant vascular plaque and inflammatory cell invasion in a chemokine receptor-dependent function (Dai et al. 2010; Liu et al. 2004). Of great interest, D. Freemont, Washington University St Louis, has demonstrated that M3 has a serpin sequence homology, but is modified such that the structure of M3 is now a chemokine-modulating protein.

17.4.6.2 Baculoviral Serpin

A serpin ortholog has recently been detected in baculovirus (group II alphabaculovirus) representing the first serpin expressed outside of the *Poxviridae* (poxvirus family) (Rohrmann et al. 2013). The baculoviral gene HESP018 has approximately 34 % amino acid sequence homology to serpins from host insects suggesting the gene was captured from an infected insect. One of the main antiviral responses in insects involves melanization which involves a serine protease cascade which catalyzes phenol oxidation. Oxidized phenols then polymerize and form melanin which encapsulates foreign bodies. Although the role of the baculoviral serpin is not known, it may interfere with host defenses against viral invasion in Lepidoptera.

17.4.7 Viral Serpins and the Virome

Natural serine protease inhibitors (serpins) are now reported to detect intruding microbes with attendant activation of cellular immune responses, and serpins are now predicted to represent part of the host innate immune response to viral infections. These innate immune responses include lectins, soluble CD14, defensins, and antimicrobial peptides. This early inflammatory response includes the mononuclear cells, neutrophils, monocytes, and T lymphocytes as well as the endothelial cell barrier all of which cell activity may be modified by serpins targeting various protease pathways. We would postulate that viruses have adapted these serpin anti-inflammatory mechanism to provide their own defense system, not only against the host immune response but also to prevent infection and invasion by competing viruses (Norman et al. 2015; Virgin 2014).

This serpin-mediated soliciting of host responses to pathogens is reported for human immunodeficiency virus (HIV), herpes simplex virus (HSV), and hepatitis C virus (HCV)-infected cells. Recent work has reported a role for AT in early responses to herpes simplex virus (HSV). AT reduced mortality in an abrasion model in mice after HSV infection. As noted above, the coagulation cascade is now more and more recognized as part of an early innate immune response. AT (SERPINC1) has also been reported to reduce viral titers in vitro and to reduce mortality by 20 % in mouse-adapted H1N1 infections, further promising inhibitory activity for a range of viral infections. Local epithelial mucosal surface serpins also decrease susceptibility to HIV infections. SERPINA3 abrogates activity of the serine proteases cathepsin G and elastase that enhance macrophage susceptibility to HIV infection and promotes pro-inflammatory cytokine expression, with recruitment of HIV target cells, including T cells.

Serpins demonstrate primarily indirect antiviral mechanisms, through interference with host proteases. Serpin A1/alpha-1 antitrypsin (AAT, SERPINA1) however binds and inhibits HIV. Interaction of AAT, with its target protease, leads to cleavage of a C-terminal fragment, designated C-36. C-36 is an inhibitor for gp120. This fragment has also recently been identified as a gp41 fusion protein inhibitor and is under investigation in a phase I/II clinical trial as antiretroviral therapy. AAT (SERPINA1) also interferes with viral protein late processing, through interference with gp120 formation and processing of p55 (gal) to p24 (Aboud et al. 2014). AAT and Serpin C1 (AT) both interfere with HIV infection through obstruction of innate immunity and nuclear factor κB (NFκB) activity. Mutations in the AAT gene, the most abundant circulating serpin, are associated with increased risk of HIV infections (Aboud et al. 2014). The serpin B clade is also believed to protect against HIV ingress into the female genital tract. SERPINB1 inhibits elastase-like and chymotrypsin-like proteases and cathepsin G and is believed to control inflammation. SLPI is a major HIV-inhibiting component, present in mucosal secretions and in serum. SLPI has inhibitory activity against neutrophil serine proteases, e.g., cathepsin G and neutrophil elastase, while also inhibiting chymotrypsin and trypsin. Yellow fever virus also modifies serine protease and serpin activity (Woodson et al. 2013).

Thus, one might have predicted that viruses would develop their own serpins, which would not only subvert the host antimicrobial defenses, but which might also undermine invasion by competing microbes, specifically viruses, thus providing an advantage for the virus in establishing an infection, invading a host, proliferating, and spreading throughout the infected host. Serpins are also modified by selected bacterial infections.

Our recent finding that the myxoma viral serpin Serp-1 can improve mortality in severe and lethal MHV68 infections in mice with associated reductions in detected viral load (antigen for MHV68) would support a role for virus-derived serpins in modifying infections by unrelated viral species (Chen et al. 2013b). One small study assessed plaque growth in a cholesterol-fed swine model after swine pox infection and atheromatous plaque sizes were reduced by the swine pox infection (Shimamura et al. 2012). Given the very marked potency of the virus-derived serpins in many animal models tested to date, we would suggest that the evolution of viral serpins over many millions of years may provide a new source of antiviral agents with exceptional efficacy.

17.4.8 Summation and Overview of the Viral Serpins and Their Therapeutic Potential

Like a kaleidoscope of butterflies, the range of virus-derived serpin families may be much more varied than has been previously known. Scientists have speculated that the viral serpins may both represent mammalian genes that have been borrowed or mimicked by the virus or that the viral serpins have evolved in parallel with mammalian genes. Conversely, viruses may have been the original source for some mammalian serpins. There is an extraordinary potential for the adaption of these viral serpins themselves, either as therapeutic proteins or as a basis for drug design to mimic and/or target the same protease pathways. One can also envision the modulation of host immune responses through altered viral populations that inhabit the mammalian host, the so-termed virome (Lucas et al. 2014b; Norman et al. 2015; Virgin 2014). Thus, treatment with a serpin or related agent may alter the host symbiotic infestation of the biome to modify the mammalian immune response and provide an enhanced protection against other invading pathogens, viral, bacterial, or parasitic. What is currently understood about serpin immune regulation may thus represent only a small portion for the future discovery and the true range of serpin functions.

References

Aboud L, Ball TB, Tjernlund A, Burgener A (2014) The role of serpin and cystatin antiproteases in mucosal innate immunity and their defense against HIV. Am J Reprod Immunol 71:12–23

Alexander JM, Nelson CA, van Berkel V, Lau EK, Studts JM, Brett TJ, Speck SH, Handel TM, Virgin HW, Fremont DH (2002) Structural basis of chemokine sequestration by a herpesvirus decoy receptor. Cell 111:343–356

Ali AN, Turner PC, Brooks MA, Moyer RW (1994) The SPI-1 gene of rabbitpox virus determines host range and is required for hemorrhagic pock formation. Virology 202:305–314

Bedard EL, Jiang J, Arp J, Qian H, Wang H, Guan H, Liu L, Parry N, Kim P, Garcia B, Li X, Macaulay C, McFadden G, Lucas A, Zhong R (2006) Prevention of chronic renal allograft rejection by SERP-1 protein. Transplantation 81:908–914

Bennett PC, Silverman SH, Gill PS, Lip GY (2009) Peripheral arterial disease and Virchow's triad. Thromb Haemost 101:1032–1040

Bird CH, Christensen ME, Mangan MS, Prakash MD, Sedelies KA, Smyth MJ, Harper I, Waterhouse NJ, Bird PI (2014) The granzyme B-Serpinb9 axis controls the fate of lymphocytes after lysosomal stress. Cell Death Differ 21:876–887

Bot I, von der Thusen JH, Donners MM, Lucas A, Fekkes ML, de Jager SC, Kuiper J, Daemen MJ, van Berkel TJ, Heeneman S, Biessen EA (2003) Serine protease inhibitor Serp-1 strongly impairs atherosclerotic lesion formation and induces a stable plaque phenotype in ApoE-/- mice. Circ Res 93:464–471

Bradshaw CB, Davis RL, Shrimpton AE, Holohan PD, Rea CB, Fieglin D, Kent P, Collins GH (2001) Cognitive deficits associated with a recently reported familial neurodegenerative disease: familial encephalopathy with neuroserpin inclusion bodies. Arch Neurol 58:1429–1434

Brahn E, Lee S, Lucas A, McFadden G, Macaulay C (2014) Suppression of collagen-induced arthritis with a serine proteinase inhibitor (serpin) derived from myxoma virus. Clin Immunol 153:254–263

Brooks MA, Ali AN, Turner PC, Moyer RW (1995) A rabbitpox virus serpin gene controls host range by inhibiting apoptosis in restrictive cells. J Virol 69:7688–7698

Carmeliet P, Moons L, Lijnen R, Baes M, Lemaitre V, Tipping P, Drew A, Eeckhout Y, Shapiro S, Lupu F, Collen D (1997) Urokinase-generated plasmin activates matrix metalloproteinases during aneurysm formation. Nat Genet 17:439–444

Carrell R, Travis J (1985) Alpha-1-antitrypsin and the serpins - variation and countervariation. Trends Biochem Sci 10:20–24

Chen D, Dorling A (2009) Critical roles for thrombin in acute and chronic inflammation. J Thromb Haemost 7(Suppl 1):122–126

Chen D, Carpenter A, Abrahams J, Chambers RC, Lechler RI, McVey JH, Dorling A (2008) Protease-activated receptor 1 activation is necessary for monocyte chemoattractant protein 1-dependent leukocyte recruitment in vivo. J Exp Med 205:1739–1746

Chen H, Davids JA, Zheng D, Bryant M, Bot I, van Berckel TJ, Biessen E, Pepine C, Ryman K, Progulski-Fox A, Kesavalu L, Moyer R, McFadden G, Lucas A (2013a) The serpin solution; targeting thrombotic and thrombolytic serine proteases in inflammation. Cardiovasc Hematol Disord Drug Targets 13:99–110

Chen H, Zheng D, Abbott J, Liu L, Bartee MY, Long M, Davids J, Williams J, Feldmann H, Strong J, Grau KR, Tibbetts S, Macaulay C, McFadden G, Thoburn R, Lomas DA, Spinale FG, Virgin HW, Lucas A (2013b) Myxomavirus-derived serpin prolongs survival and reduces inflammation and hemorrhage in an unrelated lethal mouse viral infection. Antimicrob Agents Chemother 57:4114–4127

Chen H, Zheng D, Ambadapadi S, Davids J, Ryden S, Samy H, Bartee M, Sobel E, Dai E, Liu L, Macaulay C, Yachnis A, Weyand C, Thoburn R, Lucas A (2015) Serpin treatment suppresses inflammatory vascular lesions in temporal artery implants (TAI) from patients with giant cell arteritis. PLoS One 10, e0115482

Christov A, Korol RM, Dai E, Liu L, Guan H, Bernards MA, Cavers PB, Susko D, Lucas A (2005) In vivo optical analysis of quantitative changes in collagen and elastin during arterial remodeling. Photochem Photobiol 81:457–466

Chu AJ (2010) Blood coagulation as an intrinsic pathway for proinflammation: a mini review. Inflamm Allergy Drug Targets 9:32–44

Chuang SY, Yang CH, Chou CC, Chiang YP, Chuang TH, Hsu LC (2013) TLR-induced PAI-2 expression suppresses IL-1beta processing via increasing autophagy and NLRP3 degradation. Proc Natl Acad Sci USA 110:16079–16084

Coller BS, Owen J, Jesty J, Horowitz D, Reitman MJ, Spear J, Yeh T, Comp PC (1987) Deficiency of plasma protein S, protein C, or antithrombin III and arterial thrombosis. Arteriosclerosis 7:456–462

Conard J, Samama MM (1986) Inhibitors of coagulation, atherosclerosis, and arterial thrombosis. Semin Thromb Hemost 12:87–90

Dai E, Guan H, Liu L, Little S, McFadden G, Vaziri S, Cao H, Ivanova IA, Bocksch L, Lucas A (2003) Serp-1, a viral anti-inflammatory serpin, regulates cellular serine proteinase and serpin responses to vascular injury. J Biol Chem 278:18563–18572

Dai E, Viswanathan K, Sun YM, Li X, Liu LY, Togonu-Bickersteth B, Richardson J, Macaulay C, Nash P, Turner P, Nazarian SH, Moyer R, McFadden G, Lucas AR (2006) Identification of myxomaviral serpin reactive site loop sequences that regulate innate immune responses. J Biol Chem 281:8041–8050

Dai E, Liu LY, Wang H, McIvor D, Sun YM, Macaulay C, King E, Munuswamy-Ramanujam G, Bartee MY, Williams J, Davids J, Charo I, McFadden G, Esko JD, Lucas AR (2010) Inhibition of chemokine-glycosaminoglycan interactions in donor tissue reduces mouse allograft vasculopathy and transplant rejection. PLoS One 5, e10510

Davids JA, Dai E, Chen H, Bartee MY, Liu L, Fortune A, Moyer R, McFadden G, Lucas AR (2014) Viral anti-inflammatory proteins target diverging immune pathways with converging effects on arterial dilatation, plaque, and apoptosis. Eur J Inflamm 12:131–145

Davis RL, Shrimpton AE, Holohan PD, Bradshaw C, Feiglin D, Collins GH, Sonderegger P, Kinter J, Becker LM, Lacbawan F, Krasnewich D, Muenke M, Lawrence DA, Yerby MS, Shaw CM, Gooptu B, Elliott PR, Finch JT, Carrell RW, Lomas DA (1999a) Familial dementia caused by polymerization of mutant neuroserpin. Nature 401:376–379

Davis RL, Holohan PD, Shrimpton AE, Tatum AH, Daucher J, Collins GH, Todd R, Bradshaw C, Kent P, Feiglin D, Rosenbaum A, Yerby MS, Shaw CM, Lacbawan F, Lawrence DA (1999b) Familial encephalopathy with neuroserpin inclusion bodies. Am J Pathol 155:1901–1913

de Waard V, Arkenbout EK, Carmeliet P, Lindner V, Pannekoek H (2002) Plasminogen activator inhibitor 1 and vitronectin protect against stenosis in a murine carotid artery ligation model. Arterioscler Thromb Vasc Biol 22:1978–1983

Devy L, Blacher S, Grignet-Debrus C, Bajou K, Masson V, Gerard RD, Gils A, Carmeliet G, Carmeliet P, Declerck PJ, Noel A, Foidart JM (2002) The pro- or antiangiogenic effect of plasminogen activator inhibitor 1 is dose dependent. FASEB J 16:147–154

Esmon CT (2005) The interactions between inflammation and coagulation. Br J Haematol 131:417–430

Farlow J, Ichou MA, Huggins J, Ibrahim S (2010) Comparative whole genome sequence analysis of wild-type and cidofovir-resistant monkeypoxvirus. Virol J 7:110

Fleetwood AJ, Achuthan A, Schultz H, Nansen A, Almholt K, Usher P, Hamilton JA (2014) Urokinase plasminogen activator is a central regulator of macrophage three-dimensional invasion, matrix degradation, and adhesion. J Immunol 192:3540–3547

Francis SE, Ersoy RA, Ahn JW, Atwell BJ, Roberts TH (2012) Serpins in rice: protein sequence analysis, phylogeny and gene expression during development. BMC Genomics 13:449

Fu Q, Cheng J, Gao Y, Zhang Y, Chen X, Xie J (2014) Protease-activated receptor 4: a critical participator in inflammatory response. Inflammation 38(2):886–95

Gelderblom M, Neumann M, Ludewig P, Bernreuther C, Krasemann S, Arunachalam P, Gerloff C, Glatzel M, Magnus T (2013) Deficiency in serine protease inhibitor neuroserpin exacerbates ischemic brain injury by increased postischemic inflammation. PLoS One 8, e63118

Gong Y, Hart E, Shchurin A, Hoover-Plow J (2008) Inflammatory macrophage migration requires MMP-9 activation by plasminogen in mice. J Clin Invest 118:3012–3024

Hiebert PR, Granville DJ (2012) Granzyme B in injury, inflammation, and repair. Trends Mol Med 18:732–741

Hoffmann JN, Wiedermann CJ, Juers M, Ostermann H, Kienast J, Briegel J, Strauss R, Warren BL, Opal SM, KyberSept Investigators (2006) Benefit/risk profile of high-dose antithrombin in patients with severe sepsis treated with and without concomitant heparin. Thromb Haemost 95:850–856

Hu JH, Touch P, Zhang J, Wei H, Liu S, Lund IK, Hoyer-Hansen G, Dichek DA (2015) Reduction of mouse atherosclerosis by urokinase inhibition or with a limited-spectrum matrix metalloproteinase inhibitor. Cardiovasc Res 105(3):372–382

Huang M, Gong Y, Grondolsky J, Hoover-Plow J (2014) Lp(a)/apo(a) modulate MMP-9 activation and neutrophil cytokines in vivo in inflammation to regulate leukocyte recruitment. Am J Pathol 184:1503–1517

Huntington JA, Read RJ, Carrell RW (2000) Structure of a serpin-protease complex shows inhibition by deformation. Nature 407:923–926

Irving JA, Pike RN, Lesk AM, Whisstock JC (2000) Phylogeny of the serpin superfamily: implications of patterns of amino acid conservation for structure and function. Genome Res 10:1845–1864

Ishiguro K, Kojima T, Kadomatsu K, Nakayama Y, Takagi A, Suzuki M, Takeda N, Ito M, Yamamoto K, Matsushita T, Kusugami K, Muramatsu T, Saito H (2000) Complete antithrombin deficiency in mice results in embryonic lethality. J Clin Invest 106:873–878

Iwanaga S, Kawabata S (1998) Evolution and phylogeny of defense molecules associated with innate immunity in horseshoe crab. Front Biosci 3:D973–D984

Janciauskiene S (2001) Conformational properties of serine proteinase inhibitors (serpins) confer multiple pathophysiological roles. Biochim Biophys Acta 1535:221–235

Janciauskiene SM, Bals R, Koczulla R, Vogelmeier C, Kohnlein T, Welte T (2011) The discovery of alpha1-antitrypsin and its role in health and disease. Respir Med 105:1129–1139

Jiang J, Arp J, Kubelik D, Zassoko R, Liu W, Wise Y, Macaulay C, Garcia B, McFadden G, Lucas AR, Wang H (2007) Induction of indefinite cardiac allograft survival correlates with toll-like receptor 2 and 4 downregulation after serine protease inhibitor-1 (Serp-1) treatment. Transplantation 84:1158–1167

Kanse SM, Chavakis T, Al-Fakhri N, Hersemeyer K, Monard D, Preissner KT (2004) Reciprocal regulation of urokinase receptor (CD87)-mediated cell adhesion by plasminogen activator inhibitor-1 and protease nexin-1. J Cell Sci 117:477–485

Kettle S, Blake NW, Law KM, Smith GL (1995) Vaccinia virus serpins B13R (SPI-2) and B22R (SPI-1) encode M(r) 38.5 and 40K, intracellular polypeptides that do not affect virus virulence in a murine intranasal model. Virology 206:136–147

Khan MS, Singh P, Azhar A, Naseem A, Rashid Q, Kabir MA, Jairajpuri MA (2011) Serpin inhibition mechanism: a delicate balance between native metastable state and polymerization. J Amino Acids 2011:606797

Khanaree C, Chairatvit K, Roytrakul S, Wongnoppavich A (2013) Reactive center loop moiety is essential for the maspin activity on cellular invasion and ubiquitin-proteasome level. Oncol Res 20:427–435

Komiyama T, Ray CA, Pickup DJ, Howard AD, Thornberry NA, Peterson EP, Salvesen G (1994) Inhibition of interleukin-1 beta converting enzyme by the cowpox virus serpin CrmA. An example of cross-class inhibition. J Biol Chem 269:19331–19337

Law KM, Smith GL (1992) A vaccinia serine protease inhibitor which prevents virus-induced cell fusion. J General Virol 73(Pt 3):549–557

Law RH, Zhang Q, McGowan S, Buckle AM, Silverman GA, Wong W, Rosado CJ, Langendorf CG, Pike RN, Bird PI, Whisstock JC (2006) An overview of the serpin superfamily. Genome Biol 7:216

Lawrence DA, Olson ST, Muhammad S, Day DE, Kvassman JO, Ginsburg D, Shore JD (2000) Partitioning of serpin-proteinase reactions between stable inhibition and substrate cleavage is regulated by the rate of serpin reactive center loop insertion into beta-sheet A. J Biol Chem 275:5839–5844

Li X, Schneider H, Peters A, Macaulay C, King E, Sun Y, Liu L, Dai E, Davids JA, McFadden G, Lucas A (2008) Heparin alters viral serpin, serp-1, anti-thrombolytic activity to anti-thrombotic activity. Open Biochem J 2:6–15

Liu L, Dai E, Miller L, Seet B, Lalani A, Macauley C, Li X, Virgin HW, Bunce C, Turner P, Moyer R, McFadden G, Lucas A (2004) Viral chemokine-binding proteins inhibit inflammatory responses and aortic allograft transplant vasculopathy in rat models. Transplantation 77:1652–1660

Liu L, Dai E, Kidwai B, Davids J, Macaulay C, McFadden G, Lucas AR (2013) Comparative analysis of plaque growth after arterial stent implant with anti-inflammatory Chemokine and serine protease inhibitor treatment. J Clin Exp Cardiol 4:274

Liu Y, Hou F, He S, Qian Z, Wang X, Mao A, Sun C, Liu X (2014) Identification, characterization and functional analysis of a serine protease inhibitor (Lvserpin) from the Pacific white shrimp, Litopenaeus vannamei. Dev Comp Immunol 43:35–46

Lomas DA, Carrell RW (2002) Serpinopathies and the conformational dementias. Nat Rev Genet 3:759–768

Lomas DA, Evans DL, Upton C, McFadden G, Carrell RW (1993) Inhibition of plasmin, urokinase, tissue plasminogen activator, and C1S by a myxoma virus serine proteinase inhibitor. J Biol Chem 268:516–521

Lomas DA, Belorgey D, Mallya M, Miranda E, Kinghorn KJ, Sharp LK, Phillips RL, Page R, Robertson AS, Crowther DC (2005) Molecular mousetraps and the serpinopathies. Biochem Soc Trans 33:321–330

Lopez D, Lin L, Monaghan JR, Cogle CR, Bova FJ, Maden M, Scott EW (2014) Mapping hematopoiesis in a fully regenerative vertebrate: the axolotl. Blood 124:1232–1241

Lowe GD (2003) Virchow's triad revisited: abnormal flow. Pathophysiol Haemost Thromb 33:455–457

Lucas A, McFadden G (2004) Secreted immunomodulatory viral proteins as novel biotherapeutics. J Immunol 173:4765–4774

Lucas A, Liu L, Macen J, Nash P, Dai E, Stewart M, Graham K, Etches W, Boshkov L, Nation PN, Humen D, Hobman ML, McFadden G (1996) Virus-encoded serine proteinase inhibitor SERP-1 inhibits atherosclerotic plaque development after balloon angioplasty. Circulation 94:2890–2900

Lucas A, Dai E, Liu L, Guan H, Nash P, McFadden G, Miller L (2000) Transplant vasculopathy: viral anti-inflammatory serpin regulation of atherogenesis. J Heart Lung Transplant 19:1029–1038

Lucas A, Liu L, Dai E, Bot I, Viswanathan K, Munuswamy-Ramunujam G, Davids JA, Bartee MY, Richardson J, Christov A, Wang H, Macaulay C, Poznansky M, Zhong R, Miller L, Biessen E, Richardson M, Sullivan C, Moyer R, Hatton M, Lomas DA, McFadden G (2009) The serpin saga; development of a new class of virus derived anti-inflammatory protein immunotherapeutics. Adv Exp Med Biol 666:132–156

Lucas AR, Verma RK, Dai E, Liu L, Chen H, Kesavalu S, Rivera M, Velsko I, Ambadapadi S, Chukkapalli S, Kesavalu L (2014a) Myxomavirus anti-inflammatory chemokine binding protein reduces the increased plaque growth induced by chronic Porphyromonas gingivalis oral infection after balloon angioplasty aortic injury in mice. PLoS One 9, e111353

Lucas A, Ambadapadi S, Zheng D, Chen H, Lakshmyya K, Progulske-Fox A, Krafft A, Bryant M, O'Regan S, Moyer R, McFadden G, Jobin C, Rand K (2014b) The virome – viral ghost companion, virus wars. J Clin Exp Cardiol 5, e136

MacNeill AL, Turner PC, Moyer RW (2006) Mutation of the Myxoma virus SERP2 P1-site to prevent proteinase inhibition causes apoptosis in cultured RK-13 cells and attenuates disease in rabbits, but mutation to alter specificity causes apoptosis without reducing virulence. Virology 356:12–22

Maden M (2009) Molecular embryology: methods and protocols. In: Sharpe P, Mason I (eds) Methods in molecular biology. Humana Press, Totowa, NJ

Maksymowych WP, Nation N, Nash P, Macen J, Lucas A, McFadden G, Russell AS (1996) Amelioration of antigen induced arthritis in rabbits treated with a secreted viral serine proteinase inhibitor. J Rheumatol 23:878–882

Marszal E, Shrake A (2006) Serpin crystal structure and serpin polymer structure. Arch Biochem Biophys 453:123–129

Masters SL (2013) Specific inflammasomes in complex diseases. Clin Immunol 147:223–228

McFadden G, Graham K, Ellison K, Barry M, Macen J, Schreiber M, Mossman K, Nash P, Lalani A, Everett H (1995) Interruption of cytokine networks by poxviruses: lessons from myxoma virus. J Leukoc Biol 57:731–738

Melo-Silva CR, Tscharke DC, Lobigs M, Koskinen A, Wong YC, Buller RM, Mullbacher A, Regner M (2011) The ectromelia virus SPI-2 protein causes lethal mousepox by preventing NK cell responses. J Virol 85:11170–11182

Messud-Petit F, Gelfi J, Delverdier M, Amardeilh MF, Py R, Sutter G, Bertagnoli S (1998) Serp2, an inhibitor of the interleukin-1beta-converting enzyme, is critical in the pathobiology of myxoma virus. J Virol 72:7830–7839

Miller LW, Dai E, Nash P, Liu L, Icton C, Klironomos D, Fan L, Nation PN, Zhong R, McFadden G, Lucas A (2000) Inhibition of transplant vasculopathy in a rat aortic allograft model after infusion of anti-inflammatory viral serpin. Circulation 101:1598–1605

Miranda E, Lomas DA (2006) Neuroserpin: a serpin to think about. Cell Mol Life Sci 63:709–722

Moon KB, Turner PC, Moyer RW (1999) SPI-1-dependent host range of rabbitpox virus and complex formation with cathepsin G is associated with serpin motifs. J Virol 73:8999–9010

Motal MC, Klaus DA, Lebherz-Eichinger D, Tudor B, Hamp T, Wiegele M, Seemann R, Krenn CG, Roth GA (2015) Increased plasma vaspin concentration in patients with sepsis: an exploratory examination. Biochem Med 25:90–96

Munuswamy-Ramanujam G, Dai E, Liu L, Shnabel M, Sun YM, Bartee M, Lomas DA, Lucas AR (2010) Neuroserpin, a thrombolytic serine protease inhibitor (serpin), blocks transplant vasculopathy with associated modification of T-helper cell subsets. Thromb Haemost 103:545–555

Nagai N, De Mol M, Lijnen HR, Carmeliet P, Collen D (1999) Role of plasminogen system components in focal cerebral ischemic infarction: a gene targeting and gene transfer study in mice. Circulation 99:2440–2444

Nash P, Lucas A, McFadden G (1997) SERP-1, a poxvirus-encoded serpin, is expressed as a secreted glycoprotein that inhibits the inflammatory response to myxoma virus infection. Adv Exp Med Biol 425:195–205

Nash P, Whitty A, Handwerker J, Macen J, McFadden G (1998) Inhibitory specificity of the anti-inflammatory myxoma virus serpin, SERP-1. J Biol Chem 273:20982–20991

Nathaniel R, MacNeill AL, Wang YX, Turner PC, Moyer RW (2004) Cowpox virus CrmA, Myxoma virus SERP2 and baculovirus P35 are not functionally interchangeable caspase inhibitors in poxvirus infections. J General Virol 85:1267–1278

Norman JM, Handley SA, Baldridge MT, Droit L, Liu CY, Keller BC, Kambal A, Monaco CL, Zhao G, Fleshner P, Stappenbeck TS, McGovern DP, Keshavarzian A, Mutlu EA, Sauk J, Gevers D, Xavier RJ, Wang D, Parkes M, Virgin HW (2015) Disease-specific alterations in the enteric virome in inflammatory bowel disease. Cell 160:447–460

Olson ST, Richard B, Izaguirre G, Schedin-Weiss S, Gettins PG (2010) Molecular mechanisms of antithrombin-heparin regulation of blood clotting proteinases. A paradigm for understanding proteinase regulation by serpin family protein proteinase inhibitors. Biochimie 92:1587–1596

Onda M, Belorgey D, Sharp LK, Lomas DA (2005) Latent S49P neuroserpin forms polymers in the dementia familial encephalopathy with neuroserpin inclusion bodies. J Biol Chem 280:13735–13741

Patston PA, Church FC, Olson ST (2004) Serpin-ligand interactions. Methods 32:93–109

Rashid Q, Kapil C, Singh P, Kumari V, Jairajpuri MA (2015) Understanding the specificity of serpin-protease complexes through interface analysis. J Biomol Struct Dyn 33:1352–1362

Richardson J, Viswanathan K, Lucas A (2006) Serpins, the vasculature, and viral therapeutics. Front Biosci 11:1042–1056

Rohrmann GF, Erlandson MA, Theilmann DA (2013) The genome of a baculovirus isolated from Hemileuca sp. encodes a serpin ortholog. Virus Genes 47:357–364

Ross R, Glomset J, Harker L (1977) Response to injury and atherogenesis. Am J Pathol 86:675–684

Ross R, Faggiotto A, Bowen-Pope D, Raines E (1984) The role of endothelial injury and platelet and macrophage interactions in atherosclerosis. Circulation 70:III77–III82

Schroder WA, Major LD, Le TT, Gardner J, Sweet MJ, Janciauskiene S, Suhrbier A (2014) Tumor cell-expressed SerpinB2 is present on microparticles and inhibits metastasis. Cancer Med 3:500–513

Seet BT, Johnston JB, Brunetti CR, Barrett JW, Everett H, Cameron C, Sypula J, Nazarian SH, Lucas A, McFadden G (2003) Poxviruses and immune evasion. Annu Rev Immunol 21:377–423

Seifert AW, Monaghan JR, Voss SR, Maden M (2012) Skin regeneration in adult axolotls: a blueprint for scar-free healing in vertebrates. PLoS One 7, e32875

Shi HS, Yang LP, Wei W, Su XQ, Li XP, Li M, Luo ST, Zhang HL, Lu L, Mao YQ, Kan B, Yang L (2013) Systemically administered liposome-encapsulated Ad-PEDF potentiates the anti-cancer effects in mouse lung metastasis melanoma. J Transl Med 11:86

Shimamura T, Jeng D, Lucas A, Essani K (2012) Suppression of neointimal hyperplasia following angioplasty-induced vascular injury in pigs infected with swinepox virus. Open Virol J 6:91–96

Shisler JL, Isaacs SN, Moss B (1999) Vaccinia virus serpin-1 deletion mutant exhibits a host range defect characterized by low levels of intermediate and late mRNAs. Virology 262:298–311

Silverman GA, Bird PI, Carrell RW, Church FC, Coughlin PB, Gettins PG, Irving JA, Lomas DA, Luke CJ, Moyer RW, Pemberton PA, Remold-O'Donnell E, Salvesen GS, Travis J, Whisstock JC (2001) The serpins are an expanding superfamily of structurally similar but functionally diverse proteins. Evolution, mechanism of inhibition, novel functions, and a revised nomenclature. J Biol Chem 276:33293–33296

Stocks BB, Sarkar A, Wintrode PL, Konermann L (2012) Early hydrophobic collapse of alpha(1)-antitrypsin facilitates formation of a metastable state: insights from oxidative labeling and mass spectrometry. J Mol Biol 423:789–799

Tardif JC, L'Allier PL, Gregoire J, Ibrahim R, McFadden G, Kostuk W, Knudtson M, Labinaz M, Waksman R, Pepine CJ, Macaulay C, Guertin MC, Lucas A (2010) A randomized controlled, phase 2 trial of the viral serpin Serp-1 in patients with acute coronary syndromes undergoing percutaneous coronary intervention. Circ Cardiovasc Interv 3:543–548

Thompson JP, Turner PC, Ali AN, Crenshaw BC, Moyer RW (1993) The effects of serpin gene mutations on the distinctive pathobiology of cowpox and rabbitpox virus following intranasal inoculation of Balb/c mice. Virology 197:328–338

Turner PC, Moyer RW (1992) An orthopoxvirus serpinlike gene controls the ability of infected cells to fuse. J Virol 66:2076–2085

Turner PC, Moyer RW (1995) Orthopoxvirus fusion inhibitor glycoprotein SPI-3 (open reading frame K2L) contains motifs characteristic of serine proteinase inhibitors that are not required for control of cell fusion. J Virol 69:5978–5987

Turner PC, Moyer RW (2001) Serpins enable poxviruses to evade immune defenses. Am Soc Microbiol News 67:201–209

Turner PC, Moyer RW (2008) The vaccinia virus fusion inhibitor proteins SPI-3 (K2) and HA (A56) expressed by infected cells reduce the entry of superinfecting virus. Virology 380:226–233

Turner PC, Sancho MC, Thoennes SR, Caputo A, Bleackley RC, Moyer RW (1999) Myxoma virus Serp2 is a weak inhibitor of granzyme B and interleukin-1beta-converting enzyme in vitro and unlike CrmA cannot block apoptosis in cowpox virus-infected cells. J Virol 73:6394–6404

Turner PC, Baquero MT, Yuan S, Thoennes SR, Moyer RW (2000) The cowpox virus serpin SPI-3 complexes with and inhibits urokinase-type and tissue-type plasminogen activators and plasmin. Virology 272:267–280

Upton C, Macen JL, Wishart DS, McFadden G (1990) Myxoma virus and malignant rabbit fibroma virus encode a serpin-like protein important for virus virulence. Virology 179:618–631

van Berkel V, Barrett J, Tiffany HL, Fremont DH, Murphy PM, McFadden G, Speck SH, Virgin HI (2000) Identification of a gammaherpesvirus selective chemokine binding protein that inhibits chemokine action. J Virol 74:6741–6747

Virgin HW (2014) The virome in mammalian physiology and disease. Cell 157:142–150

Viswanathan K, Liu L, Vaziri S, Dai E, Richardson J, Togonu-Bickersteth B, Vatsya P, Christov A, Lucas AR (2006) Myxoma viral serpin, Serp-1, a unique interceptor of coagulation and innate immune pathways. Thromb Haemost 95:499–510

Viswanathan K, Richardson J, Togonu-Bickersteth B, Dai E, Liu L, Vatsya P, Sun YM, Yu J, Munuswamy-Ramanujam G, Baker H, Lucas AR (2009) Myxoma viral serpin, Serp-1, inhibits human monocyte adhesion through regulation of actin-binding protein filamin B. J Leukoc Biol 85:418–426

Viswanathan K, Bot I, Liu L, Dai E, Turner PC, Togonu-Bickersteth B, Richardson J, Davids JA, Williams JM, Bartee MY, Chen H, van Berkel TJ, Biessen EA, Moyer RW, Lucas AR (2012) Viral cross-class serpin inhibits vascular inflammation and T lymphocyte fratricide; a study in rodent models in vivo and human cell lines in vitro. PLoS One 7, e44694

Wang YX, Turner PC, Ness TL, Moon KB, Schoeb TR, Moyer RW (2000) The cowpox virus SPI-3 and myxoma virus SERP1 serpins are not functionally interchangeable despite their similar proteinase inhibition profiles in vitro. Virology 272:281–292

Wang H, Jiang J, McFadden G, Garcia B, Lucas AR, Zhong R (2003) Serp-1, a viral anti-inflammatory serpin, attenuates acute xenograft rejection in a rat-to-mouse cardiac transplantation model. Xenotransplant 10:506

Wen H, Ting JP, O'Neill LA (2012) A role for the NLRP3 inflammasome in metabolic diseases--did Warburg miss inflammation? Nat Immunol 13:352–357

Wensink AC, Hack CE, Bovenschen N (2015) Granzymes regulate proinflammatory cytokine responses. J Immunol 194:491–497

Whisstock JC, Bottomley SP (2006) Molecular gymnastics: serpin structure, folding and misfolding. Curr Opin Struct Biol 16:761–768

Woodson SE, Freiberg AN, Holbrook MR (2013) Coagulation factors, fibrinogen and plasminogen activator inhibitor-1, are differentially regulated by yellow fever virus infection of hepatocytes. Virus Res 175:155–159

Yamasaki M, Sendall TJ, Harris LE, Lewis GM, Huntington JA (2010) Loop-sheet mechanism of serpin polymerization tested by reactive center loop mutations. J Biol Chem 285:30752–30758

Ye S, Cech AL, Belmares R, Bergstrom RC, Tong Y, Corey DR, Kanost MR, Goldsmith EJ (2001) The structure of a Michaelis serpin-protease complex. Nat Struct Biol 8:979–983

Zalai CV, Nash P, Dai E, Liu L, Lucas AR (2002) The potential role of serine proteinase inhibitors for the prevention of plaque rupture. In: Brown DL (ed) Cardiovascular plaque rupture. Marcel Dekker, New York, pp 447–482

Zetterstrom R (2009) The 1908 Nobel Prize--discovery of the basic principles of innate and acquired immunity. Acta Paediatr 98:1066–1069

Zheng D, Chen H, Bartee MY, Williams J, Davids JA, Lomas DA, McFadden G, Lucas AR (2013) Myxomaviral anti-inflammatory serpin reduces myeloid – derived suppressor cells and human pancreatic cancer cell growth in mice. J Cancer Sci Therapy 5:291–299

Zhou J, Sun XY, Fernando GJ, Frazer IH (1992) The vaccinia virus K2L gene encodes a serine protease inhibitor which inhibits cell-cell fusion. Virology 189:678–686

Zhou Y, Geng Y, Yang J, Zhao C, Harrison TJ, Wang Y (2014) Hepatitis E virus open reading frame 3 protein interacts with porcine liver-specific plasminogen and alpha2-antiplasmin. J Med Virol 86:487–495

Index

M. Geiger et al. (eds.), *The Serpin Family*, DOI 10.1007/978-3-319-22711-5

MIX
Papier aus verantwortungsvollen Quellen
Paper from responsible sources
FSC® C105338

If you have any concerns about our products,
you can contact us on
ProductSafety@springernature.com

In case Publisher is established outside the EU,
the EU authorized representative is:
Springer Nature Customer Service Center GmbH
Europaplatz 3, 69115 Heidelberg, Germany

Printed by Libri Plureos GmbH
in Hamburg, Germany